ASM全埋点开发实战

王灼洲 张伟◎著

人民邮电出版社
北京

图书在版编目（CIP）数据

ASM全埋点开发实战 / 王灼洲，张伟著. -- 北京 : 人民邮电出版社，2023.7
ISBN 978-7-115-61261-8

Ⅰ. ①A… Ⅱ. ①王… ②张… Ⅲ. ①移动终端—应用程序—程序设计 Ⅳ. ①TN929.53

中国国家版本馆CIP数据核字(2023)第038070号

内容提要

本书由业内知名团队神策数据的专业人士编写，结合实战案例，深入浅出地介绍了 ASM 技术和 Android 全埋点技术。

作者从神策数据服务超过 2000 家客户的经历中，发现了行业用户对全埋点技术的迫切需求。本书针对这一点，详细、客观地阐述了 ASM 在 Android 全埋点中的应用，涵盖各种真实商业场景，并清晰地讲解其技术原理和实现步骤，以帮助用户利用好全埋点技术的特长和优势。

本书作为一本技术参考书，特别适合非专业开发工程师在日常工作中使用。

◆ 著　　　　王灼洲　张　伟
　责任编辑　张天怡
　责任印制　陈　犇
◆ 人民邮电出版社出版发行　　北京市丰台区成寿寺路 11 号
　邮编　100164　　电子邮件　315@ptpress.com.cn
　网址　https://www.ptpress.com.cn
　三河市祥达印刷包装有限公司印刷
◆ 开本：787×1092　1/16
　印张：25　　　　2023 年 7 月第 1 版
　字数：627 千字　　　　2023 年 7 月河北第 1 次印刷

定价：99.80 元

读者服务热线：(010)81055410　印装质量热线：(010)81055316
反盗版热线：(010)81055315
广告经营许可证：京东市监广登字 20170147 号

序 | PREFACE

在移动应用开发中，埋点技术是不可或缺的一环。通过对用户行为的合规监控和分析，开发者可以更好地了解用户需求，优化应用性能，提升用户体验。但是，传统的手动埋点方式存在诸多问题：开发成本高、维护成本大、埋点覆盖不全等。因此，寻找一种更加高效和自动化的埋点方式成了开发者的共同需求。

神策数据作为一家大数据分析与营销科技服务提供商，在用户行为数据埋点、传输、存储和分析等领域积累了多年经验。灼洲与他所负责的团队，一直负责各个客户端与服务端数据采集 SDK 的研发工作，并且为客户提供完整的数据采集方案，以及解决客户在数据采集过程中碰到的各种疑难杂症。在整个研发与服务客户的过程中，灼洲团队积累了丰富的经验，并且通过采集 SDK 开源、持续举办各种技术沙龙，为技术社区做出贡献。

灼洲通过之前编写的《Android全埋点解决方案》和《iOS 全埋点解决方案》两本书，对于Android和 iOS 应用的自动化埋点技术进行了全面且深入的介绍。而本书则是在这一基础之上，更进一步讲述关于Android应用自动化埋点技术的实用指南。本书深入浅出地介绍了如何使用ASM技术对Android应用进行全埋点，从而降低埋点的成本和代价。

ASM技术作为一种基于Java字节码的操作工具，可以在不改变原有代码逻辑的情况下，对字节码进行修改和增强。这使得ASM技术成为一种理想的自动化埋点工具。本书从基础的ASM技术原理入手，详细介绍了ASM在Android应用中的应用方式，同时，本书还介绍了ASM在实际项目中的应用案例，让读者更好地理解ASM技术的应用场景和优势。

除了介绍ASM技术，本书还涵盖了其他与自动化埋点相关的知识，如常见的埋点方案、埋点数据分析等。通过全面的介绍，读者不仅可以掌握ASM技术的实现方法，还可以对自动化埋点的相关知识进行全面了解。本书不仅适合Android开发者阅读，对于任何对自动化埋点感兴趣的读者来说都是一本值得阅读的图书。

神策数据联合创始人＆CTO 曹犟

前言 | INTRODUCTION

为什么要写这本书

大数据时代已经到来，数据采集变得愈加重要，企业对这方面的需求也越来越多。我的第一本书《Android全埋点解决方案》自发布后，收到了许多读者的反馈，有的与我分享他们的阅读感悟，有的说这本书改变了他们的职业生涯，有的与我探讨具体技术。其中，有不少读者跟我探讨了在实际编写Android插件时遇到的各种问题，主要原因是读者对 Gradle 和 ASM还不够熟悉。

ASM确实是入手门槛比较高的一项技术，它更加偏向底层，需要我们对底层涉及的技术有所了解。ASM的主要用途是操作字节码，它的应用非常广，例如Java JDK、Groovy编译器、Gradle、AGP中都需要用到它。

神策数据深耕埋点技术多年，在此方面积累了大量的经验，同时神策数据也是开源领域的践行者，为此，我专门编写本书，将沉淀的知识毫无保留地分享出来，希望能够推动该领域进一步向上发展。本书结构设计合理，充分考虑读者渐进式接受知识的方式，内容翔实，即便是初学者也能够掌握。

读者对象

- 初级、中级、高级水平的Android开发工程师。
- Java工程师、Java架构师。
- 对Gradle和ASM技术感兴趣的读者。
- 对埋点技术和数据行为分析感兴趣的读者。

勘误和支持

为了方便读者更好地学习ASM技术，读者可以从https://github.com/sensorsdata/asm-book获取本书案例源码。

由于作者的水平有限，以及技术不断更新和迭代，书中难免会出现一些不恰到的地方，恳请读者批评指正。联系邮箱：zhangtianyi@ptpress.com.cn。

致谢

感谢神策数据创始人团队桑文锋、曹犟、付力力、刘耀洲在工作中对我的指导和帮助。感谢神策数据开源社区每一位充满活力和共享精神的朋友们。

谨以此书献给大数据行业的关注者和建设者！

王灼洲

2023年5月

目录 | CONTENTS

7. ASM 实现全埋点——基础部分

8. ASM 实现全埋点——进阶部分

9. ASM 实践分享和未来展望

1. Gradle插件介绍

ASM 是一个通用的 Java 字节码操作和分析框架，其应用非常广泛，例如 CGLib 动态代理的实现、Java 8 Lambda 功能的实现、字节码插桩等。目前国内有不少介绍 ASM 的资料，不过这些资料只对 ASM 应用程序接口（Application Programming Interface，API）进行了简单介绍。读者如果参考这些资料来学习 ASM 框架，会发现学起来很费劲，通常只能“照葫芦画瓢”地抄一些代码，而且容易出错，出错的时候也不知道原因。产生这种现象的原因是 ASM 的入门门槛比较高，学习曲线很陡，要想学好 ASM 框架必须要具备一些基础知识。

本书是一本介绍 ASM 框架的专业书。本书结构设计合理，从基础知识开始讲解，先介绍基础知识再介绍 ASM 框架，最后介绍 ASM 框架在全埋点中的实际应用，知识点层层深入，最终形成知识闭环，即使完全不了解 ASM 的读者也能够较轻松地掌握这门技术。读者在学完本书中的知识后，可以发挥自己的奇思妙想，用 ASM 实现一些看起来很酷、很神奇的功能。

本书详细介绍了 ASM 在 Android 全埋点中的应用，而要实现全埋点功能，必须对 Gradle 插件有所了解。本章将介绍 Gradle 相关知识，以及如何在 Android 中定义一个 Gradle 插件。

1.1 什么是Gradle插件

Gradle 是一款强大的构建工具，不仅能构建 Java/Android 项目，还可以构建 C++ 项目、JavaScript 项目和 Swift 项目。对于 Gradle，接触过 Android 开发的读者应该比较熟悉。那什么是 Gradle 插件呢？我们知道Android app工程中的build.gradle文件开头都会有一段apply plugin: 'com.android.application'代码，它的作用就是使用“com.android.application”这个插件。那么 Gradle 插件到底是怎么定义的呢？

Gradle 插件就是 Gradle 工具提供的一种可以将用户的代码逻辑与他人分享的方式。为了使读者能更好地理解后续章节的内容，我们先简单介绍一些 Gradle 的基础知识。

1.2 Gradle基础知识

1.2.1 学习前提

首先按照 Gradle 官网的教程下载并安装 Gradle，安装好以后在控制台中运行如下命令：

```
$ gradle help
```

这段命令的意思是运行 Gradle 中的 help 任务。另外，Gradle 脚本通常使用 Groovy 语言来编写，因此读者需要花一点儿时间学习 Groovy，包括 Groovy 基本语法、闭包、领域特定语言（Domain Specific Language，DSL）等知识。Groovy 兼容 Java，熟悉 Java 的读者可以很快上手。

与 Groovy 相关的教程请参考其官网。

1.2.2 Gradle项目结构

接下来介绍 Gradle 项目结构。此处创建一个名为 Chapter1_00 的 Android 项目。选择 Android 项目的原因是本书面向的读者主要是 Android 开发商，当然，读者可以使用 gradle init 来创建其他类型（如 Java、C++）的项目。下面是项目的基本结构，我们以此来介绍 Gradle 项目的组成。

```
Chapter1_00
├── app(3)
│    ├── build.gradle(4)
│    └── src
├── build.gradle(2)
├── gradle
│    └── wrapper(6)
│         ├── gradle-wrapper.jar (7)
│         └── gradle-wrapper.properties(8)
├── gradle.properties (5)
├── gradlew(9)
├── gradlew.bat(10)
├── local.properties(11)
└── settings.gradle(1)
```

接下来按照上面标注的序号，着重对前 6 个进行介绍。

（1）settings.gradle： 该文件在 Gradle 项目中是可选的。如果一个 Gradle 工程中有多个子项目，那么必须要有 settings.gradle 文件，它的配置决定哪些项目会参与构建，例如：

```
include ':app'
rootProject.name = "Chapter1_01"
```

其中 include 表示使同级目录中的 app 这个模块参与构建，rootProject.name 表示当前项目的名称。

（2）build.gradle： 表示根项目的构建文件，每一个 build.gradle 文件都对应一个 Project（项目）对象。根项目中的 build.gradle 对应的 Project 对象，我们称为 root project；其他模块 / 项目我们称为 sub project，例如这里的 app 目录下的 build.gradle 就是 sub project。为了方便后续的内容介绍，我们约定用 app/build.gradle 来表示 app 项目的 build.gradle 构建文件，用 root project/build.gradle 来表示根项目的 build.gradle 构建文件。根项目的 build.gradle 用于做一些全局的配置，例如：

```
// Top-level build file where you can add configuration options common to all sub-
projects/modules
buildscript {
      repositories {
          google()
          jcenter()
      }
      dependencies {
          classpath "com.android.tools.build:gradle:4.1.2"
          // 动态添加
          // NOTE: Do not place your application dependencies here; they belong
          // in the individual module build.gradle files
      }
}

allprojects {
      repositories {
          google()
          jcenter()
      }
}
```

说明如下。

① buildscript{} 用于配置 Gradle 插件需要的依赖，其中，repositories{} 用于声明插件所在的仓库地址，dependencies{} 用于声明依赖的插件。

② allprojects{} 用于对所有子项目进行配置，其中 repositories{} 用于声明子项目添加依赖包的仓库地址。

（3）app： 子项目目录。

（4）app/build.gradle： 子项目对应的构建文件。

（5）gradle.properties： 用于配置 Gradle 运行时相关的值，也可以在其中定义一些其他键值对，配置的值可以在 build.gradle 或 Project 中使用。

（6）gradle/wrapper： 这个目录下有 gradle-wrapper.jar 和 gradle-wrapper.properties 两个文件，这两个文件是给 gradlew 和 gradlew.bat 脚本使用的。这两个脚本的作用都是运行 Gradle 命令，不同点是 gradlew 脚本运用于 macOS/Linux 平台，而 gradlew.bat 脚本运用于 Windows 平台。gradlew 是 Gradle Wrapper 的缩写。相较于直接运行 Gradle 命令，gradlew 脚本在运行命令的时候会根据 gradlew-wrapper.properties 中的配置运行指定版本的 Gradle。如果指定版本的 Gradle 不存在，会执行 gradlew-wrapper.jar 包中的逻辑去下载。这么做的好处是可以保证项目组中不同的成员都使用同样版本的 Gradle 进行构建。

再来看看 gradle-wrapper.properties 中的内容：

```
#Sat Mar 06 11:46:33 CST 2021
distributionBase=GRADLE_USER_HOME
distributionPath=wrapper/dists
zipStoreBase=GRADLE_USER_HOME
zipStorePath=wrapper/dists
distributionUrl=https\://services.gradle.org/distributions/gradle-6.5-all.zip
```

说明如下。

① distributionBase：下载的 gradle*.zip 解压后所在位置的父目录。

② distributionPath：distributionBase 确定了父目录，distributionBase + distributionPath 的组合就能确定 Gradle 解包后的存放位置。

③ zipStoreBase：Gradle 压缩包下载后存储的父目录。

④ zipStorePath：zipStoreBase 确定了父目录，zipStoreBase + zipStorePath 的组合就能确定 Gradle 压缩包的存放位置。

⑤ distributionUrl：Gradle 指定版本的压缩包下载地址，上面的 gradle-6.5-all.zip 中的 all 表示会下载文件。当我们想要查看 API 的时候很有帮助，如果只想下载可执行文件，将 all 改成 bin 即可。

以上就是 Gradle 项目的典型结构。

1.2.3 生命周期

Gradle 生命周期具有 3 个不同的阶段。

- **初始（initialization）阶段**。Gradle 支持单个或多个项目构建，Gradle 在初始阶段，会查找 settings.gradle 文件并根据其中的配置决定让哪些项目参与构建，并且为每一个项目创建一个 Project 对象。对于 settings.gradle 文件，Gradle 会首先在项目中查找。如果存在此文件，就按照多项目构建来执行；如果在当前目录中没有找到此文件，就从父目录中寻找，如果父目录中没有此文件，就将此项目作为单个项目进行构建。

- **配置(configuration)阶段**。在这个阶段，将配置Project对象，即执行各项目下的build.gradle脚本，构造任务(task)的依赖关系并执行任务中的配置代码。
- **执行(execution)阶段**。按照顺序执行任务中定义的动作(action)，例如在Terminal中执行help任务。

```
$ ./gradlew help
```

1.2.4 Project API介绍

在Gradle构建体系中，build.gradle和Project对象是一对一的关系，每个build.gradle文件都对应一个Project对象。我们在build.gradle文件中调用的方法相当于调用Project类中的方法，例如rootProject/build.gradle中使用的buildscript{}，它实际上调用的是Project类中的buildscript(Closure configureClosure)方法。

下面列出了Project类中的部分常用方法。

- **Project getProject()**：获取当前的Project对象。
- **File getProjectDir()**：获取当前项目的目录。
- **Set<Project> getAllprojects()**：获取当前项目和子项目的Project集合。
- **Map<String,Project> getChildProjects()**：获取子项目，此键值对中键是项目名(例如Chapter1_00中的项目app)，值是Project实例。
- **File getBuildDir()**：获取当前项目的build文件夹。
- **void setBuildDir(File path)**：设置编译目录。
- **File getBuildFile()**：获取build.gradle文件。
- **ScriptHandler getBuildscript()**：获取build.gradle文件中对应的buildscript()方法中创建的值，例如rootProject/build.gradle中定义的内容。

```
buildscript {
      repositories {
          google()
          jcenter()
      }
      dependencies {
          classpath "com.android.tools.build:gradle:4.1.2"
      }
}
```

我们可以根据获取到的ScriptHandler对象添加和修改内容。

- **void buildscript(Closure configureClosure)**：设置buildscript()中的内容，我们也可以通过getBuildScript()方法进行操作。
- **DependencyHandler getDependencies()**：获取build.gradle文件中对应的dependencies()方法中创建的值，例如1.2.2小节中序号为4的build.gradle中定义的内容。

```
//build.gradle
dependencies {
      implementation 'androidx.appcompat:appcompat:1.2.0'
```

```
    implementation 'com.google.android.material:material:1.3.0'
    implementation 'androidx.constraintlayout:constraintlayout:2.0.4'
    testImplementation 'junit:junit:4.+'
    androidTestImplementation 'androidx.test.ext:junit:1.1.2'
    androidTestImplementation 'androidx.test.espresso:espresso-core:3.3.0'
}
```

当然，我们可以根据返回值动态地添加项目依赖。

- **Gradle getGradle()**：获取当前项目所属的Gradle对象。Gradle对象中提供了生命周期相关的回调。
- **ExtensionContainer getExtensions()**：为当前项目添加扩展对象，这对我们后面介绍的插件编写很有用。例如1.2.2小节的build.gradle文件中定义的android { ... }DSL语句，它的内容就是用在“com.android.application”中的。
- **String getName()**：获取当前项目名。
- **Task task(String name)**：创建一个指定名字的Task，将其绑定到Project上并返回该Task对象。
- **Task task(String name, Closure configureClosure)**：创建一个指定名字的Task，将其绑定到Project上，然后运行configureClosure，最后返回该Task对象。
- **Task task(Map<String,?> args, String name)**：创建Task，与Task相关的配置可以放在Map中，其中Map支持如下类型。

  type：Task的类型，默认值是DefaultTask。

  overwrite：是否替换已经存在的Task，需要和type配合使用，默认值是false。

  dependsOn：依赖的Task名字，默认值是[ ]。

  action：添加到Task的Action或者闭包，默认值是null（表示“空”）。

  description：Task的描述，默认值是null。

  group：Task所属的分组，默认值是null。

- **TaskContainer getTasks()**：获取当前Project的所有Task，这里是一个TaskContainer对象，通过它可以对Task进行增删改操作。
- **Map<Project,Set<Task>> getAllTasks(boolean recursive)**：获取所有项目中的Task，包括子项目的Task。
- **void setProperty(String name, Object value)**：此方法会查找有没有对应的属性，如果没有就会抛出异常，如果有就为其设置新的值。例如可以通过为Project设置extra、extension、gradle.properties等方式设置属性。
- **Map<String,?> getProperties()**：获取当前Project的所有属性。
- **void setVersion(Object version)**：设置当前项目的版本号。
- **Object getVersion()**：获取当前项目的版本号。
- **void setGroup()**：设置项目的组织信息，通常使用类似“com.sensorsdata”这种公众域名字符串来代替。Version和Group在Maven打包的时候需要用到。
- **Object getGroup()**：获取项目组织信息。
- **void defaultTasks(String...defaultTasks)**：设置默认Task，默认Task会在执行其他Task时自动执行。

- **<T> Iterable<T> configure(Iterable<T> objects, Action<? super T> configureAction)**：对给定的对象执行给定的配置动作。
- **void afterEvaluate(Closure closure)**：这是Project生命周期中提供的钩子函数，是在项目评估完成后的回调。另外，Task、Gradle都提供了类似的钩子函数，关于生命周期回调可以参考1.2.6小节。
- **void beforeEvaluate(Closure closure)**：这是Project生命周期中提供的钩子函数，是在项目评估前的回调。另外，Task、Gradle都提供了类似的钩子函数。
- **void apply(Closure closure)**：引入插件或脚本，例如apply plugin: 'com.android.application'。
- **void apply(Map<String,?> options)**：引入插件或脚本，例如apply plugin: 'com.android.application'。
- **PluginContainer getPlugins()**：获取包含所有插件的容器对象。

1.2.5 Gradle任务介绍

Gradle 的核心模型是一个以任务为工作单元的有向无环图（Directed Acyclic Graph，DAG），这意味着一次构建实际上是配置一些任务并将它们按照 DAG 的关系联系在一起。例如图 1-1 描述了这种关系，左边是抽象图，右边是具体的依赖关系图。

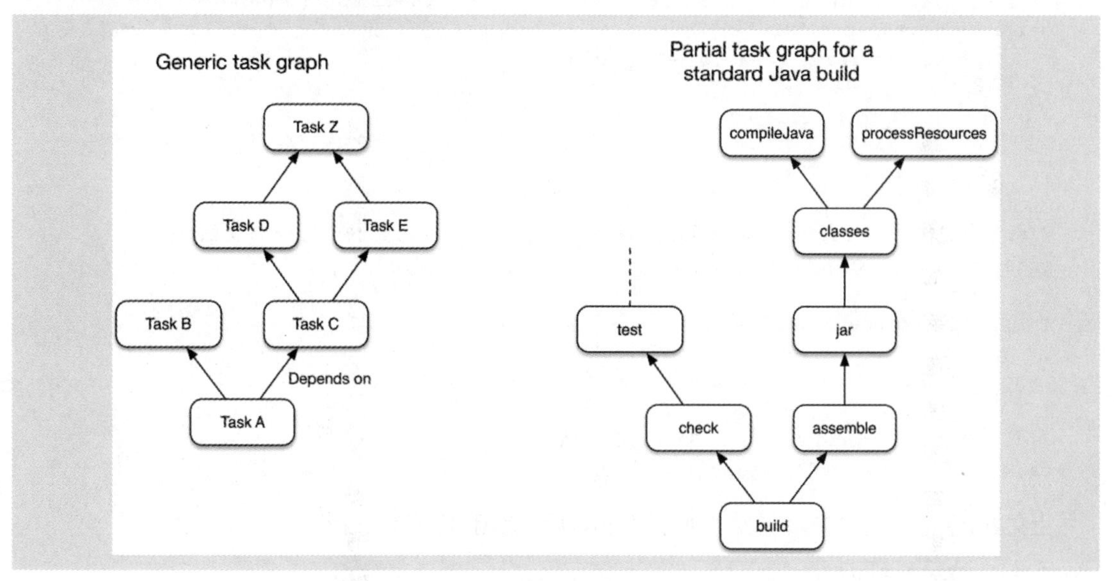

图 1-1　Gradle Task DAG（摘自 Gradle 官网）

当我们使用 Gradle 编译项目的时候，实际上是执行一个个 Task，例如图 1-1 所示：任务之间使用 DAG 的方式相互依赖，当执行构建任务时会按照 DAG 中的依赖关系先执行 check、assemble 任务，而执行 assemble 任务的时候又会先执行 jar 任务，按照这样的规律执行任务。

一个任务由如下几个部分组成。

- **Action**：任务执行片段，可以定义多个，当任务执行时会触发Action中的逻辑。
- **Input**：供Action消费的资源，资源可以是values、files和directories。

- **Output**：Action执行后输出的产物，产物可以是values、files和directories，可以供其他任务消费。

1. 创建任务的方式

有多种方式可以创建任务，相关接口都集中在 Project 和 TaskContainer 这两个类中。参考前面介绍的 Project 中的 API，下面的代码展示了在 build.gradle 脚本文件中创建任务的方式。

```
//1. 对应：Project#task(String name)
task("style1")

//2. 对应：Project#task(String name, Closure configureClosure)
task("style2"){
      println("create task style2")
}

//3. 对应：Project#task(Map<String,?> args, String name)
task(["description":"create task style3"],"style3"){
      println("create task style3")
}

//4. 先获取 TaskContainer，然后通过 TaskContainer#create() 方法来创建任务
getTasks().create("style4"){
      println("create task style4")
}

//5. 先获取 TaskContainer，然后通过 TaskContainer#register() 方法来创建任务
// 建议使用这种方式来创建任务，使用懒加载的方式，在使用的时候才真正调用闭包中的值
getTasks().register("style5"){
      println("create task style5")
      doLast{
          println("!23123")
      }
}

//6. 先创建一个 DefaultTask 的子类，然后将类型传给 TaskContainer#create() 方法
class MyTask1 extends DefaultTask{
      @TaskAction
      void doSomething(){
          println("create task style6")
      }
}
// 这里的 tasks == getTasks，是 Groovy 的语法特性
tasks.create("style6", MyTask1.class)
```

上文演示了创建任务的基本方式，创建完任务以后，使用如下命令来运行。

```
# 运行 style6 这个 Task
$./gradlew style6
# Windows 中使用 gradlew.bat 这个命令
```

输出结果如下。

```
$ ./gradlew style6
Starting a Gradle Daemon (subsequent builds will be faster)

> Configure project :app  # 配置阶段
create task style2
create task style3
create task style4
```

```
> Task :app:style6        # 执行阶段
create task style6

BUILD SUCCESSFUL in 11s
1 actionable task: 1 executed
```

通过输出结果可以看到，当执行 style6 任务的时候，其他任务闭包中的代码也执行了，这是因为在 Gradle 生命周期的配置阶段中会执行各项目的 build.gradle 脚本并创建和执行任务中的配置代码。另外，我们注意到 style5 任务的闭包中的代码没有被执行，因为通过 register() 创建的任务是懒加载的，只有在使用的时候才会执行，这样效率更高，也是官方推荐的做法。

2. 任务运作

任务运作是任务在执行的时候运行的代码，例如运行如下代码。

```
Task task1 = task("task1"){
      println("task1 configure")
      doLast{
          println("task1 action==>doLast")
      }
}

//运行任务 task1
$./gradlew task1
```

输出结果如下。

```
> Configure project :app
task1 configure

> Task :app:task1
task1 action==>doLast

BUILD SUCCESSFUL in 1s
1 actionable task: 1 executed
```

代码中 doLast 闭包的作用就是给任务添加动作，动作会在任务执行的时候运行。表 1-1 列出了任务中与动作相关的方法。

表1-1 任务中与动作相关的方法

返回值类型	方法	描述
Task	doFirst(Closure action)	将动作添加到动作列表的头部
Task	doFirst(String actionName,Action<? superTask> action)	将动作添加到动作列表的头部，actionName 可用于日志输出
Task	doFirst(Action<? super Task> action)	将动作添加到动作列表的头部
Task	doLast(Closure action)	将动作添加到动作列表的尾部
Task	doLast(String actionName, Action<? super Task> action)	将动作添加到动作列表的尾部
Task	doLast(Action<? super Task> action)	将动作添加到动作列表的尾部
List<Action<? super Task>>	getActions()	获取所有的动作
void	setActions(List<Action<? super Task>> actions)	设置动作集合，会删除已经添加的动作

任务内部维护了一个 List<InputChangesAwareTaskAction> actions 列表，doFirst 表示添加到列

表的头部，doLast 表示添加到列表的尾部。除了上面提供的方法外，Gradle 还提供了 @TaskAction 注解，用于添加动作，例如下面这种用法。

```
class MyTask extends DefaultTask{
      @TaskAction
      void action1(){
          println("action1")
      }

      @TaskAction
      void action2(){
          println("action2")
      }
}

Task task = tasks.create("myTask", MyTask)
task.doFirst {
      println("do first")
}
task.doLast {
      println("do last")
}
```

运行以后的输出结果如下。

```
$ ./gradlew myTask
> Task :app:myTask
do first
action1
action2
do last
```

3. 查找任务

在实际使用中，可能经常要查找一些任务，找到后对其进行配置，例如为其添加依赖、动作等。Gradle 提供了一些查找任务的方法，这些方法主要集中在 TaskContainer 类中，相关方法如表 1-2 所示。

表1-2　查找任务的方法

返回值类型	方法	描述
Task	findByPath(String path)	根据路径来查找任务，例如“:app:myTask”，意思是查找 app 项目中名字是 myTask 的任务，如果找不到则返回 null
Task	getByPath(String path)	根据路径来查找任务，如果找不到则抛出 UnknownTaskException
T	findByName(String name)	通过名字来查找任务，如果找不到则返回 null
T	getByName(String name)	通过名字来查找任务，如果找不到则抛出 UnknownTaskException
TaskProvider<T>	named(String name)	查找任务，这里返回的是 TaskProvider，可以通过 get() 方法来获取任务

下面是使用示例。

```
getTasks().named("myTask").get().doLast {
      println("last2")
}

getTasks().findByName("myTask").doLast {
```

```
        println("last3")
}

getTasks().getByPath(":app:myTask").doLast {
        println("last4")
}
```

4. 任务的依赖方式

前面讲过任务是按照 DAG 方式依赖的，当我们单击“AndroidStudio → Build → Make Project”的时候，实际上是执行 Gradle 的一系列任务。

```
Executing tasks: [:app:assembleDebug] in project /Users/wangzhuozhou/Documents/work/
others/book-asm/Chapter1_00

> Task :app:preBuild UP-TO-DATE
> Task :app:preDebugBuild UP-TO-DATE
> Task :app:compileDebugAidl NO-SOURCE
> Task :app:compileDebugRenderscript NO-SOURCE
> Task :app:generateDebugBuildConfig UP-TO-DATE
> Task :app:javaPreCompileDebug UP-TO-DATE
...
> Task :app:mergeDexDebug UP-TO-DATE
> Task :app:mergeDebugJniLibFolders UP-TO-DATE
> Task :app:mergeDebugNativeLibs UP-TO-DATE
> Task :app:stripDebugDebugSymbols NO-SOURCE
> Task :app:validateSigningDebug UP-TO-DATE
> Task :app:packageDebug UP-TO-DATE
> Task :app:assembleDebug UP-TO-DATE

BUILD SUCCESSFUL in 2s
25 actionable tasks: 25 up-to-date

Build Analyzer results available
```

可以看到上面执行了 :app:assembleDebug 任务，同时执行了很多其他的任务，这是因为任务间会相互依赖，被依赖的任务也会执行，任务间的依赖按照 DAG 规则进行。可以通过如下命令来查看任务在执行时，有哪些依赖的任务会执行。

```
$ ./gradlew taskName --dry-run
```

也可以使用一些第三方的插件来更加直观地显示任务依赖树。

任务接口中也提供了 dependsOn(Object...) 或 setDependsOn(Iterable) 方法来设置依赖。

```
task("lib1"){
      doLast {
          println("lib1 executed!")
      }
}

task("lib2"){
      doLast {
          println("lib2 executed!")
      }
}

task("lib3"){
```

```
        doLast {
            println("lib3 executed!")
        }
        dependsOn("lib2", "lib1")
}
```

上面的代码定义了 3 个任务，其中 lib3 依赖 lib2、lib1。运行 lib3，输出结果如下。

```
$ ./gradlew lib3

> Task :app:lib1
lib1 executed!

> Task :app:lib2
lib2 executed!

> Task :app:lib3
lib3 executed!
```

可以看到运行 lib3 后，其依赖的任务先运行了，最后运行 lib3。另外，还需注意到依赖的任务的运行顺序与通过 dependsOn() 方法添加的顺序不一致，所以 dependsOn() 方法并不能保证依赖的任务的运行顺序。其实还有一种非常重要的依赖类型，我们将在 1.2.10 小节对其进行介绍。

5. 任务排序

Gradle 中并没有真正的任务排序的功能，这是由 Gradle 的设计思想决定的，即用户不用关注任务链上的任务连接方式、不用关注任务链上有哪些任务，而只用声明在一个给定的任务运行之前什么任务应该被运行。Gradle 为任务排序制定了两条规则——mustRunAfter 和 shouldRunAfter，对应着任务的如下两个方法。

- Task mustRunAfter(Object... paths)；
- TaskDependency shouldRunAfter(Object... paths)。

这两个方法的使用方式如下。

（1）mustRunAfter 的使用方式如下。

```
def taskA = tasks.register('taskA') {
        doLast {
            println 'taskA'
        }
}
def taskB = tasks.register('taskB') {
        doLast {
            println 'taskB'
        }
}
taskB.configure {
        mustRunAfter taskA
}
```

执行上面的 taskA、taskB。

```
$ ./gradlew -q taskB taskA
taskA
taskB
```

（2）shouldRunAfter 的使用方式如下。

```
def taskA = tasks.register('taskA') {
```

```
        doLast {
            println 'taskA'
        }
}
def taskB = tasks.register('taskB') {
        doLast {
            println 'taskB'
        }
}
taskB.configure {
        shouldRunAfter taskA
}
```

执行上面的 taskA、taskB。

```
$ ./gradlew -q taskB taskA
taskA
taskB
```

这里要注意，通过上面这两种方式确定的任务执行顺序并不代表任务之间创建了依赖关系。例如单独执行上面的taskB并不会执行taskA。对于shouldRunAfter()，我们还需要注意一点，其可能并不起作用，前面介绍了 Gradle的任务是一个 DAG，假如破坏了这种结构，shouldRunAfter() 将不起作用。

1.2.6 生命周期回调

前面介绍了 Gradle 生命周期的 3 个阶段，对此 Gradle 提供了相应的生命周期回调方法，接下来分别介绍 Project、Gradle、Task 这 3 个类中设置的回调方法。

1. Project 类

Project 类提供的回调方法如下。

```
// 在 Project 配置前调用
void beforeEvaluate(Closure closure)
// 在 Project 配置后调用
void afterEvaluate(Closure closure)
```

这里要注意 beforeEvaluate() 需要在父 Project 类中调用，如果在当前的 build.gradle 中调用，当执行到 beforeEvaluate() 方法的时候实际上评估已经结束了，所以不会回调。除了在父 Project 类中配置外，还可以在 Gradle 中配置，后面会介绍。

2. Gradle 类

Gradle 类中提供的回调方法非常多，与 Project 类也有一些相似的回调方法。

```
// 需要在父类 Project 或者 settings.gradle 中设置才会生效
void beforeProject(Closure closure)

// 在 Project 配置后调用
void afterProject(Closure closure)

// 构建开始前调用
void buildStarted(Closure closure)

// 构建结束后调用
void buildFinished(Closure closure)
```

```
// 所有 Project 配置完成后调用
void projectsEvaluated(Closure closure)

// 当 settings.gradle 中引入的所有 Project 都被创建好后调用，只在该文件设置才会生效
void projectsLoaded(Closure closure)

//settings.gradle 配置完后调用，只对 settings.gradle 设置生效
void settingsEvaluated(Closure closure)
```

除了上面这些提供的方法外，Gradle 还提供了一些 Listener，通过添加支持的 Listener 也能取代上面介绍的方法。总之，Gradle 既提供了一些单独的闭包回调方法，也提供了一些 Listener 来实现相同的效果。

```
/**
   * Adds the given listener to this build. The listener may implement any of the
given listener interface
   *
   * <ul>
   * <li>{@link org.gradle.BuildListener}
   * <li>{@link org.gradle.api.execution.TaskExecutionGraphListener}
   * <li>{@link org.gradle.api.ProjectEvaluationListener}
   * <li>{@link org.gradle.api.execution.TaskExecutionListener}
   * <li>{@link org.gradle.api.execution.TaskActionListener}
   * <li>{@link org.gradle.api.logging.StandardOutputListener}
   * <li>{@link org.gradle.api.tasks.testing.TestListener}
   * <li>{@link org.gradle.api.tasks.testing.TestOutputListener}
   * <li>{@link org.gradle.api.artifacts.DependencyResolutionListener}
   * </ul>
   *
   * @param listener The listener to add. Does nothing if this listener has already
been added.
   */
void addListener(Object listener);
```

3. Task 类

Gradle 配置阶段结束后，任务会被构建成一个 DAG，这决定了任务的执行顺序，同样，Task 类也提供了对应的回调方法:

```
// 任务执行前调用
void afterTask(Closure closure)
// 任务执行后调用
void beforeTask(Closure closure)
// 任务准备好后调用
void whenReady(Closure closure)
```

下面我们针对上面介绍的这些回调方法编写一个例子来说明，其中回调方法主要集中在 settings.gradle 中。读者也可以在 app/build.gradle 中定义一个任务，然后观察任务的执行顺序。settings.gradle 中的代码如下。

```
//include ':lib'
include ':app'
rootProject.name = "Chapter1_00"

//called after project evaluated
getGradle().afterProject {
```

```
    println("setting: after project===${it}")
}

//called before project evaluated
getGradle().beforeProject {
    println("setting: before project===${it}")
}

getGradle().addBuildListener(new BuildListener() {
    @Override
    void buildStarted(Gradle gradle) {
        println("setting: buildStarted===${gradle}")
    }

    @Override
    void settingsEvaluated(Settings settings) {
        println("setting: settingsEvaluated===${settings}")
    }

    @Override
    void projectsLoaded(Gradle gradle) {
        println("setting: projectsLoaded===${gradle}")
    }

    @Override
    void projectsEvaluated(Gradle gradle) {
        println("setting: projectsEvaluated===${gradle}")
    }

    @Override
    void buildFinished(BuildResult result) {
        println("setting: buildFinished===${result}")
    }
})

getGradle().addListener(new TaskExecutionGraphListener() {
    //TaskGraph 构成以后
    @Override
    void graphPopulated(TaskExecutionGraph graph) {
        println("setting: graphPopulated===${graph}")
    }
})

getGradle().addListener(new ProjectEvaluationListener(){
    @Override
    void beforeEvaluate(Project project) {
        println("setting: beforeEvaluate===${project}")
    }

    @Override
    void afterEvaluate(Project project, ProjectState state) {
        println("setting: afterEvaluate===${project}")
    }
})

getGradle().addListener(new TaskExecutionListener(){
```

```
        @Override
        void beforeExecute(Task task) {
            println("setting: task beforeExecute")
        }

        @Override
        void afterExecute(Task task, TaskState state) {
            println("setting: task afterExecute")
        }
})

getGradle().addListener(new TaskActionListener(){

        @Override
        void beforeActions(Task task) {
            println("setting: task beforeActions")
        }

        @Override
        void afterActions(Task task) {
            println("setting: task afterActions")
        }
})

getGradle().beforeSettings {
        println("setting: beforeSettings")
}

getGradle().buildStarted {
        println("setting: buildStarted2")
}
getGradle().buildFinished {
        println("setting: buildFinished2")
}
getGradle().projectsLoaded {
        println("setting: projectsLoaded22")
}

getGradle().getTaskGraph().addTaskExecutionGraphListener(new TaskExecutionGraphListener() {
        @Override
        void graphPopulated(TaskExecutionGraph graph) {

        }
})

getGradle().getTaskGraph().addTaskExecutionListener(new TaskExecutionListener() {
        @Override
        void beforeExecute(Task task) {

        }

        @Override
        void afterExecute(Task task, TaskState state) {

        }
})
```

```
getGradle().getTaskGraph().whenReady {
      println("setting : whenReady")
}

getGradle().getTaskGraph().beforeTask {
      println("setting : beforeTask2")
}
```

最后我们用一张图将 Gradle 生命周期的不同阶段对应的回调方法串联起来，如图 1-2 所示。

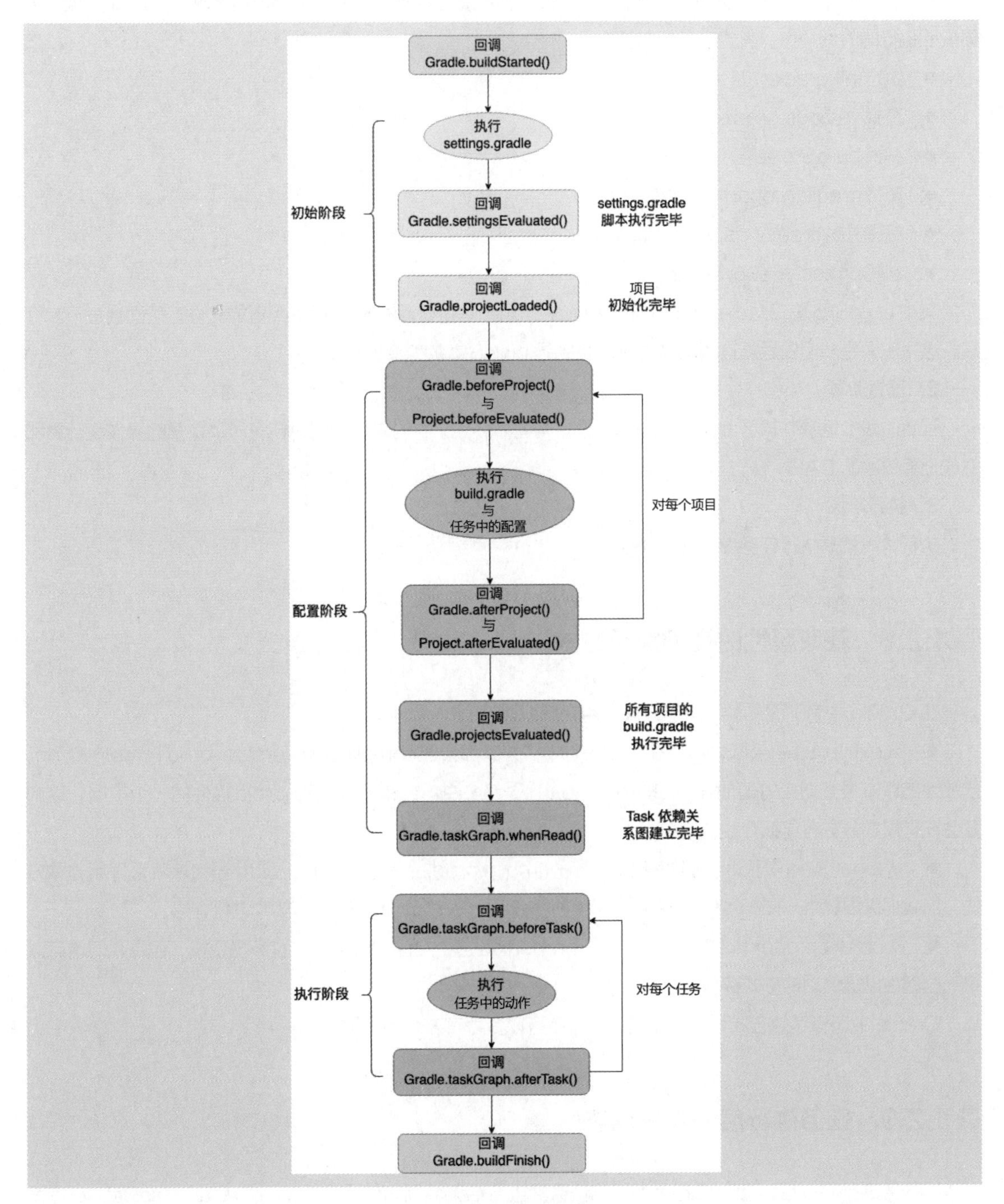

图 1-2　Gradle 生命周期的回调方法

1.2.7 Gradle执行流程

前文介绍了 Gradle 的生命周期回调，本小节将补充生命周期各阶段的一些细节。

1. 初始阶段

前面我们介绍初始阶段会执行 setting.gradle，并根据 setting.gradle 中的内容构建 Project 对象。如果按照执行的时间顺序，其内容包括如下。

- 执行init.gradle；
- 查找settings.gradle；
- 编译buildSrc目录；
- 解析gradle.properties内容；
- 编译并执行settings.gradle；
- 创建project和subproject。

其中 buildSrc 是 Gradle Plugin 的一种写法，gradle.properties 是环境配置的文件，后面会进一步介绍，这里主要是要明白运行这两项的时机，特别是 buildSrc 的编译时机。

2. 配置阶段

配置阶段会编译并执行 build.gradle 文件，构建 Project 中的内容，同时还有一项重要的操作是计算和生成任务依赖的 DAG。

3. 执行阶段

执行阶段就是执行任务中定义的动作。

1.2.8 获取属性的几种常见方式

在 Gradle 中有下面几种常见的方式可用来获取属性。

- 从extra属性中获取，例如在Android中，大家都会把compileSdkVersion、buildToolsVersion、依赖库的版本号等对应的值放在rootProject.ext{}语句块中，它实际上是Android插件的一个扩展，这部分会在扩展部分进行详细介绍。
- 从extension中获取，例如在app工程的build.gradle中定义的android{}语句块就是采用这种方式。当我们需要提供一些配置信息的时候可以使用extension，这部分在扩展部分进行介绍。
- 从gradle.properties中获取，gradle.properties这个文件一般用来设置环境变量，但是我们也可以将一些属性以键值对的形式存放在此文件中。

Gradle 中还有其他方式可以获取属性，读者可以参考官方介绍了解。

1.2.9 任务执行后的几种状态

当任务执行完以后，会在控制台上显示任务的执行结果，例如使用 ./gradlew assembleDebug 来编译 App debug 包的时候，任务执行结果如图 1-3 所示。

图 1-3 Gradlew assembleDebug 任务执行结果

这里的 UP-TO-DATE、NO-SOURCE 表示任务执行后的状态结果。Gradle 中的任务执行结果状态如表 1-3 所示。

表1-3 Gradle中的任务执行结果状态

任务执行结果状态	描述	前提
no label 或者 EXECUTED	任务执行了动作	● 任务有动作并且执行了 ● 任务没有动作但有一些依赖，这些依赖都被执行了
UP-TO-DATE	任务的输出没有变化	● 任务有输入和输出，这些输入和输出没有变化 ● 任务有动作，但是任务告诉Gradle，其输出没有变化 ● 任务没有动作但是有一些依赖，这些依赖都是UP-TO-DATE、SKIPPED 或者 FROM-CACHE ● 任务没有动作也没有依赖
FROM-CACHE	任务的输出可以从之前执行的缓存中找到	● 任务在构建缓存中可以找到相应的输出
SKIPPED	任务有动作，但是没有执行	● 任务在命令行中被声明为跳过不执行 ● 任务的 onlyIf 判断返回 false
NO-SOURCE	任务不需要去执行动作	● 任务有输入和输出，但是没有来源（Source）

1.2.10 增量构建

增量构建是 Gradle 中很重要的一个知识点，试想一下在编译的时候，当你输入的文件没有发生变化，意味着输出也不会有变化，那就不需要再执行编译，直接将上一次的编译结果返回即可；而当只修改了部分文件时，也只需要单独编译发生变化的文件即可，这就是增量编译。

前面介绍任务由输入、输出和动作组成，任务的输入和输出作为增量构建的一部分，Gradle 会检查上一次构建依赖的输入和输出是否有变化，如果没有变化，Gradle 认为该任务是最新的，就会跳过任务的执行，任务的结果状态是 UP-TO-DATE。需要注意的是，在 Gradle 中至少要有一个任务输出，否则增量构建将无法工作。表 1-4 所示为 Gradle 提供的与增量相关的注解。

表1-4　Gradle提供的与增量相关的注解

注解	类型	描述
@Input	Serializable	可序列化的输入对象
@InputFile	File	单个输入文件（不包括文件夹）
@InputDirectory	File	单个输入文件夹（不包括文件）
@InputFiles	Iterable<File>	输入文件或文件夹的集合
@Classpath	Iterable<File>	Java 路径的输入文件和文件夹的集合
@CompileClasspath	Iterable<File>	编译 Java 路径的输入文件和文件夹的集合
@OutputFile	File	单一输出文件
@OutputDirectory	File	单一输出文件夹
@OutputFiles	Map<String, File> 或 Iterable<File>	输出文件的集合或者映射
@OutputDirectories	Map<String, File> 或 Iterable<File>	输出文件夹的集合或者映射
@Destroys	File 或 Iterable<File>	指定此任务销毁的一个文件或文件集合。请注意，任务可以定义输入 / 输出或可销毁对象，但不能同时定义两者
@LocalState	File 或 Iterable<File>	本地任务状态，当任务从缓存恢复的时候，这些文件将会被移除
@Nested	任意自定义类型	嵌套属性
@Console	任意类型	辅助属性，指出修饰属性不为输入或者输出属性
@Internal	任意类型	内部使用属性，指出修饰属性不为输入或者输出属性
@ReplacedBy	任意类型	指示该属性已被另一个替换，作为输入或输出被忽略
@SkipWhenEmpty	File	和 @InputFiles / @InputDirectory 配合使用，告诉 Gradle 如果相应的文件或目录为空，以及使用此注释声明的所有其他输入文件为空，则跳过任务，任务的状态是 NO-SOURCE
@Incremental	Provider<FileSystemLocation> 或 FileCollection	和 @InputFiles / @InputDirectory 配合使用，告诉 Gradle 跟踪对文件属性的更改。可以通过 InputChanges.getFileChanges() 查询更改
@Optional	任意类型	可选属性
@PathSensitive	File	文件属性的类型

让我们看看如何使用表 1-4 中的注解来创建增量任务。

```
class IncrementalTask extends DefaultTask {

    @OutputFile // 用在 Property() 方法上
    File outputDir= getProject().file(".proguard-rules.pro")

    @OutputFile // 用在 get() 方法上
    File getOutputFile(){
        return getProject().file(".proguard-rules.pro")
    }

    @TaskAction
    void doAction(){
```

```
            println("do something")
        }
    }

    getTasks().create("testIncremental", IncrementalTask)
```

注意，Gradle 提供的与增量相关的注解要么用在 Property() 方法上，要么用在 get() 方法上。运行 testIncremental 任务两次，当第二次运行的时候会看到如下输出。

```
$./gradlew testIncremental

BUILD SUCCESSFUL in 840ms
1 actionable task: 1 up-to-date
```

可以看到 up-to-date 关键字，表示输入和输出没有变化，没有执行动作中的内容直接返回了结果，这就是增量构建的一个简单应用。

前面介绍任务依赖的知识时，还提到一种依赖，就是 Gradle 可以根据任务的输入和输出来推断依赖关系，例如下面的这段代码。

```
class MyTask1 extends DefaultTask {
      @OutputFiles
      FileCollection outputDir

      @TaskAction
      void processTemplates() {

      }
}

class MyTask2 extends DefaultTask {
      @InputFiles
      FileCollection inputFileTmp

      @TaskAction
      void t() {
          println("task2")
      }
}

def task1 =  tasks.create('task1', MyTask1) {
      println("task1")
      doLast{
          println("task1 doLast")
      }
      outputDir =
          files(layout.buildDirectory.getAsFile().get().getAbsolutePath()+"/log.txt")
}

tasks.create("task2", MyTask2){
      inputFileTmp = files(task1.outputs)
}
```

上面的代码是通过输入和输出的依赖关系来确定两个任务之间的依赖，当执行任务 task2 的时候，task1 也会一起执行，这其实就是第 2 章介绍的 Android Gradle Plugin Transform 对应的 TransformTask 的关键原理。

1.3 插件类型

前文介绍了什么是 Gradle 插件，以及有关 Gradle 的一些基础知识，本节将正式介绍 Gradle 插件的用法，首先来认识 Plugin<T> 接口。

```
package org.gradle.api;

/**
   * <p>A <code>Plugin</code> represents an extension to Gradle. A plugin applies
some configuration to a target object.
   * Usually, this target object is a {@link org.gradle.api.Project}, but plugins
can be applied to any type of
   * objects.</p>
   *
   * @param <T> The type of object which this plugin can configure.
   */
public interface Plugin<T> {
      /**
       * Apply this plugin to the given target object.
       *
       * @param target The target object
       */
      void apply(T target);
}
```

可以看到插件的这个接口只有一个 apply() 方法，通常这里的泛型 T 基本上都是 org.gradle.api.Project 类型。本书中所指的插件均指实现了此接口的插件，需要注意的是 apply() 方法中的代码会在 Gradle 配置阶段执行。插件的使用也比较简单，例如 apply plugin: 'com.android.application'，这里实际上调用的是 Project.apply(Map<String, ?> options) 方法。

根据插件代码所在位置的不同，Gradle 提供了 3 种创建插件的方式。

- 直接在构建脚本（build.gradle）中定义插件，我们称之为脚本插件；
- 创建一个名为buildSrc的模块（module），并在其中定义插件，我们称之为buildSrc插件；
- 在一个单独的项目中定义插件，我们称之为单独项目插件。

插件的实现可以使用任何基于 Java 虚拟机（Java Virtual Machine，JVM）的编程语言，例如 Java、Kotlin、Groovy。通常实现插件的时候推荐使用 Java 或 Kotlin，相对于 Groovy，这两种语言是静态类型的，在开发的过程中借助集成开发环境（Intergrated Development Environment，IDE）的语法检测会方便很多。

接下来依次介绍这 3 种类型的插件。

1.3.1 脚本插件

脚本插件是指在构建脚本中定义的插件。这么做的好处是插件会被自动编译并且添加到构建的 classpath 中，用户不需要做其他操作；坏处是只能在当前的构建脚本中使用，其他的构建脚本无法使用。

下面创建一个 Android 项目并在 app/build.gradle 中实现脚本插件。

第一步：创建一个新的 Android 项目，名称为 Chapter1_01。

第二步：在 app/build.gradle 文件中创建插件的实现类。

```
class MyScriptPlugin implements Plugin<Project> {

      @Override
      void apply(Project target) {
          println("My First Plugin")
          target.task("getBuildDir") {
              println("get build dir at configure phase") // 配置阶段输出
              doLast { // 为任务添加动作
                  println("build dir: ${buildDir}")
              }
          }
      }
}
```

上述代码在 apply() 方法中输出了一条日志，并创建了一个名为 getBuildDir 的任务，该任务输出编译目录。

第三步：引用插件。

在 app/build.gradle 中引用插件。

```
// 使用我们定义的插件
apply plugin: MyScriptPlugin
```

第四步：执行 getBuildDir 任务进行测试。

```
$ ./gradlew getBuildDir

> Configure project :app
My First Plugin
get build dir at configure phase

> Task :app:getBuildDir
build dir: /Users/wangzhuozhou/Documents/work/others/book-asm/Chapter1_01/app/build
```

可以看到在配置阶段输出了“My First Plugin”并且输出了 build 目录。

完整的 app/build.gradle 文件内容如下。

```
plugins{
      id 'com.android.application'
}
apply plugin: MyScriptPlugin // 使用插件

android {
      ...
}

dependencies {
      ...
}

class MyScriptPlugin implements Plugin<Project> {

      @Override
      void apply(Project target) {
```

```
            println("My First Plugin")
            target.task("getBuildDir") {
                println("get build dir at configure phase") // 配置阶段输出
                doLast { // 为任务添加动作
                    println("build dir: ${target.buildDir}")
                }
            }
        }
}
```

1.3.2 buildSrc插件

buildSrc 插件指的是在项目中定义一个名为 buildSrc 的模块，在模块中实现插件逻辑。这种插件的优点是项目中的所有模块都可以使用、Gradle 负责编译并将其加入构建环境中，缺点是不方便单独维护和发布。下面介绍如何将脚本插件转换成 buildSrc 插件。

第一步：创建一个新的 Android 项目，名称为 Chapter1_02。

第二步：创建一个新的模块，类型选择“Java Library”，名称为 buildSrc。在 Android Studio 中的完整操作路径是: File → New → New Module → Java Library → Next → Finish。在创建完 buildSrc Library 以后，Gradle 会重新编译，Android Studio 会报图 1-4 所示的错误。

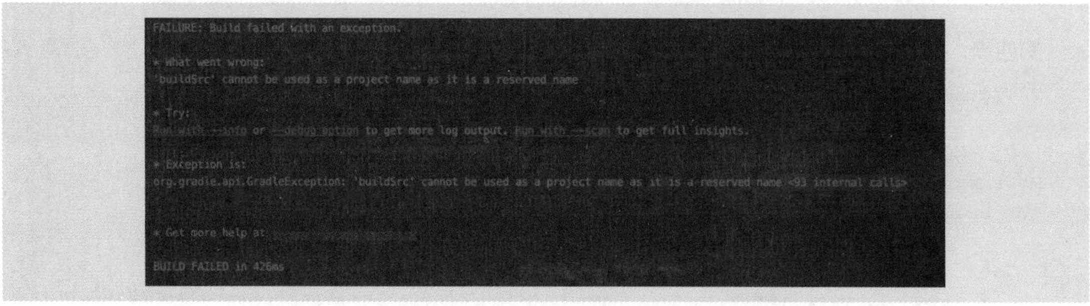

图 1-4 编译报错

这是因为 buildSrc 是一个保留名字，不能用于项目名，只需将 settings.gradle 文件中的 include ':buildSrc' 配置删除即可。

第三步：创建插件类，首先在 buildSrc 模块中创建 MyBuildSrcPlugin.java 类来实现 Plugin 接口，项目结构和示例代码如图 1-5 所示。

注意，这里使用的语言是 Java，也可以使用 Groovy 或 Kotlin 来实现。

第四步：在 app/build.gradle 文件中使用插件。

```
apply plugin: cn.sensorsdata.asmbook.plugin.MyBuildScrPlugin
```

第五步：执行 getBuildDir 任务进行测试，可以看到控制台中输出了插件的内容。

```
$ ./gradlew getBuildDir

> Configure project :app
My Second Plugin

> Task :app:getBuildDir
build dir: /Users/wangzhuozhou/Documents/work/others/book-asm/Chapter1_02/app/build
```

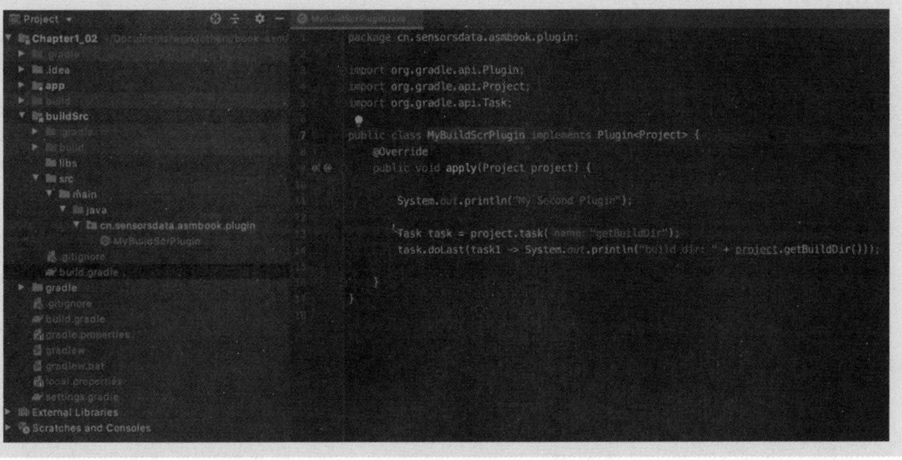

图 1-5　项目结构和示例代码

本小节介绍了 buildSrc 插件，使用的时候只需要在模块中进行引入即可，这种插件通常适合项目内部使用。

1.3.3　单独项目插件

相对于脚本插件和 buildSrc 插件，单独项目插件要复杂一些，可以将其打成插件包并发布出去，发布后更方便他人使用。接下来介绍如何创建单独项目插件。

第一步：创建一个新的 Android 项目，名称为 Chapter1_03。

第二步：创建一个新的模块，类型选择“Java Library”，名称为 myPlugin。在 Android Studio 中的完整操作路径是：File → New → New Module → Java Library → Next → Finish。

第三步：修改 myPlugin/build.gradle 文件，将依赖的 java-library 插件改成依赖 java 插件，并依赖 Gradle API，修改后的内容如下。

```
apply plugin:'java' //依赖 java 插件
apply plugin:'maven-publish' //依赖 maven-publish 插件

java {
      sourceCompatibility = JavaVersion.VERSION_1_8
      targetCompatibility = JavaVersion.VERSION_1_8
}

dependencies {
      implementation gradleApi() //依赖 Gradle API，才能使用 Plugin 这个接口
}

//组织或者公司名称
group="cn.sensorsdata.asmbook.plugin"

//版本号
version='1.0.0'

//模块名称
archivesBaseName='myPlugin'
```

```
// 打包 Maven
// 使用命令 ./gradlew publish 或者 publish{PubName}PublicationTo{RepoName}Repository 形式
        的命令
publishing {
    publications {
        myLibrary(MavenPublication) {
            from components.java
        }
    }

    repositories {
        maven {
          url '../maven-repo'
        }
    }
}
```

publishing{ } 语句块用于配置 Maven 仓库的版本信息，此处的配置是将插件发布到本地。Maven 仓库发布还需要指定如下信息。

- group：指groupId，一般为组织或者公司名称。
- version：库的版本。
- archivesBaseName：指artifactId，是项目名或者模块名。

第四步： 编写插件。

在 myPlugin 模块中创建插件实现类。

```
public class MyStandalonePlugin implements Plugin<Project> {

    @Override
    public void apply(Project project) {
        System.out.println("My Third Plugin");

        Task task = project.task("getBuildDir");
           task.doLast(task1 -> System.out.println("build dir: " + project.
getBuildDir()));
    }
}
```

第五步： 编译插件。

单击图 1-6 所示的 publish 任务即可编译插件并发布到本地 Maven 仓库。

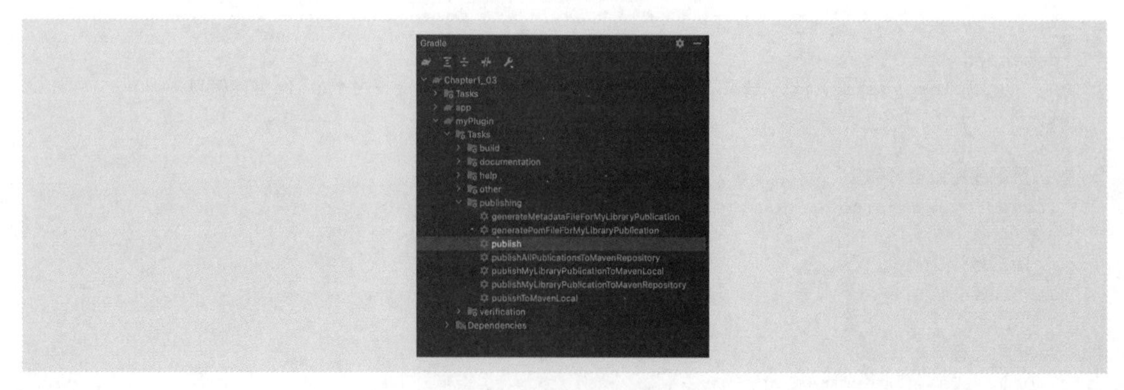

图 1-6　publish 任务

也可以在终端中执行 publish 任务。

```
$ ./gradlew publish
```

该任务的作用是打包 Maven 并将其发布到本地。此时，在项目的根目录下会多一个 maven-repo 目录，目录中的文件即构建的插件，如图 1-7 所示。

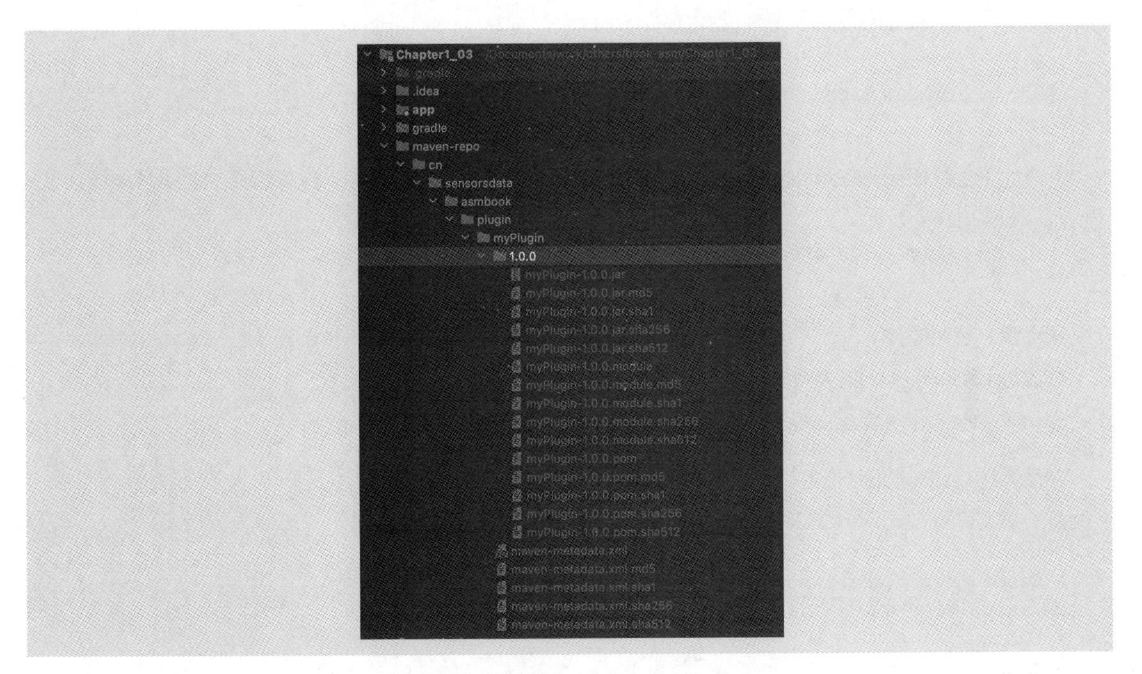

图 1-7 maven-repo 目录

第六步：添加对插件的依赖。修改项目根目录下的 rootProject/build.gradle 文件，添加对插件的依赖。

```
// Top-level build file where you can add configuration options common to all sub-
projects/modules.
buildscript {
      repositories {
          google()
          mavenCentral()
          maven {
              url(uri("maven-repo"))
          }
      }
      dependencies {
          classpath "com.android.tools.build:gradle:4.1.2"
          classpath 'cn.sensorsdata.asmbook.plugin:myPlugin:1.0.0' //依赖插件
          // NOTE: Do not place your application dependencies here; they belong
          // in the individual module build.gradle files
      }
}

allprojects {
      repositories {
          google()
          mavenCentral()
      }
}
```

```
task clean(type: Delete) {
      delete rootProject.buildDir
}
```

下面这段代码是声明在哪里寻找插件，因为是在 rootProject/build.gradle 文件中声明的，所以这里的意思是在 build.gradle 同级目录中寻找 maven-repo。

```
maven{
      url(uri('maven-repo'))
}
```

另外，下面这段代码表示添加插件的依赖（即将库添加到 classpath 中，这样编译的时候才能找到）。

```
dependencies {
      classpath 'cn.sensorsdata.asmbook.plugin:myPlugin:1.0.0'
}
```

第七步：使用插件。

修改 app/build.gradle 文件，添加引用插件的代码。

```
apply plugin: cn.sensorsdata.asmbook.myplugin.MyStandalonePlugin
```

然后运行 getBuildDir 任务。

```
$ ./gradlew getBuildDir

> Configure project :app
My Third Plugin

> Task :app:getBuildDir
build dir: /Users/wangzhuozhou/Documents/work/others/book-asm/Chapter1_03/app/build
```

可以看到插件正常运行了。你可能注意到引用插件时需要填写完整的包名和类名，字符串比较长，难以书写和记忆。我们也可以给插件起一个精简且有意义的名字，比如“testplugin”，下面介绍如何操作。

1.3.4 单独项目插件优化

首先在 myPlugin/src/main 目录下创建 resources/META-INF.gradle-plugins 目录，再在 gradle-plugins 目录下新建一个属性配置文件 testplugin.properties，这里的 testplugin 就是最终的插件名，最后在 testplugin.properties 文件中添加如下内容。

```
implementation-class=cn.sensorsdata.asmbook.myplugin.MyStandalonePlugin
```

效果如图 1-8 所示。

上述内容声明了插件的实现类，重新编译插件，这样就可以通过如下方式依赖插件。

```
apply plugin: 'testplugin'
```

再次运行 getBuildDir 任务，可以看到输出了预期的结果。

```
$ ./gradlew getBuildDir

> Configure project :app
```

```
My Third Plugin

> Task :app:getBuildDir
build dir: /Users/wangzhuozhou/Documents/work/others/book-asm/Chapter1_03/app/build
```

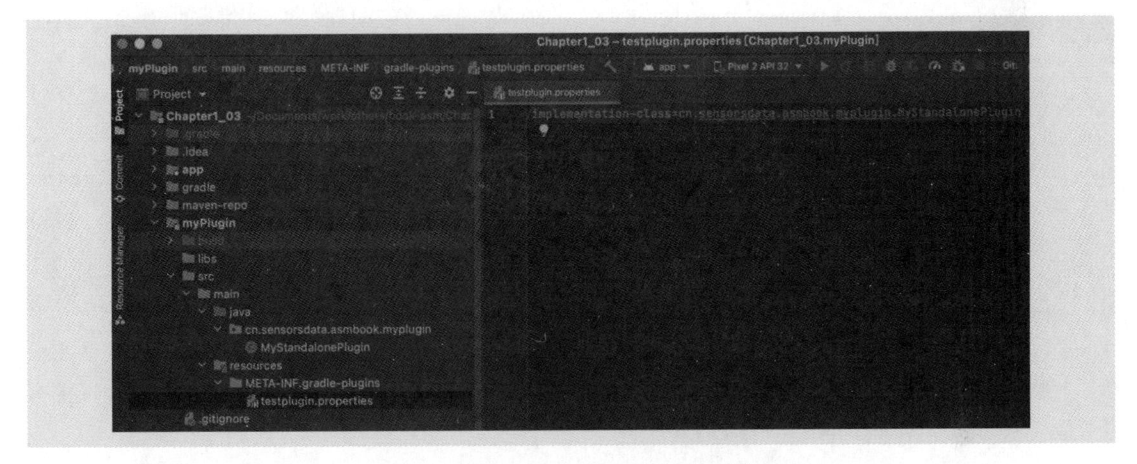

图 1-8　定义插件 META-INF 信息后的效果

以上就是一个单独项目插件的标准做法。可以看到定义一个单独项目插件的操作步骤很多，那么有没有方式可以简化操作步骤呢？答案是有，Gradle 官方提供了 java-gradle-plugin 插件，该插件具有如下作用。

- 添加java插件，这样就不需要添加apply plugin: 'java'这段代码；
- 自动依赖Gradle API，这样就不需要添加implementation gradleApi() 这段代码；
- 自动生成resources/META-INF.gradle-plugins目录和properties文件，这样就不需要手动添加；
- 打包时自动生成插件标识产物（plugin marker atifacts）,1.3.5小节会详细介绍；
- 配合maven-publish插件生成对应的Maven包和Gradle Plugin Portal包。

接下来继续对插件进行优化。复制一份 Chapter1_03 项目，将其副本的名称改为 Chapter1_03_better。然后修改 myPlugin/build.gradle 文件，添加 java-gradle-plugin 插件。修改后的内容如下。

```
plugins{
    id 'java-gradle-plugin'
    id 'maven-publish'
}

java {
    sourceCompatibility = JavaVersion.VERSION_1_8
    targetCompatibility = JavaVersion.VERSION_1_8
}

// 不再需要依赖 Gradle API
dependencies {
    implementation gradleApi()
}

// 组织或者公司名称
group="cn.sensorsdata.asmbook.plugin"

// 版本号
```

```
version='1.0.0'

//模块名称
archivesBaseName='myPlugin'

//Gradle 插件配置
gradlePlugin {
      plugins {
          myPluginB {
              id = 'mytestPlugin'
                    implementationClass = 'cn.sensorsdata.asmbook.myplugin.
MyStandalonePlugin'
          }
      }
}

//打包
publishing {
      //可以不需要
      publications {
          myLibrary(MavenPublication) {
              from components.java
          }
      }

      repositories {
          maven {
              url '../maven-repo'
          }
      }
}
```

引入了 java-gradle-plugin 后就可以删除 Gradle API 的依赖与 resources/META-INF 目录中的配置。在上述配置中新添加了 gradlePlugin{ } 语句块，它的作用是配置插件发布信息，其中的 id 项定义了插件的唯一标识，implementationClass 项定义了插件的实现类。这一段配置相当于创建了 resources/META-INF.gradle-plugins/mytestPlugin.property 文件，文件内容是 implementation-class=cn.sensorsdata.asmbook.myplugin.MyStandalonePlugin。

运行 ./gradlew publish 命令来打包，可以看到生成了 maven-repo，其内容如图 1-9 所示。

可以看到 maven-repo 目录下有两个子目录，其中 cn 这个子目录，与之前发布的包结构是一致的，子目录 mytestPlugin 与 cn 不一样，它的结构符合插件标识产物的结构，这样我们就可以使用 Gradle 提供的 plugins{} DSL 来声明插件，例如：

```
plugins{
      id 'com.android.application'
      id 'mytestPlugin' version '1.0.0' //试用此种方式来使用插件
}

android {
      ...
}
```

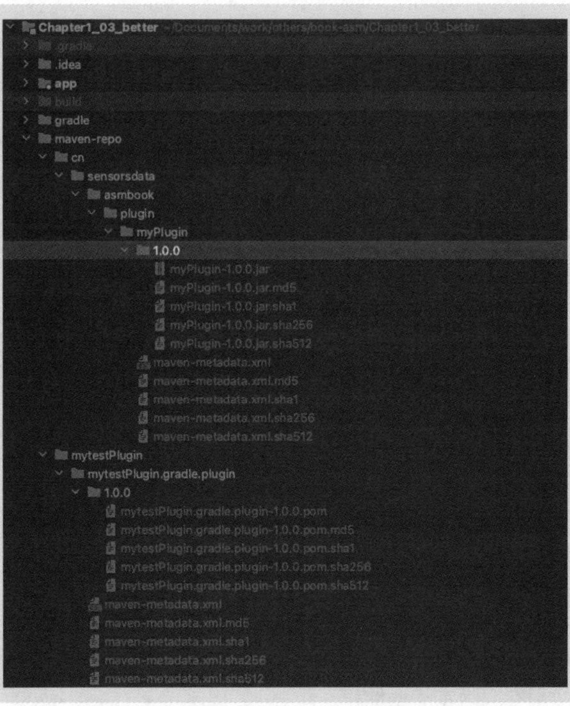

图 1-9　maven-repo 的内容

关于 plugins{} 会在 1.3.5 小节详细介绍。根据 myPlugin/build.gradle 文件中的 gradlePlugin{ } 语句块的作用，新生成插件的 id 是 mytestplugin，这一点可以通过反编译 myPlugin-1.0.0.jar 看到，图 1-10 所示为反编译的结果。

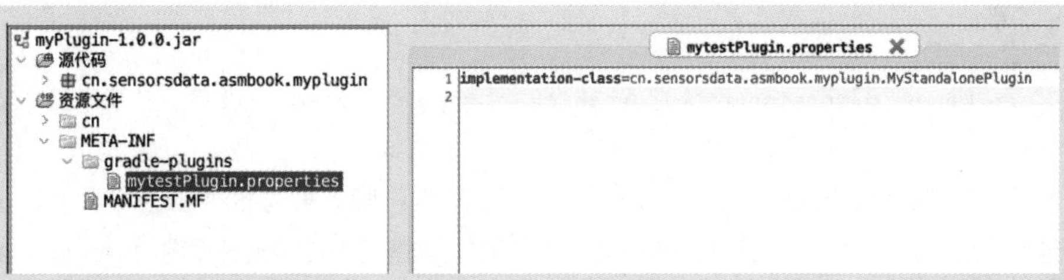

图 1-10　反编译结果

对此需要调整 app/build.gradle，修改后的结果如下。

```
apply plugin: 'com.android.application'
apply plugin: 'mytestPlugin'
```

当然也可以使用下面的方式来声明插件。

```
plugins{
      id 'com.android.application'
      id 'mytestPlugin'
}
```

以上就是优化后的插件，接下来介绍 plugins{} DSL 的用法。

1.3.5 插件使用方式

前文简单介绍了插件的使用方式，本小节对此做更进一步的介绍。Gradle 在使用插件时需要执行两步操作。

- 识别插件；
- 应用插件。

其中，识别插件的意思是从插件 JAR 包中找到对应的版本并将其添加到构建环境中；应用插件意味着具体执行 Plugin.apply(T) 方法。为能识别插件，需要在 rootProject/build.gradle 的 buildscript 块中引入插件，再在 module/build.gradle 中应用插件。Gradle 为简化识别插件和应用插件的步骤，引入了 plugins{} DSL 语句块，将两个步骤放在一个步骤中，例如使用下面的代码引入插件。

```
plugins {
      id 'com.jfrog.bintray' version '1.8.5'
}
```

相比于“传统”的 apply() 方法，plugins{} DSL 不仅简化了使用方式，还有如下优点。

（1）优化了插件的加载速度；

（2）不同的项目可以使用不同版本的插件；

（3）良好的 IDE 支持。

例如对于同一个插件，不同的模块可以使用不同的版本插件，这个对于传统的 apply() 方法是无法实现的。不过使用 plugins{} 也有一些限制，plugins{} 语句块必须在 build.gradle 文件的顶部声明，而 apply() 可以在任何地方使用；再有 plugins{} 的语句块有严格的语法限制，如果不满足语法就会产生编译器报错，其语法格式如下。

```
plugins {
      id«plugin id»                                              (1)
      id«plugin id»version«plugin version» [apply«false»]        (2)
}
```

其中（1）表示应用核心插件（例如 java 插件）或者对构建脚本来说插件已经可用；（2）表示应用插件并指定某个版本，apply«false» 表示引入插件但不使用。假设项目中有一个名为 module-a 的模块，rootProject/build.gradle 文件中有如下内容。

```
plugins {
      id 'com.example.hello' version '1.0.0' apply false
}
```

上述代码表示在 rootProject/build.gradle 中引入 com.example.hello 插件，根据加载的顺序，此时插件对 module-a 来说已经可用。module-a 中可按照如下方式引入插件，即不需要指定版本号。

```
plugins {
      id 'com.example.hello'
}
```

除了上述两个限制外，plugins{} 在使用上还有一个限制，前面提到 plugins{} 将识别插件和应用插件集中在一个步骤中，这是有前提的，即它需要满足 Gradle Plugin Portal（Gradle 官方插件平台）的格式标准才能如此简单地使用，如果插件托管在 Maven 上，仍然需要在 rootProject/build.gradle 的 buildscript

块中声明插件信息。

Gradle Plugin Portal 要求的插件信息称为插件标识产物，它的格式如下。

```
{plugin.id}:{plugin.id}.gradle.plugin:{plugin.version}
```

Gradle 加载 plugins{} 语句块中的插件时会按照上述格式去定位插件。另外，如果插件已经上传到 Gradle Plugin Portal，就可以在 plugins{} 中直接声明插件；如果插件在本地，还需要在 settings.gradle 中声明插件的位置，这与传统的 apply() 需要在 rootProject/build.gradle 中声明 repositories 是不同的。下面使用具体的例子来演示带有插件标识产物的插件在本地的用法。

复制项目 Chapter1_03_better，并将其副本的项目名修改为 Chapter1_03_best。使用 java-gradle-plugin 和 maven-publish 插件，java-gradle-plugin 插件会在使用 maven-publish 发布版本时生成符合插件标识产物的插件，运行 ./gradlew publish 命令，其输出结果如图 1-11 所示。

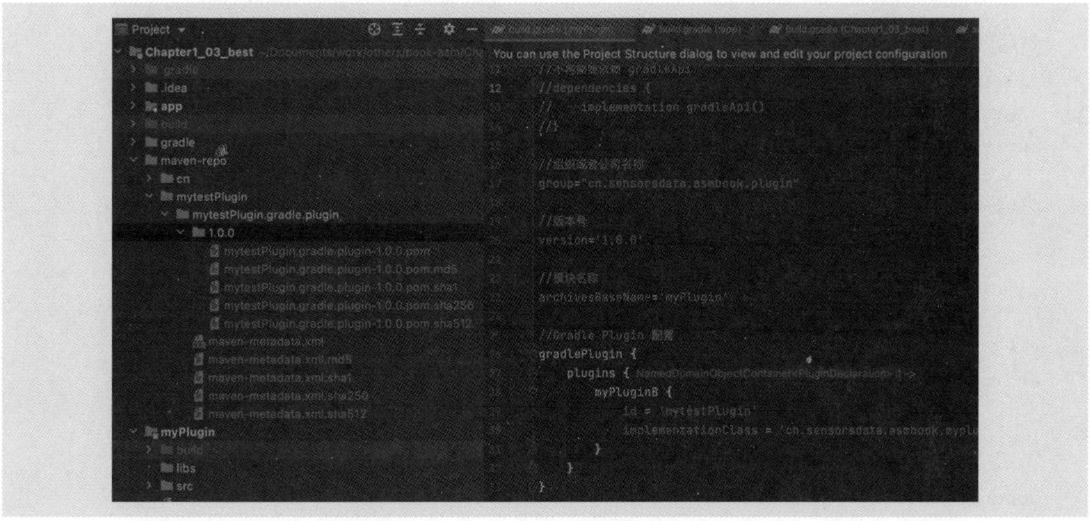

图 1-11　./gradlew publish 命令的输出结果

可以看到 maven-repo 中 mytestPlugin 的目录结构符合插件标识产物的格式。接下来就是引入插件，首先删除 rootProject/build.gradle 中传统的插件依赖方式，修改后的结果如下。

```
buildscript {
      repositories {
          google()
          mavenCentral()
          // 删除传统的插件依赖方式
          maven {
              url(uri("maven-repo"))
          }
      }
      dependencies {
          classpath "com.android.tools.build:gradle:4.1.2"
          classpath 'cn.sensorsdata.asmbook.plugin:myPlugin:1.0.0'// 删除传统的插件依赖方式
          // NOTE: Do not place your application dependencies here; they belong
          // in the individual module build.gradle files
      }
}
...
```

接着在 settings.gradle 文件中使用 pluginManagement{} 语句块声明插件仓库，修改后的结果如下。

```
// 声明依赖关系
pluginManagement {
      plugins {
      }
      repositories {
          maven {
              url(uri("maven-repo"))
          }
      }
}

include ':myPlugin'
include ':app'
rootProject.name = "Chapter1_03_best"
```

然后就可以在 app/build.gradle 中使用插件，代码如下。

```
// 传统方式
apply plugin: 'com.android.application'
apply plugin: 'mytestPlugin'

plugins{
      id 'com.android.application'
      id 'mytestPlugin' version '1.0.0' // 使用插件
}


android {
     ...
}

dependencies {
      ...
}
```

本小节主要介绍了 plugins{} DSL 的使用方式，关于如何将插件放到 Gradle Plugin Portal 平台上，会在 1.6.1 小节详细介绍。

1.3.6 小结

本小节我们从简单到复杂依次介绍了脚本插件、buildSrc 插件以及单独项目插件，关于这 3 种创建自定义插件的方式，我们做了总结，如表 1-5 所示。

表1-5 不同插件的总结

创建插件方式	总结
直接在构建文件 build.gradle 中自定义插件	简单直接，不方便在当前项目内的其他构建文件引用
创建 buildSrc 模块自定义插件	适用于项目内模块引用，并且不需要对插件进行发布
创建 Gradle Plugin 项目自定义插件	方便分享和传播，需要将插件单独发布

1.4 Gradle扩展

1.4.1 什么是扩展

关于扩展，首先来看看 Android 项目 app/build.gradle 中这段非常经典的代码。

```
android {
      compileSdkVersion 30
      buildToolsVersion "30.0.3"

      defaultConfig {
          applicationId "cn.sensorsdata.asmbook.chapter1_04"
          minSdkVersion 16
          targetSdkVersion 30
          versionCode 1
          versionName "1.0"

          testInstrumentationRunner "androidx.test.runner.AndroidJUnitRunner"
      }

      buildTypes {
          release {
              minifyEnabled false
              proguardFiles getDefaultProguardFile(
                  'proguard-android-optimize.txt'), 'proguard-rules.pro'
          }
      }
      compileOptions {
          sourceCompatibility JavaVersion.VERSION_1_8
          targetCompatibility JavaVersion.VERSION_1_8
      }
}
```

相信 Android 开发者对这段配置很熟悉，但读者可能会有如下疑问：

Project类中并没有android{}这样的API，为何能这样配置？

其实，上面的这段配置用的是 Gradle 的扩展机制，简称扩展。它的作用是为插件提供配置，赋予插件特定的功能。

1.4.2 ExtensionContainer API介绍

要想自定义扩展，必须先了解 ExtensionContainer。这是因为一般情况下，我们都是通过 ExtensionContainer 来管理扩展的，可以通过 Project 的 getExtensions() 方法获取当前 Project 的 ExtensionContainer 对象。Project 中此方法的定义如下。

```
/**
 * Allows adding DSL extensions to the project. Useful for plugin authors.
```

```
   *
   * @return Returned instance allows adding DSL extensions to the project
   */
@Override
ExtensionContainer getExtensions();
```

ExtensionContainer 中常用的 API 如下。

```
/**
   * Allows adding 'namespaced' DSL extensions to a target object.
   */
@HasInternalProtocol
public interface ExtensionContainer {

      /**
       * 在容器中添加一个新的扩展
       *
       * @param publicType 扩展向外暴露的类型
       * @param name 扩展的名字，例如上面提到的“android”，可以是任意字符串，
       * 但需要确保唯一性，否则会有异常
       * @param extension 实现了 publicType 类型的扩展实例
       * @throws IllegalArgumentException 如果该名字的扩展已经存在，就会抛出异常
       */
      <T> void add(Class<T> publicType, String name, T extension);

      /**
       * 在容器中添加一个新的扩展
       *
       * @param name 要创建的扩展的名字，可以是任意字符串，
       * 但需要确保唯一性，否则会有异常
       * @param Extension 实例
       * @throws IllegalArgumentException 如果该名字的扩展已经存在就会抛出异常
       */
      void add(String name, Object extension);

      /**
       * 创建一个扩展并添加到容器中
       *
       * @param <T> 扩展对外暴露的类型
       * @param publicType 扩展对外暴露的类型
       * @param name 要创建的扩展的名字，可以是任意字符串，
       * 但需要确保唯一性，否则会有异常
       * @param instanceType 此扩展的实例类型
       * @param constructionArguments 构造扩展实例时需要的构造参数
       * @return 返回创建的扩展实例
       * @throws IllegalArgumentException 如果该名字的扩展已经存在就会抛出异常
       */
      <T> T create(Class<T> publicType, String name,
            Class<? extends T> instanceType,Object...constructionArguments);

      /**
       * 创建一个扩展并添加到容器中
       *
       * @param name 要创建的扩展的名字，可以是任意字符串，
       * 但需要确保唯一性，否则会有异常
```

```
     * @param type 此扩展的实例类型
     * @param constructionArguments 构造扩展实例时需要的构造参数
     * @return 返回创建的扩展实例
     * @throws IllegalArgumentException 如果该名字的扩展已经存在就会抛出异常
     */
    <T> T create(String name, Class<T> type, Object... constructionArguments);

    /**
     * 根据扩展类型获取实例，如果找不到会抛出异常
     *
     * @param type 扩展类型
     * @return 返回扩展，不会为 null
     * @throws UnknownDomainObjectException 如果根据扩展类型找不到扩展就会抛出异常
     */
    <T> T getByType(Class<T> type) throws UnknownDomainObjectException;

    /**
     * 根据扩展类型获取实例，如果找不到会返回 null，注意与 getByType 的区别
     *
     * @param type 扩展类型
     * @return extension or null
     */
    @Nullable
    <T> T findByType(Class<T> type);

    /**
     * 根据扩展的名字获取实例，如果找不到会抛出异常
     *
     * @param name 扩展的名字
     * @return 返回扩展，不会为 null
     * @throws UnknownDomainObjectException 如果找不到会抛出异常
     */
   Object getByName(String name) throws UnknownDomainObjectException;

    /**
     * 根据扩展的名字获取实例，如果找不到返回 null，注意与 getByName 的区别
     *
     * @param name 扩展的名字
     * @return extension or null
     */
    @Nullable
    Object findByName(String name);

    /**
     * 从当前扩展容器中获取 extra 属性扩展，就是我们在 build.gradle 中定义的 ext{}
     *
     * 这个扩展对应的名字是 ext
     */
    ExtraPropertiesExtension getExtraProperties();
}
```

上面列出了 ExtensionContainer 中常用的 API，下面我们用实例来展示具体的使用方式。

1.4.3 创建扩展

扩展的使用大概有 3 个步骤：

（1）定义扩展类；

（2）编译插件创建扩展；

（3）在 Gradle 脚本中配置扩展。

首先创建一个 Android 项目 Chapter1_04，为了简单地运行和测试，我们直接在 app/build.gradle 文件中进行代码演示。

第一步：在 app/build.gradle 中创建 ServerNode 类。

```
class ServerNode{
      String address
      int cpuCount

      @Override
      public String toString() {
          return "ServerNode{" +
                  "address='" + address + '\'' +
                  ", cpuCount=" + cpuCount +
                  '}';
      }
}
```

第二步：创建脚本插件并通过 apply() 方法使用它。

```
class MyExtensionTestPlugin implements Plugin<Project>{

      @Override
      void apply(Project target) {
          println("My Extension Test Plugin")
           // 通过 ExtensionContainer 创建扩展实例
          ServerNode serverNode = target.getExtensions()
                                          .create("serverNode",ServerNode)
          // 定义一个任务获取结果
          target.task("getResult"){
              doLast {
                  println("result is: $serverNode")
              }
          }
      }
}

apply plugin:MyExtensionTestPlugin
```

上面这段代码创建了 MyExtensionTestPlugin 插件，然后在 apply() 方法中创建了一个名称是 serverNode、类型是 ServerNode 并且不带构造函数的扩展实例，接着又定义了一个 getResult 任务用于获取配置的结果。

第三步：在 app/build.gradle 脚本里配置扩展。

```
serverNode{
      address = "SensorsData HeFei"
      cpuCount = 16
}
```

这里的 serverNode 需要与 target.getExtensions().create("serverNode",ServerNode) 中 create() 方法使用的名称保持一致。

第四步：运行 getResult 任务。

```
$ ./gradlew getResult

> Configure project :app
My Extension Test Plugin

> Task :app:getResult
result is: ServerNode{address='SensorsData HeFei', cpuCount=16}
```

可以看到，我们正确获取了配置内容中的值。

我们还可以给 ServerNode 添加默认值，具体有两种方式。一种是在定义成员变量的时候直接设置默认值，例如：

```
class ServerNode {
      String address = "SensorsData HeFei"
      int cpuCount = 16
...
}
```

另一种是调用 create() 方法的时候为其设置默认值。如果要在调用 create() 方法时，为其设置默认值，需要为 ServerNode 添加构造函数。

```
class ServerNode {
      String address
      int cpuCount

      //添加了一个构造函数
      ServerNode(String address, int cpuCount) {
          this.address = address
          this.cpuCount = cpuCount
      }

      @Override
      public String toString() {
          ...
      }
}
```

对应的 create() 调用方式的修改如下。

```
class MyExtensionTestPlugin implements Plugin<Project> {

      @Override
      void apply(Project target) {
          println("My Extension Test Plugin ")
          //添加构造函数
          ServerNode serverNode = target.getExtensions()
              .create("serverNode", ServerNode, "SensorsData Beijing", 36)
```

```
            println("configure phase 's result: $serverNode")
            target.task("getResult") {
                doLast {
                    println("result is: $serverNode")
                }
            }
        }
}
```

保持 app/build.gradle 中 serverNode 的配置不变，运行 getResult 任务的输出结果如下。

```
$ ./gradlew getResult

> Configure project :app
My Extension Test Plugin
configure phase 's result: ServerNode{address='SensorsData Beijing', cpuCount=36}
// 输出默认值

> Task :app:getResult
result is: ServerNode{address='SensorsData HeFei', cpuCount=16}
```

通过结果可以看到输出了默认值。上述代码是对创建扩展方法 <T> T create(String name, Class<T> type, Object... constructionArguments) 的介绍。

1.4.4 添加和查找扩展

1. 添加扩展

添加扩展有如下两种方法。

```
void add(String name, Object extension);
<T> void add(Class<T> publicType, String name, T extension);
```

对于扩展来说，create() 和 add() 的区别是，create() 是创建并添加扩展，add() 是添加已有的扩展实例。下面使用 add() 方法来添加 ServerNode 扩展。

```
ServerNode addServerNode = new ServerNode("SensorsData HeFei by Add", 4)
target.getExtensions().add("addServerNode", addServerNode)
target.task("getAddResult"){
      doLast {
          ServerNode tmpServerNode =
              target.getExtensions().findByName("addServerNode")
          println("add result is: $tmpServerNode")
      }
}
```

上面这段代码首先创建了一个 ServerNode 实例，然后通过 add() 方法添加到 ExtensionContainer 中，最后定义了一个任务查找添加的扩展（该任务通过 ExtensionContainer 提供的 findByName API 来查找 ServerNode）。

接下来在 app/build.gradle 中添加 addServerNode 配置。

```
addServerNode{
      cpuCount = 8
}
```

最后运行 getAddResult 命令，输出结果如下。

```
add result is: ServerNode{address='SensorsData HeFei by Add', cpuCount=8}
```

2. 查找扩展

上一个例子使用 findByName() 方法来查找已存在的扩展，另外还有如下几个方法可以查找扩展。

```
<T> T getByType(Class<T> type) throws UnknownDomainObjectException;
<T> T findByType(Class<T> type);
Object getByName(String name) throws UnknownDomainObjectException;
Object findByName(String name);
```

这些方法的主要区别在于它们的返回值是否可以为 null 或者抛出异常，在此不做详细介绍。至此我们已经了解了扩展的基本用法，完整的示例代码如下。

```
plugins {
      id 'com.android.application'
}

android {
      ...
}

dependencies {
      ...
}

class ServerNode {
      String address
      int cpuCount

      ServerNode(String address, int cpuCount) {
          this.address = address
          this.cpuCount = cpuCount
      }

      @Override
      public String toString() {
          return "ServerNode{" +
                  "address='" + address + '\'' +
                  ", cpuCount=" + cpuCount +
                  '}';
      }
}

class MyExtensionTestPlugin implements Plugin<Project> {

      @Override
      void apply(Project target) {
          println("My Extension Test Plugin ")

          // 示例 1: 创建扩展
          ServerNode serverNode = target.getExtensions()
                   .create("serverNode", ServerNode, "SensorsData Beijing", 36)
          println("1.configure phase 's result: $serverNode")
          target.task("getResult") {
              doLast {
                  println("result is: $serverNode")
```

```
                }
            }

            //示例 2: 添加扩展
            ServerNode addServerNode = new ServerNode("SensorsData HeFei by Add", 4)
            target.getExtensions().add("addServerNode", addServerNode)
            target.task("getAddResult"){
                doLast {
                    ServerNode tmpServerNode =
                            target.getExtensions().findByName("addServerNode")
                    println("add result is: $tmpServerNode")
                }
            }

        }
}

apply plugin: MyExtensionTestPlugin

serverNode {
        address = "SensorsData HeFei"
        cpuCount = 16
}

addServerNode{
        cpuCount = 8
}
```

1.4.5 扩展嵌套

什么是扩展嵌套呢？扩展嵌套就是扩展里嵌套扩展，例如下面的配置。

```
server {
        message = "Server Config Info"
        defaultConfig {
            address = "hefei"
            ip = "11.11.11.11"
            cpuCount = 8
        }
}
```

server 里面嵌套了 defaultConfig 配置。对于这个配置，我们该如何创建对应的扩展呢？

首先创建对应的 Extension Bean，代码如下。

```
class ServerExtension {
        String message
        ServerNode defaultConfig = new ServerNode()

        void defaultConfig(Action<ServerNode> action) {
            action.execute(defaultConfig)
        }

        @Override
        public String toString() {
            return "ServerExtension{" +
                    "message='" + message + '\'' +
```

```
                    ", defaultConfig=" + defaultConfig +
                    '}';
        }
}

class ServerNode {
        String address
        String ip
        int cpuCount

        @Override
        public String toString() {
            return "ServerNode{" +
                    "address='" + address + '\'' +
                    ", ip='" + ip + '\'' +
                    ", cpuCount=" + cpuCount +
                    '}';
        }
}
```

其中，最关键的是下面这个方法的定义。

```
void defaultConfig(Action<ServerNode> action) {
        action.execute(defaultConfig)
}
```

上面的例子中的 defaultConfig 配置实际上就是调用了这个方法。定义好这样的 Extension Bean 之后，其他的操作跟前面创建扩展的操作是一样的，这里就不赘述了。

1.4.6 NamedDomainObjectContainer

1.4.5 小节介绍的扩展嵌套，只是嵌套的一种方式，也是最常见的方式，其实还有一种更复杂的嵌套，即 NamedDomainObjectContainer。观察 android 中的这段配置。

```
android {
     ...

        buildTypes {
            release {
                minifyEnabled false
                    proguardFiles getDefaultProguardFile('proguard-android-optimize.
txt'), 'proguard-rules.pro'
            }

            debug {
                minifyEnabled true
                zipAlignEnabled true
            }

            preVersion {
                applicationIdSuffix ".pre"
            }
        }
}
```

可以看到 buildTypes 中定义的配置项，有我们熟悉的 release 和 debug，除此之外还能够添加自定义的 preVersion 这个配置项，类似地可以添加任意多个自定义的配置项，有点像往列表（list）中添加数据一样。那这个功能是如何实现的呢？

android 这个配置中的 buildTypes 是定义在 Android Gradle 插件的 BaseExtension 类中（注意，为方便演示代码，本书使用的是 3.2.0 版本，不同版本可能有所变化）的，查看其定义。

```
public abstract class BaseExtension implements AndroidConfig {

      ...

      private final NamedDomainObjectContainer<BuildType> buildTypes;

      ...

      /**
       * Encapsulates all build type configurations for this project.
       *
       * <p>For more information about the properties you can configure in
       * this block, see {@link
       * BuildType}.
       */
      public void buildTypes(
          Action<? super NamedDomainObjectContainer<BuildType>> action) {
          checkWritability();
          action.execute(buildTypes);
      }

      ...
```

从源码很容易看出，buildTypes 的类型是 NamedDomainObjectContainer<BuildType>，其中的泛型 BuildType 是对应的 Extension Bean；对应的 buildTypes(Action) 方法与 1.4.5 小节的扩展嵌套一样，也使用 Action，只不过泛型发生了变化。

当新建一个 Android 项目时，Android Studio 会默认添加 release 和 debug 这两个配置项，实际上还可以增加其他的配置项，使配置更加丰富。因此可以根据不同场景或者需求定义不同的配置项，每个不同的配置项都会生成一个新的 BuildType 配置。这些都需要借助 NamedDomainObjectContainer 来实现。

那究竟什么是 NamedDomainObjectContainer 呢？

它直译过来就是命名领域对象容器，主要用途是：通过 DSL 创建指定类型的对象实例。

再来看一下官方对它的说明：

```
/**
  * <p>A named domain object container is a specialization of
  * {@link NamedDomainObjectSet} that adds the ability to create
  * instances of the element type.</p>
  *
  * <p>Implementations may use different strategies for creating
  * new object instances.</p>
  *
  * <p>Note that a container is an implementation of {@link java.util.SortedSet},
  * which means that the container is guaranteed
  * to only contain elements with unique names within this container.
  * Furthermore, items are ordered by their name.</p>
  *
```

```
    * <p>You can create an instance of this type using the factory method
    * {@link org.gradle.api.model.ObjectFactory#domainObjectContainer(Class)}.</p>
    *
    * @param <T> The type of objects in this container.
    */
public interface NamedDomainObjectContainer<T> extends NamedDomainObjectSet<T>,
Configurable<NamedDomainObjectContainer<T>>
```

使用 NamedDomainObjectContainer<T> 有两个限制。

- 指定的泛型T必须有一个public构造函数，并且必须带有一个name字符串参数；
- 它实现了SortedSet接口，要求所有类型对象的name属性必须是唯一的，在内部会使用name属性进行排序。

那如何创建 NamedDomainObjectContainer 呢?

Project 类提供了如下几种方式来创建 NamedDomainObjectContainer。

```
/**
    * 为指定的类型创建一个容器，指定的类型必须要有一个构造函数，构造函数中需要有一个 name
    * 字符串参数，并且这个对象必须暴露它的 name 字段
    *
    * @param type 这个容器包含的对象的类型
    * @return 返回创建的容器
    */
<T> NamedDomainObjectContainer<T> container(Class<T> type);

/**
    * 为指定的类型创建一个容器，指定的类型必须要有一个构造函数，构造函数中需要有一个 name
    * 字符串参数，并且这个对象必须暴露它的 name 字段
    *
    * @param type 这个容器包含的对象的类型
    * @param factory 自定义实现 T 的方式
    * @return 返回创建的容器
    */
<T> NamedDomainObjectContainer<T> container(Class<T> type, NamedDomainObjectFactory
<T> factory);

/**
    * 为指定的类型创建一个容器，指定的类型必须要有一个构造函数，构造函数中需要有一个 name
    * 字符串参数，并且这个对象必须暴露它的 name 字段。其中的闭包参数用于创建 T 实例
    * 这个闭包需要有一个 name 参数
    *
    * <p>All objects <b>MUST</b> expose their name as a bean property named "name".
    * The name must be constant for the life of the object.</p>
    *
    * @param type 这个容器包含的对象的类型
    * @param factoryClosure 通过闭包方式来创建实例
    * @return 返回创建的容器
    */
<T> NamedDomainObjectContainer<T> container(Class<T> type,
                                            Closure factoryClosure);
```

接下来通过一个具体的例子来说明 NamedDomainObjectContainer 如何使用。

第一步：创建一个 Android 项目 Chapter1_05。

第二步：在 app/build.gradle 中添加相关的 Extension Bean。

```
class ServerExtension {
      String message
      ServerNode defaultConfig = new ServerNode()
      NamedDomainObjectContainer<ServerNode> nodesContainer

      // 通过构造函数将 Project 对象传入
      ServerExtension(Project project) {
          // 调用 project.container() 方法创建 NamedDomainObjectContainer
          nodesContainer = project.container(ServerNode)
      }

      // 默认节点配置
      void defaultConfig(Action<ServerNode> action) {
          action.execute(defaultConfig)
      }

      // 其他节点配置，注意 NamedDomainObjectContainer 的泛型
      void otherConfig(Action<NamedDomainObjectContainer<ServerNode>> action) {
          action.execute(nodesContainer)
      }

      @Override
      public String toString() {
          return "ServerExtension{" +
                  "message='" + message + '\'' +
                  ", defaultConfig=" + defaultConfig +
                  '}';
      }
}

class ServerNode {
      String name
      String address
      String ip
      int cpuCount

      ServerNode(String name) {
          this.name = name
      }

      @Override
      public String toString() {
          return "ServerNode{" +
                  "name='" + name + '\'' +
                  ", address='" + address + '\'' +
                  ", ip='" + ip + '\'' +
                  ", cpuCount=" + cpuCount +
                  '}';
      }
}
```

上面的代码中，需要关注 ServerExtension(Project project) 这个构造方法，这里需要将 Project 作为参数传入，然后调用 project.container() 方法创建 ServerNode 类型的 NamedDomainObjectContainer。

第三步：创建插件，并在插件中创建扩展的实例。

```
class MyExtensionTestPlugin implements Plugin<Project> {

      @Override
      void apply(Project target) {
          println("My NamedDomainObjectContainer Test Plugin")
          //创建扩展，传入 ServerExtension 构造参数的值 target
          ServerExtension serverNode =
              target.getExtensions().create("server", ServerExtension, target)
          target.task("getResult") {
              doLast {
                  println("result is: $serverNode")

                  println("\nshow all other configs:")
                  //遍历容器中所有的 ServerNode 配置
                  serverNode.nodesContainer.all {
                      println(it)
                  }

                  println("\nshow shanghai config in other configs:")
                  // 获取指定的配置
                  ServerNode node =
                      serverNode.nodesContainer.findByName("shanghai")
                  println(node)
              }
          }

      }
}

apply plugin: MyExtensionTestPlugin
```

上面的代码在 app/build.gradle 文件中创建了一个插件，并在插件中创建了对应的扩展，以及创建了一个 getResult 任务来获取 ServerNode 配置。代码中展示了两种获取 ServerNode 配置的方式，一种是通过 all() 方法获取所有的配置，一种是通过 findByName() 方法获取指定的配置。

第四步：添加 ServerExtension 配置。

```
server {
      message = "Server Config Info"
      defaultConfig {
          address = "hefei"
          ip = "11.11.11.11"
          cpuCount = 8
      }

      otherConfig {
          shanghai {
              address = "shanghai base"
              ip = "22.22.22.22"
              cpuCount = 8
          }

          chengdu {
              address = "chengdu base"
              ip = "122.122.122.133"
              cpuCount = 8
          }
```

```
    }
}
```

注意观察 otherConfig 中的两个配置项。

第五步：运行 getResult 任务。

```
$ ./gradlew getResult

> Configure project :app
My NamedDomainObjectContainer Test Plugin

> Task :app:getResult
result is: ServerExtension{message='Server Config Info', defaultConfig=ServerNode
{name='null', address='hefei', ip='11.11.11.11', cpuCount=8}}

show all other configs:
ServerNode{name='chengdu', address='chengdu base', ip='122.122.122.133', cpuCount=8}
ServerNode{name='shanghai', address='shanghai base', ip='22.22.22.22', cpuCount=8}

show shanghai config in other configs:
ServerNode{name='shanghai', address='shanghai base', ip='22.22.22.22', cpuCount=8}
```

注意观察这里使用 all() 遍历输出的结果，我们定义的配置项顺序是 shanghai→chengdu，但是输出结果的顺序是 chengdu→shanghai，这就是我们前面提到的使用 NamedDomainObjectContainer 有限制的原因：实现了 SortedSet 接口，会对 name 参数进行排序。

关于 NamedDomainObjectContainer 的完整代码如下。

```
plugins {
    id 'com.android.application'
}

android {
    ...
}

dependencies {
    ...
}

class ServerExtension {
    String message
    ServerNode defaultConfig = new ServerNode()
    NamedDomainObjectContainer<ServerNode> nodesContainer

    // 通过构造方法将 Project 对象传入
    ServerExtension(Project project) {
        // 调用 project.container() 方法创建 NamedDomainObjectContainer
        nodesContainer = project.container(ServerNode)
    }

    // 默认节点配置
    void defaultConfig(Action<ServerNode> action) {
        action.execute(defaultConfig)
    }

    // 其他节点配置，注意 NamedDomainObjectContainer 的泛型
```

```
    void otherConfig(Action<NamedDomainObjectContainer<ServerNode>> action) {
        action.execute(nodesContainer)
    }

    @Override
    public String toString() {
        return "ServerExtension{" +
                "message='" + message + '\'' +
                ", defaultConfig=" + defaultConfig +
                '}';
    }
}


class ServerNode {
    String name
    String address
    String ip
    int cpuCount

    ServerNode(String name) {
        this.name = name
    }

    @Override
    public String toString() {
        return "ServerNode{" +
                "name='" + name + '\'' +
                ", address='" + address + '\'' +
                ", ip='" + ip + '\'' +
                ", cpuCount=" + cpuCount +
                '}';
    }
}

class MyExtensionTestPlugin implements Plugin<Project> {

    @Override
    void apply(Project target) {
        println("My NamedDomainObjectContainer Test Plugin")
        //创建扩展，传入 ServerExtension 构造参数的值 target
        ServerExtension serverNode =
            target.getExtensions().create("server", ServerExtension, target)
        target.task("getResult") {
            doLast {
                println("result is: $serverNode")

                println("\nshow all other configs:")
                //遍历容器中所有的 ServerNode 配置
                serverNode.nodesContainer.all {
                    println(it)
                }

                println("\nshow shanghai config in other configs:")
                // 获取指定的配置
                ServerNode node =
```

```
                    serverNode.nodesContainer.findByName("shanghai")
                println(node)
            }
        }

    }
}

apply plugin: MyExtensionTestPlugin

server {
    message = "Server Config Info"
    defaultConfig {
        address = "hefei"
        ip = "11.11.11.11"
        cpuCount = 8
    }

    otherConfig {
        shanghai {
            address = "shanghai base"
            ip = "22.22.22.22"
            cpuCount = 8
        }

        chengdu {
            address = "chengdu base"
            ip = "122.122.122.133"
            cpuCount = 8
        }
    }
}
```

上面在第三步中介绍了两种从 NamedDomainObjectContainer 中获取配置的方式，NamedDomainObjectContainer 中还提供了如下几种常用方式，对应的方法如下，比较简单，在此不多做介绍。

```
// 找不到返回 null，不会抛出异常
T findByName(String name);
// 找不到会抛出异常
T getByName(String name) throws UnknownDomainObjectException;
// 遍历所有
void all(Closure action);
```

1.5 综合示例

前文我们介绍了 Gradle 的基础知识，本节用一个例子来对前面的内容做综合的总结。

1.5.1 概述

本节的这个例子比较简单，目标是简化神策 Android 全埋点 SDK 的集成步骤。关于神策 Android 全

埋点，可以参考官方文档中的描述。

为方便阅读，下面简述神策 Android 全埋点 SDK 的集成步骤。

第一步： 添加插件依赖。在 rootProject/build.gradle 中添加插件依赖，示例代码如下。

```
buildscript {
   repositories {
      mavenCentral()
   }
   dependencies {
    classpath 'com.android.tools.build:gradle:3.5.3'
    // 添加神策分析 android-gradle-plugin2 依赖
    classpath 'com.sensorsdata.analytics.android:android-gradle-plugin2:3.3.4'
   }
}
```

第二步： 添加 SDK 依赖和使用插件。在 app/build.gradle 文件中使用神策插件和依赖神策 SDK，示例代码如下。

```
apply plugin: 'com.android.application'
apply plugin: 'com.sensorsdata.analytics.android' // 使用神策插件

dependencies {
     // 添加神策 SDK 库
     implementation 'com.sensorsdata.analytics.android:SensorsAnalyticsSDK:5.1.0'
}
```

可以看到，这个集成步骤相对来说还是有点麻烦的，为了进一步简化这个步骤，现提出如下要求。

（1）使用插件进一步优化集成步骤；

（2）将这个插件发布到线上仓库上供别人下载使用。

1.5.2 集成步骤

第一步： 创建一个新的 Android 项目，名称为 Chapter1_06。

第二步： 创建一个新的模块，类型选择“Java Library”，名称为 myPlugin。在 Android Studio 中完整的操作路径是：File → New → New Module → Java Library → Next → Finish。

第三步： 修改 myPlugin/build.gradle 文件，添加 java-gradle-plugin 插件以及 maven-publish 插件，并配置发布信息。修改后的内容如下。

```
plugins {
      id "java-gradle-plugin"
      id 'maven-publish'
}

java {
      sourceCompatibility = JavaVersion.VERSION_1_8
      targetCompatibility = JavaVersion.VERSION_1_8
}

repositories {
      google()
      mavenCentral()
      gradlePluginPortal()
```

```
}

dependencies {
    //添加神策分析 android-gradle-plugin2 依赖
    implementation "com.sensorsdata.analytics.android:android-gradle-plugin2:+"
    implementation "com.android.tools.build:gradle:4.1.2"
}

//组织或者公司名称
group="cn.sensorsdata.asmbook.plugin"
//版本号
version='1.0.0'

gradlePlugin {
    plugins {
        sensorsAutoPlugin {
            id = 'sensorsdata.autosdk'
            implementationClass =
                'cn.sensorsdata.asmbook.myplugin.AutoAddSensorsDataSDKPlugin'
        }
    }
}

//./gradlew publish  发布到本地
publishing {
    publications {
        myLibrary(MavenPublication) {
            from components.java
        }
    }

    repositories {
        maven {
            name = 'myPluginRepo'
            url = '../repo_maven'
        }
    }
}
```

上述 build.gradle 中添加了 implementation "com.sensorsdata.analytics.android:android-gradle-plugin2:+"，它表示使用神策插件的最新版本。

第四步： 编写 Plugin<T> 的实现类。为了能够灵活地修改神策 SDK 版本号，我们需要创建一个扩展。这个扩展很简单，只有一个字段，代码如下。

```
package cn.sensorsdata.asmbook.myplugin;

public class SDKVersionExtension {
    String version;

    @Override
    public String toString() {
        return "SDKVersionExtension{" +
                "version='" + version + '\'' +
                '}';
    }
}
```

接着创建 Plugin<T> 的实现，具体代码如下。

```
package cn.sensorsdata.asmbook.myplugin;

public class AutoAddSensorsDataSDKPlugin implements Plugin<Project> {
      // 默认下载最新的版本
      String sdkVersion = "+";

      @Override
      public void apply(Project project) {
          System.out.println("Auto Add SensorsData AutoTrack SDK");
          // 创建版本号扩展
          project.getExtensions().add("sdkVersion", SDKVersionExtension.class);//(3)
          project.afterEvaluate(project1 -> {

              Plugin saPlugin = project.getPlugins()
                      .findPlugin("com.sensorsdata.analytics.android");//(1)
              if (saPlugin == null) {
                  // 添加神策插件依赖
                  project.getPluginManager()
                          .apply("com.sensorsdata.analytics.android");//(2)
                  // 查找扩展并获取其中的版本号设置
                  Object sdkVersionExtension =
                      project1.getExtensions().findByName("sdkVersion");//(4)
                  if (sdkVersionExtension != null) {
                      SDKVersionExtension tmp =
                          (SDKVersionExtension) sdkVersionExtension;
                      if (tmp.version != null) {
                          sdkVersion = tmp.version;
                      }
                  }
                  System.out.println("====final version====" + sdkVersion);
                  // 添加对 SDK 的依赖
                  project.getDependencies().add("implementation",
       "com.sensorsdata.analytics.android:SensorsAnalyticsSDK:" + sdkVersion);//(5)
              }
          });
      }
}
```

接下来按照代码中标注的序号，对应解释代码。

（1）因为用户可能已经集成了神策插件，为了避免重复引入，所以需要查询是否已经存在神策插件。

（2）如果没有集成神策插件，此处调用 project.getPlugins().apply 方法使用神策插件。

（3）创建配置版本号的扩展，注意在 AutoAddSensorsDataSDKPlugin 类中定义了一个 sdkVersion="+" 字段。假如用户没有配置扩展，就默认使用该字段的默认值来加载最新的版本。

（4）此处通过 findByName() 方法查找对应的扩展，如果找不到会返回 null；如果找到，就将其中的 version 值赋给 sdkVersion。

（5）在确定 SDK 版本号以后，调用 project.getDependencies() 添加对 SDK 的依赖。

第五步： 编译插件。运行 ./gradlew publish 来打包插件，至此会在项目根目录下生成 repo_maven 目录，其内容如图 1-12 所示。

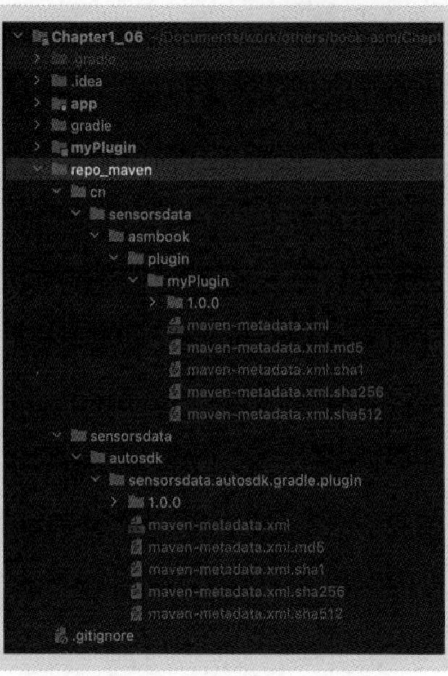

图 1-12　repo_maven 目录的内容

第六步： 使用插件。此处我们使用 plugins{} DSL 来依赖插件，因此需要先在 rootProject/settings.gradle 文件中配置本地 Maven 仓库的信息，内容如下。

```
//plugins 声明依赖关系
pluginManagement {
      plugins {
      }
      repositories {
          maven {
              url(uri("repo_maven"))
          }
          gradlePluginPortal()
          mavenCentral()
          google()
      }
}

include ':myPlugin'
include ':app'
rootProject.name = "Chapter1_06"
```

然后在 app/build.gradle 中添加插件和添加扩展配置，内容如下。

```
plugins {
      id 'com.android.application'
      id 'sensorsdata.autosdk' version '1.0.0'
}

android {
      ...
}

dependencies {
      ...
```

```
}
// 配置版本号
sdkVersion{
      version = "6.0.0"
}
```

这里我们将 SDK 版本号设置为 6.0.0。

第七步：运行项目，控制台输出结果如图 1-13 所示。

图 1-13　控制台输出结果

通过图 1-13 可以看到插件起作用了，自动集成了神策的插件和 SDK。借用这个例子对前面的内容进行总结：插件为用户提供了一个外部入口，使用户可以创建易于分发的工具，通过这个工具可以参与 Gradle 构建的流程中。

本节通过 7 个步骤创建并运行了自定义的插件，那如何将这个插件上传到公共仓库上呢？继续看 1.6 节的内容。

1.6　插件发布

通常，插件发布的平台有如下两个。

- Gradle Plugin Portal；
- Apache Maven。

1.6.1　Gradle Plugin Portal

Gradle Plugin Portal——Gradle 自己的插件平台，该平台具有操作简单、支持性良好的特点，其最大的优势是可以配合 plugins{} DSL 来依赖插件。下面介绍如何将 Chapter1_06 项目上传到 Gradle Plugin Portal 平台。首先复制一份 Chapter1_06 项目，并将其副本改名为 Chapter1_06_GPP（GPP：Gradle Plugin Portal）。

下面是具体的发布操作流程。

第一步：注册Gradle Plugin Portal平台账号。进入Gradle Plugin Portal登录页面，可以看到图1-14所示的页面效果。

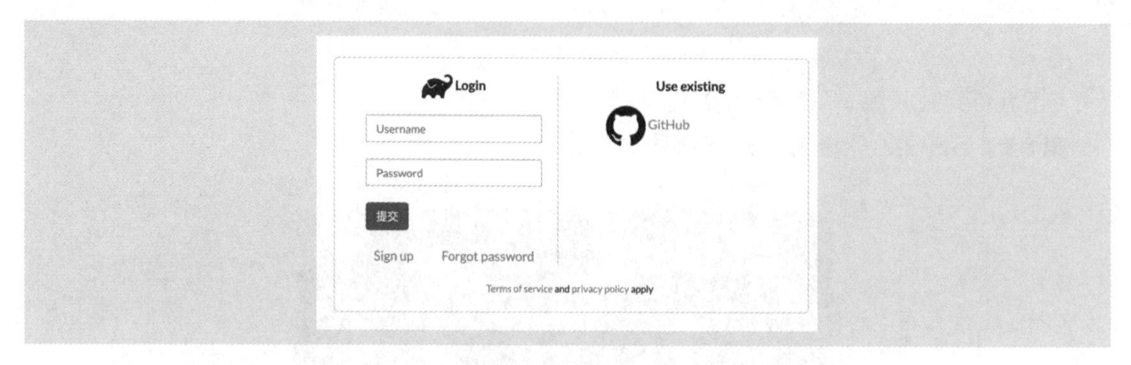

图 1-14　Gradle Plugin Portal 登录页面

用户可以使用 GitHub 账号授权登录，也可以单击“Sign Up”注册一个账号后进行登录。

第二步：获取 API Key。登录平台以后进入自己的个人账户页面，单击“API Keys”选项（见图 1-15），其中的 key 是发布插件时需要的值。

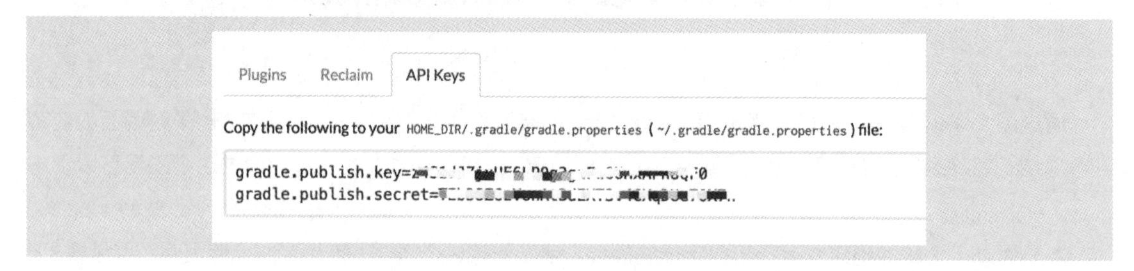

图 1-15　API Key

（1）将上述 key 配置在 ~ /.gradle/gradle.properties 文件中，这表示该配置是全局的，发布插件的时候会从该文件读取 key。

（2）将 key 放在自己项目的 gradle.properties 文件中，例如在 myPlugin 中创建一个 gradle.properties，如图 1-16 所示。

图 1-16　API Key 配置

（3）也可以在 Gradle 发布任务时以添加命令行参数的形式设置，例如：

```
$ ./gradlew publishPlugins -Pgradle.publish.key=<your-key> \
-Pgradle.publish.secret=<your-secret>
```

不管使用哪种方式都要确保 key 安全。

第三步：添加 gradle publish 插件。在 myPlugin/build.gradle 文件头部添加如下插件。

```
plugins {
      id "java-gradle-plugin"
      id 'maven-publish'
      id "com.gradle.plugin-publish" version "0.17.0"
}
```

其中 com.gradle.plugin-publish 是 Gradle 发布插件，顾名思义，其可用于发布插件。

第四步：插件发布配置。接着在 myPlugin/build.gradle 文件中添加发布插件所需要的配置。

```
group="cn.sensorsdata.asmbook.plugin" //(1) 组织或者公司名称
version='1.0.0' //(2) 版本号

gradlePlugin { //(3)java-gradle-plugin 对应的扩展，用于生成插件描述信息和插件 ID
      plugins { //(4) 添加插件的方法
        sensorsAutoPlugin { //(5) 插件名称
         id = 'cn.sensorsdata.autosdk'//(6) 插件的唯一 ID
         implementationClass =
          'cn.sensorsdata.asmbook.myplugin.AutoAddSensorsDataSDKPlugin' //(7) 实现类
        }
      }
}

pluginBundle {//(8) 配置发布到 Gradle Plugin Portal 上时的基本信息
      website = 'https://github.com/sensorsdata' //(9) 网址
      vcsUrl = 'https://github.com/sensorsdata' //(10)GitHub 项目的仓库地址
      description = 'SensorsData SDK' //(11) 项目描述
      tags = ['sensorsdata'] //(12) 项目标签

      plugins { //(13)
          sensorsAutoPlugin {//(14)
            // id is captured from java-gradle-plugin configuration
            description = 'SensorsData SDK Android Auto Plugin' //(15) 插件描述
            version = '1.0.0' //(16) 插件版本号
            tags = ['sensorsdata', 'autosdk']//(17)
            displayName = 'Plugin for SA' //(18)
          }
      }
}
```

下面按照标注的序号对代码进行详细介绍。

（1）group 是插件发布时所需的组织或者公司名称信息，这个字段在生成包和引入插件时使用。

（2）version 表示插件的版本号信息，如果插件没有设置版本号信息，则会使用这里给 Project 设置的版本号信息，可以对比序号 16 来理解。

（3）这个配置是 java-gradle-plugin 对应的扩展，用于配置插件的描述信息（包括 implementationClass、description、displayName）和 ID。

（4）plugins{} 方法就是前面介绍的 NamedDomainObjectContainer 的实现，因此可以添加多个插件，例如序号 6 所示。

（5）sensorsAutoPlugin 是自己定义的插件名称。

（6）ID 必须是唯一的，不能重复，通常使用 com.xxx.pluginName 这样的命名格式。另外注意，这里与 Chapter_06 略有不同，这里的插件 ID 与 group 都是以 cn.sensorsdata 开头的，原因是 Gradle Plugin Portal 对第三方插件的 ID 命名有这方面的要求。

（7）implementationClass 是当前插件的实现类，与前面介绍的在 myPlugin/src/main/resources/META-INF/gradle-plugins/xxxproperties 文件中添加插件的入口类类似。

（8）pluginBundle 是配置上传到 Gradle Plugin Portal 平台上的插件信息。

（9）website 是项目的网址信息。

（10）vcsUrl 是项目的仓库地址。

（11）description 是项目的描述信息。

（12）tags 是项目的标签。

（13）因为在 gradlePlugin {} 中可以配置多个插件，如果想对其中的插件做单独的配置，可以在这里添加。

（14）表示对 sensorsAutoPlugin 做单独的配置。

（15）插件的描述信息。

（16）插件的版本号信息。

（17）插件的标签信息。

（18）插件的显示名称。

第五步： 本地 Maven 配置。在发布插件之前，需要先在本地对其进行验证，确保其通过。利用 maven-publish 插件提供的发布功能，将插件发布到本地，相关配置如下。

```
publishing {
      repositories {
          maven {
              name = 'myPluginRepo'
              url = '../repo_maven'
          }
      }
}
```

然后在控制台中运行 ./gradlew publish 命令来发布，可以在项目根目录下得到一个 repo_maven 目录，进行验证即可。

第六步： 插件配置好并验证通过后，将其发布到 Gradle Plugin Portal 上，这里使用的是通过命令行添加 key 的方式来发布。

```
$ ./gradlew publishPlugins -Pgradle.publish.key=<your-key> \
-Pgradle.publish.secret=<your-secret>
```

发布成功以后可以在 Gradle Plugin Portal 的个人页面看到插件。图 1-17 表示插件正在审核中。

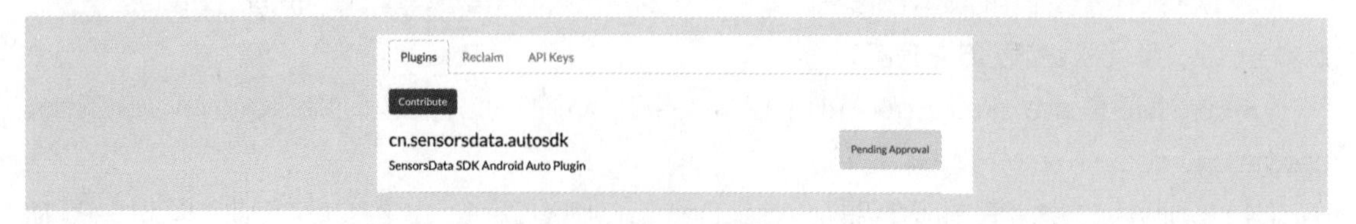

图 1-17　插件审核界面

1.6.2 Maven Central简介

在介绍如何将包发布到 Maven Central 之前，先来了解一些关于 Maven 的基础知识。

1. 什么是 Maven

Maven 是基于项目对象模型（Project Object Model，POM）的，可以通过一小段描述信息（配置）来管理项目的构建。Maven 主要作为 Java 的项目管理工具，它不仅可以用于包管理，还有许多的插件，可以支持整个项目的开发、打包、测试、部署等一系列行为，而包管理则是其核心功能。

2. 什么是仓库

Maven 既然能够管理包，自然就需要存放包的地方，这样的地方称为仓库（repositories），仓库又分为本地仓库、中央仓库（maven central）和私服仓库。本地仓库，顾名思义是指存放在本地计算机中的 Maven 仓库。Maven 会将项目中依赖的包从远端下载到本机的一个目录下管理，计算机中默认的仓库在 $user.home/.m2/repository 目录下。当通过 POM 下载依赖包时会先从本地仓库中寻找，如果找不到就到中央仓库中寻找。中央仓库包含绝大多数流行的开源 Java 构件，以及源码、作者信息、源代码控制管理（Source Control Manager，SCM）信息、许可证信息等，几乎所有开源的 Java 项目依赖的构件都可以在这里下载到。但是中央仓库在国外，速度上可能无法保证，因此一些公司为了解决这个问题会选择搭建“私服仓库”。简单来说私服仓库会将项目中的一些依赖包下载到自己的服务器上，从而解决下载速度慢的问题，而这些私服仓库中，最著名的就是 Sonatype 公司的 Nexus。同时 Sonatype 公司还提供了托管包，以及将包同步到 Maven Central 的服务，这个服务称为开源软件资源库托管（Open Source Software Repository Hosting，OSSRH）。

3. 什么是 Maven 坐标

Maven 上托管着众多的包，自然要求包满足一定的规则，以便于检索，这个规则称为“坐标”。Maven 坐标是通过 groupId、artifactId、version、packaging、classifier 这些元素来定义的。

- **groupId：** 定义当前Maven项目隶属项目、组织。每一个groupId可以对应多个项目，如图1-18所示，com.android.tools这个group对应多个项目（common、annotations等）。

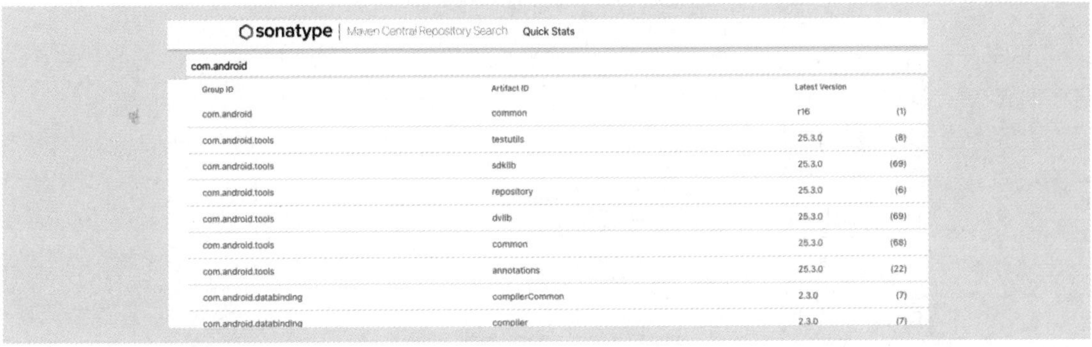

Group ID	Artifact ID	Latest Version	
com.android	common	r16	(1)
com.android.tools	testutils	25.3.0	(8)
com.android.tools	sdklib	25.3.0	(69)
com.android.tools	repository	25.3.0	(6)
com.android.tools	dvlib	25.3.0	(69)
com.android.tools	common	25.3.0	(68)
com.android.tools	annotations	25.3.0	(22)
com.android.databinding	compilerCommon	2.3.0	(7)
com.android.databinding	compiler	2.3.0	(7)

图 1-18 groupId 效果展示

- **artifactId：** 该元素定义当前实际项目中的一个模块，推荐的做法是使用实际项目名称作为artifactId值。
- **version：** 该元素定义使用构件的版本。
- **packaging：** 定义Maven项目打包的方式，打包方式通常有WAR/JAR/RAR/AAR等，默认是JAR。

- **classifier**：该元素用来帮助定义构建输出的一些附件。例如JAR包的javadoc.jar、sources.jar等内容，这些附件也有自己的坐标。

上述5个元素中，groupId、artifactId、version是必须定义的，packaging是可选的，classifier不能直接定义，需要结合插件使用。另外OSSRH对groupId的命名有一定的要求，如果你的项目托管在开源的网站上，必须满足表1-6所示的命名规则。

表1-6　groupId命名规则样例

网站	groupId样例
GitHub	io.github.myusername
GitLab	io.gitlab.myusername
Gitee	io.gitee.myusername
Bitbucket	io.bitbucket.myusername
SourceForge	io.sourceforge.myusername

1.6.3　上传到Maven Central

接下来详细介绍如何将包上传到Maven Central。

第一步：注册。首先需要注册Sonatype的Jira账号。

第二步：提交issue。注册完以后需要在Sonatype Jira上提交一个issue，详细信息如图1-19所示。

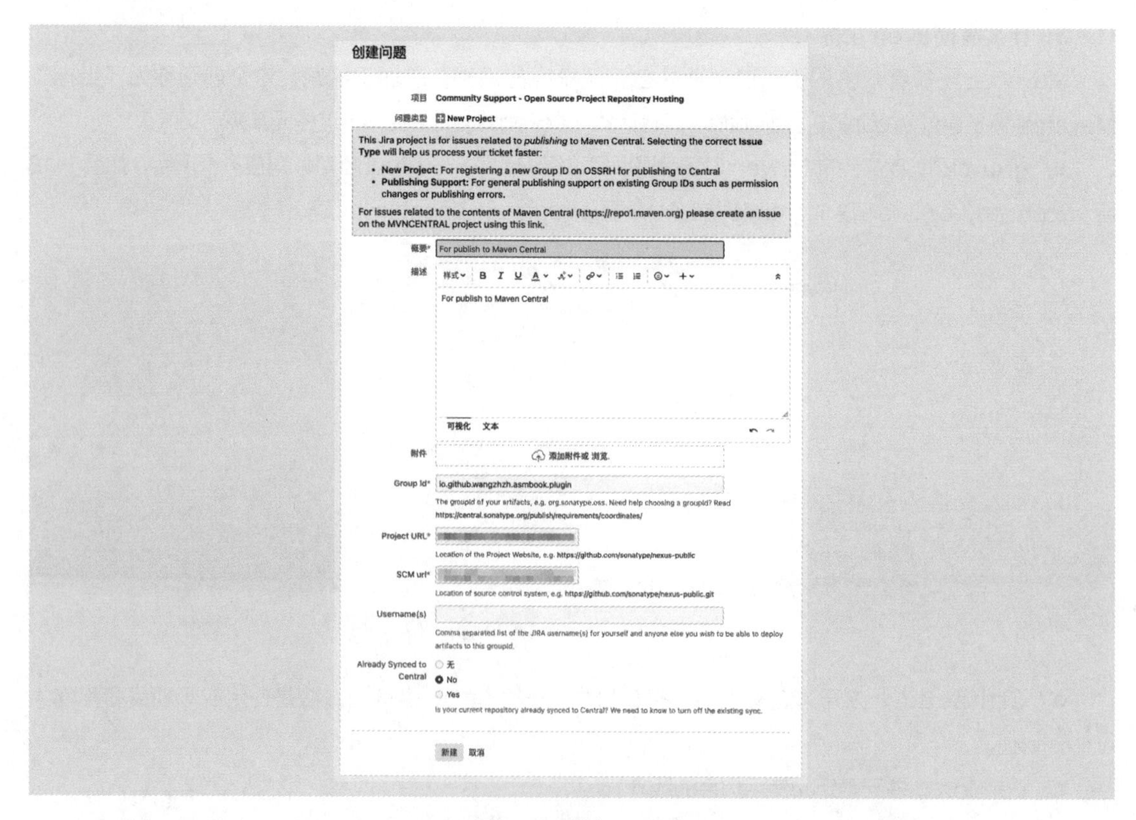

图1-19　提交issue

第三步：等待反馈。提交完 issue 后会进入问题的详情页面，Sonatype 的管理员会将所填信息中不合规的部分反馈给提交者，一定要注意刷新页面，通常 Sonatype 管理员的响应还是很及时的。下面举一个比较常见的反馈的例子。

If you do not own this domain, **you may also choose a different Group Id that reflects your project hosting following this steps.**

1. According to your project information, io.github.wangzhzh is valid and can be used. (*com.github.* Group IDs are invalid now. The only allowed groupIds for Github projects are io.github.**)

2. Create a temporary, public repository called https://github.com/wangzhzh/OSSRH-75041 to verify github account ownership.

3. **Edit this ticket** and update the Group ID field with the new GroupId, and set Status to Open.

More info: https://central.sonatype.org/publish/requirements/coordinates/

上述反馈的大致意思有两点：第一点，填写的 groupId 不合规，因为项目是放在 GitHub 上托管的，要求 groupId 必须使用 io.github.username 的形式；第二点是用户需要在自己的 GitHub 账号中创建一个 OSSRH-75041 项目，目的是确保用户自己就是该账号的所有者。按照上述提示修改完后重新提交，如果顺利，会收到如下反馈信息，表示信息验证通过了。

io.github.wangzhzh has been prepared, now user(s) curious can:Publish snapshot and release artifacts to s01.oss.sonatype.org.Have a look at this section of our official guide for deployment instructions:https://central.sonatype.org/publish/publish-guide/#deployment

第四步：准备包信息。为了确保中央仓库中包的最低质量水平，确立了一些基本要求，接下来进行简单介绍。

（1）javadoc 和 source 包。假如项目的 groupId 和 artifactId 的内容如下。

```
<groupId>com.example.applications</groupId>
<artifactId>example-application</artifactId>
<version>1.4.7</version>
```

对应的 javadoc 和 source 包的形式如下。

```
example-application-1.4.7-sources.jar
example-application-1.4.7-javadoc.jar
```

如果项目中没有 javadoc 和 source 包，需要创建一个空的 JAR 包，并且 JAR 包中需要使用 README.md 文件对原因进行说明。

（2）正确的 Maven 坐标。关于 Maven 坐标前面已做了简单的介绍，在此就不赘述。

（3）项目信息。操作者需要提供项目的一些基本信息，例如项目的名称、介绍、地址等，这些信息会在 Maven Central 中展示，方便别人了解自己的项目。下面是一个例子。

```
<name>Example Application</name>
<description>A application used as an example on how to set up pushing
    its components to the Central Repository.</description>
<url>http://www.example.com/example-application</url>
```

（4）版权信息。版权信息自然也是不可缺少的部分，对开源项目来说，最常用的有 Apache License 和 MIT License。下面是这两种版权的使用示例。

Apache License:

```
<licenses>
    <license>
      <name>The Apache License, Version 2.0</name>
      <url>http://www.apache.org/licenses/LICENSE-2.0.txt</url>
    </license>
</licenses>
```

MIT License:

```
<licenses>
    <license>
      <name>MIT License</name>
      <url>http://www.opensource.org/licenses/mit-license.php</url>
    </license>
</licenses>
```

（5）开发者信息。为了与项目关联，还需要提供开发者信息。下面是关于开发者信息的例子。

```
<developers>
      <developer>
        <name>wangzhzh</name>
        <email>xxx@sensorsdata.cn</email>
        <organization>SensorsData</organization>
        <organizationUrl>http://www.sensorsdata.cn</organizationUrl>
      </developer>
</developers>
```

（6）源码仓库信息。中央仓库中的包都是开源的，因此还需要提供源码仓库信息，方便别人能够定位到项目。根据源码托管系统的不同，这些信息也是略有不同的，如下是部分 SCM 对应的配置。

GitHub：

```
<scm>
    <connection>scm:git:git://github.com/simpligility/ossrh-demo.git</connection>
     <developerConnection>scm:git:ssh://github.com:simpligility/ossrh-demo.git
  </developerConnection>
    <url>http://github.com/simpligility/ossrh-demo/tree/master</url>
</scm>
```

SubVersion：

```
<scm>
     <connection>scm:svn:http://subversion.example.com/svn/project/trunk/
</connection>
     <developerConnection>scm:svn:https://subversion.example.com/svn/project/trunk/
</developerConnection>
    <url>http://subversion.example.com/svn/project/trunk/</url>
</scm>
```

（7）GPG 加密。为了保证数据的安全，Maven Central 要求发布的内容使用 GPG 软件进行加密[GPG 支持多种加密算法，例如 RSA（RSA 是 3 个共同发明人的姓氏首字母）、DSA（Digital Signature Algorithm，数字签名算法）等]，用户可以使用提供的公钥进行解密验证。要了解 GPG，首先要知道 PGP（Pretty Good Privacy，良好保密协议）。1991 年，程序员菲尔 · 齐默尔曼为了避开政府监视，开发了加密软件 PGP。PGP 通常用于签名、加密和解密文本、电子邮件和文件，不过 PGP 是一款商业软件，需要付费。于是在 1997 年 7 月，菲尔 · 齐默尔曼的公司 PGP Inc. 向 IETF（Internet Engineering Task Force，因特网工程任务组）提议制定一项名为 OpenPGP 的统一的标准，而 GPG（GnuPG）就是 OpenPGP 协议的一个具体实现。

接下来我们介绍关于 GPG 的一些基本用法，主要内容包括：

- 创建密钥对；
- 将公钥发往公用服务器上，提供给用户校验。

首先安装 GPG，可以从 GnuPG 网站上直接下载安装文件并安装，也可使用命令安装。以 macOS 为例，可以通过 Homebrew 安装，命令如下。

```
$ brew install gpg
```

下载完成以后，在控制台中运行如下命令查看版本信息。

```
$ gpg --version
```

输出示例如下。

```
gpg (GnuPG) 2.3.3
libgcrypt 1.9.4
Copyright (C) 2021 Free Software Foundation, Inc.
License GNU GPL-3.0-or-later <https://gnu.org/licenses/gpl.html>
This is free software: you are free to change and redistribute it.
There is NO WARRANTY, to the extent permitted by law.

Home: /Users/wangzhzh/.gnupg
支持的算法：
公钥：RSA, ELG, DSA, ECDH, ECDSA, EDDSA
密文：IDEA, 3DES, CAST5, BLOWFISH, AES, AES192, AES256, TWOFISH,
      CAMELLIA128, CAMELLIA192, CAMELLIA256
AEAD: EAX, OCB
散列：SHA1, RIPEMD160, SHA256, SHA384, SHA512, SHA224
压缩：  不压缩 , ZIP, ZLIB, BZIP2
```

通过上述输出可以看到 GPG 支持很多种加密算法，Maven Central 支持的是 RSA 算法，另外上述的 Home 路径表示 GPG 默认目录。

接下来运行如下命令创建密钥。

```
$ gpg --full-generate-key
```

运行上述命令会输出如下选项。

```
请选择您要使用的密钥类型：
    (1) RSA 和 RSA
    (2) DSA 和 Elgamal
    (3) DSA（仅用于签名）
    (4) RSA（仅用于签名）
    (9) ECC（签名和加密） *默认*
    (10) ECC（仅用于签名）
    (14) 卡中现有密钥
您的选择是？ 1
```

这里选择 RSA 加密，后续步骤按照提示操作即可。在输入密码时，记住自己输入的密码，后续会用到。按照提示的步骤操作完成后，会得到类似如下的输出。

```
pub   rsa3072 2021-11-19 [SC]
        FADC0000236CA74AAA8C101ABBE33FDCCBA14508
uid                     wangzhzh (For MavenCentral) <curious.a@qq.com>
sub   rsa3072 2021-11-19 [E]
```

其中: rsa3072 表示算法是 RSA 且长度是 3072; FADC0000236CA74AAA8C101ABBE33FDCCBA14508 为公钥，其后 8 位 CBA14508 是公钥的短写形式。使用如下命令可以显示所有的公钥。

```
$ gpg --list-keys
```

生成密钥以后就可以对文件进行加密。

```
$ gpg -ab Main.java
```

上述命令就是对文件进行加密，加密后会生成 Main.java.asc 文件。为了方便其他人验证加密结果，还需要将公钥放在服务器上。有 3 个服务器可以存放公钥，分别如下。

- keyserver.ubuntu.com；
- keys.openpgp.org；
- pgp.mit.edu。

使用如下命令将公钥信息发到服务器上。

```
$ gpg --keyserver keys.openpgp.org --send-keys FADC0000236CA74AAA8C101ABBE33FDCC
BA14508
```

发送到服务器之后，其他用户可以使用如下命令接收公钥，进而对加密文件进行校验:

```
$ gpg --keyserver keys.openpgp.org --recv-keys FADC0000236CA74AAA8C101ABBE33FDCC
BA14508
```

通常情况下，在开发中很可能有不同的成员需要使用同一个密钥来加密文件，这个时候就需要导出密钥。下面是导出密钥的方式。

```
$ gpg --export-secret-keys > ~/.gnupg/secring.gpg
```

运行上述命令，就会将对应公钥的密钥导出到 secring.gpg 文件中，这个文件后面需要用到。

第五步: 配置 Gradle。第四步介绍了在 Maven Central 上发布包需要的基本信息，第五步根据前面的准备知识介绍如何使用 Gradle 配置这些信息。首先复制 Chapter1_06_GPP 项目，将其副本改名为 Chapter1_06_Maven，然后修改 myPlugin/build.gradle 文件。修改后的结果如下。

```
plugins {
      id "java-gradle-plugin"
      id 'maven-publish'
      id 'signing'//(1)
}

java {
      sourceCompatibility = JavaVersion.VERSION_1_8
      targetCompatibility = JavaVersion.VERSION_1_8
}

repositories {
      google()
      mavenCentral()
      gradlePluginPortal()
}
```

```
dependencies {
      //添加神策分析 android-gradle-plugin2 依赖
      implementation "com.sensorsdata.analytics.android:android-gradle-plugin2:+"
      implementation "com.android.tools.build:gradle:4.1.2"
}

//组织或者公司名称
group="io.github.gvczhang.asmbook.plugin"
//版本号
version='1.0.0-SNAPSHOT'

java { //(2)
      withJavadocJar()
      withSourcesJar()
}

javadoc {//(3)
      if(JavaVersion.current().isJava9Compatible()) {
          options.addBooleanOption('html5', true)
      }
}

//./gradlew publish 命令发布到本地
publishing {

      publications {
          myLibrary(MavenPublication) {//(4)
              //meta info
              groupId = "io.github.gvczhang.asmbook.plugin"
              artifactId = 'sensorsdata.autosdk'
              version = '1.0.0-SNAPSHOT'

              from components.java

              pom {//(5)
                  name = 'SensorsDataAutoPlugin'
                   description = 'This is a plugin, that provide a easy way to integrate
SensorsData\'s SDK and plugin.'
                  url = 'https://github.com/GvcZhang/SensorsDataAutoPlugin'
                  licenses {
                      license {
                          name = 'The Apache License, Version 2.0'
                          url = 'http://www.apache.org/licenses/LICENSE-2.0.txt'
                      }
                  }
                  developers {
                     developer {
                          id = 'curious'
                          name = 'ZhangWei'
                          email = 'curious.a@qq.com'
                      }
                  }
                  scm {
                      connection = 'scm:git:https://github.com/GvcZhang/SensorsDataAutoPlugin.
      git'
```

```
                        developerConnection = 'scm:git:ssh://github.com:GvcZhang/
SensorsDataAutoPlugin.git'
                    url = 'https://github.com/GvcZhang/SensorsDataAutoPlugin'
                }
            }
        }
    }

    repositories {

        mavenCentral {//(6)
            name = 'OSSRH'
            //publish to local
            // url = '../repo_maven'


            //(7)
            //public to sonatype
                 //url = "https://s01.oss.sonatype.org/content/repositories/
snapshots" //SNAPSHOT 版本
                url = 'https://s01.oss.sonatype.org/service/local/staging/deploy/
maven2/' // 发布到 Maven 中
            credentials { // 配置 OSSRH 服务的账户密码信息
                username = findProperty("ossrhUsername")
                password = findProperty("ossrhPassword")
            }
        }
    }
}

signing {//(8)
    sign publishing.publications.myLibrary
}
```

说明如下。

- 位置（1）引入signing插件，该插件会根据配置的GPG私钥自动对包内容生成加密文件。
- 位置（2）和（3）用于配置javadoc和source包内容。
- 位置（4）和（5）定义Maven Central需要的基本信息。
- 位置（6）和（7）定义仓库的位置信息，其中位置（7）定义Sonatype的基本配置；如果想发布到本地，修改URL即可，不过发布到本地不需要credentials。
- 位置（8）是加密配置，这里表示myLibrary加密。

在正式发布之前，还需要添加 OSSRH 的账户密码以及 GPG 的密钥信息，我们选择在 myPlugin/gradle.properties 中配置，也可以选择在运行任务的时候设置属性，gradle.properties 中的内容类似如下。

```
ossrhUsername=Sonatype #JIRA 账号
ossrhPassword=Sonatype #JIRA 密码

#GPG 信息
signing.keyId=CBA14508 # 公钥短写
signing.password=test1111 # 密码
signing.secretKeyRingFile=../secring.gpg # 导出的私钥文件
```

配置好以上信息后，运行如下代码即可完成发布。

```
$ ./gradlew publishMyLibraryPublicationToOSSRHRepository
```

注意，发布到 Maven Central 上的信息不支持修改和删除，通常需要先发布 SNAPSHOT 版本，验证通过后再同步到 Maven Central 上。

1.7 插件调试

1.7.1 输出日志

输出日志是比较原始的一种查看日志的方式。例如在 Java 文件中，我们可以使用 System.out.println 来输出日志，在 Groovy 或 Kotlin 中，我们可以使用 println 来输出日志。Gradle 中也提供了输出日志的方式，我们可以通过 Project.getLogger() 获取 Logger 对象来输出日志。Gradle 中的 Logger 使用的是 slf4j。使用 Logger 的好处是，我们可以控制日志输出的级别。Gradle 的日志级别是定义在 LogLevel 类中的。

```
package org.gradle.api.logging;

/**
   * The log levels supported by Gradle.
   */
public enum LogLevel {
      DEBUG,
      INFO,
      LIFECYCLE,
      WARN,
      QUIET,
      ERROR
}
```

从枚举的定义可以看出，Gradle 日志级别总共分为 6 种，如表 1-7 所示（级别由低到高依次排列）。

表1-7　Gradle日志级别

级别	调用方法	备注
DEBUG	project.logger.debug(message)	调试信息
INFO	project.logger.info(message)	内容信息
LIFECYCLE	project.logger.lifecycle(message)	进度信息
WARN	project.logger.warn(message)	警告信息
QUIET	project.logger.quiet(message)	重要信息
ERROR	project.logger.error(message)	错误信息

默认的级别是 LIFECYCLE，即默认会输出 LIFECYCLE 及其之上级别的日志信息。那如何控制日志的输出级别呢？可以通过在 gradle 命令后面加上相应的参数来控制。

```
# 默认输出 DEBUG 及其之上级别的日志信息，即默认输出所有日志信息
./gradlew -d build
```

控制日志级别的参数如表 1-8 所示。

表1-8 控制日志级别的参数

参数	说明
无参数	LIFECYCLE及其之上级别的日志信息
-d或 --debug	DEBUG及其之上级别的日志信息
-i或 --info	INOF及其之上级别的日志信息
-w或 --warn	WARN及其之上级别的日志信息
-q或 --quiet	QUIET及其之上级别的日志信息

1.7.2 断点调试

Gradle 插件是在代码的编译器中运行的，所以调试 Android 应用程序（简称应用，缩写为 App 或 app）的方法不再适用于 Android Gradle。下面我们介绍通过断点的方式调试 Gradle 插件。

第一步： 添加断点。

这一步比较简单，单击代码行左侧部分即可添加 / 删除断点，如图 1-20 所示。

```
@Override
public void apply(Project project) {
    System.out.println("Auto Add SensorsData AutoTrack SDK");

    //判断是否已添加了插件，如果没有就对其进行依赖
    Plugin saPlugin = project.getPlugins().findPlugin("com.sensorsdata.analytics.android");
    if (saPlugin == null) {
        //添加神策插件依赖
        project.getPlugins().apply("com.sensorsdata.analytics.android");
        //创建版本号扩展
        project.getExtensions().add("sdkVersion", SDKVersionExtension.class);
        //在 Project 评估结束后，获取扩展信息
        project.afterEvaluate(project1 -> {
            //查找扩展并获取其中的版本号设置
            Object sdkVersionExtension = project1.getExtensions().findByName("sdkVersion");
            if (sdkVersionExtension != null) {
                SDKVersionExtension tmp = (SDKVersionExtension) sdkVersionExtension;
                if (tmp.version != null) {
                    sdkVersion = tmp.version;
                }
            }
            System.out.println("====final==" + sdkVersion);
            //添加对 SDK 的依赖
            project.getDependencies().add("implementation", "com.sensorsdata.analytics.android:Se
        });
    }
}
```

图 1-20 添加断点

第二步： 配置 Run/Debug Configurations。

（1）打开 Edit Configurations，操作方式如图 1-21 所示。

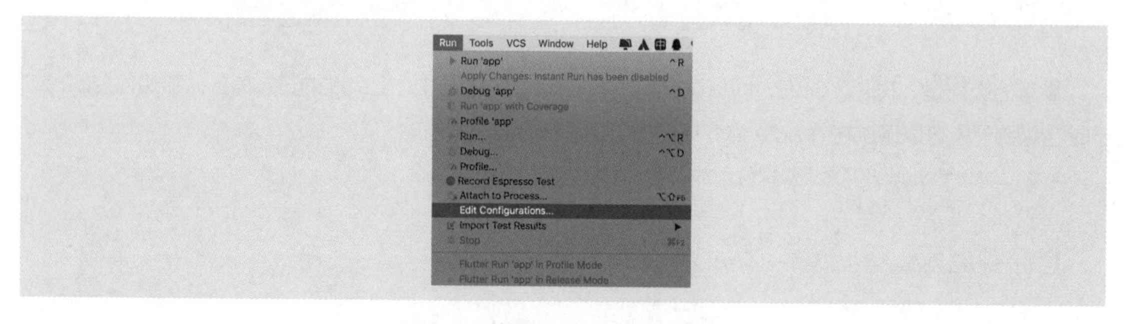

图 1-21 打开 Edit Configurations

（2）建立远程调试任务。

单击 Run/Debug Configurations 左上角的加号，然后选择 Remote，如图 1-22 所示。

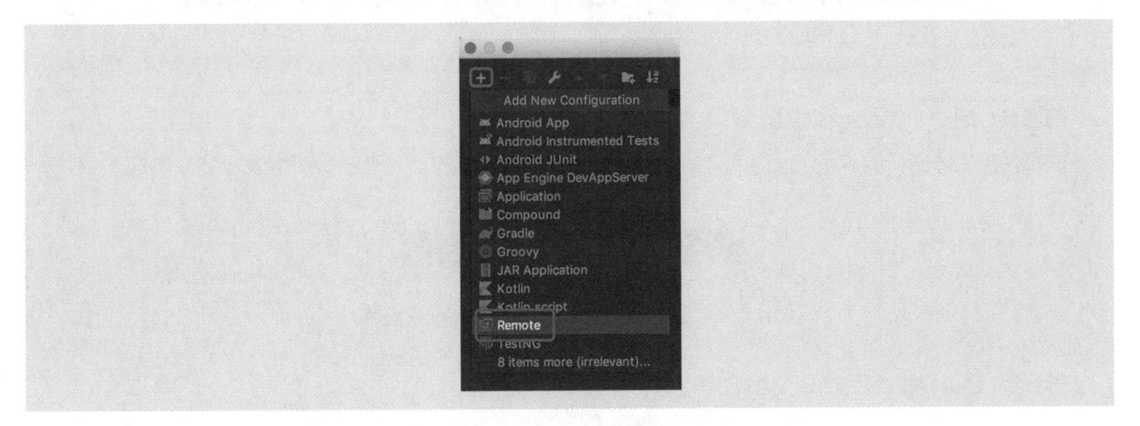

图 1-22　选择 Remote

（3）然后不需要做任何修改，直接单击 OK 按钮即可，如图 1-23 所示。

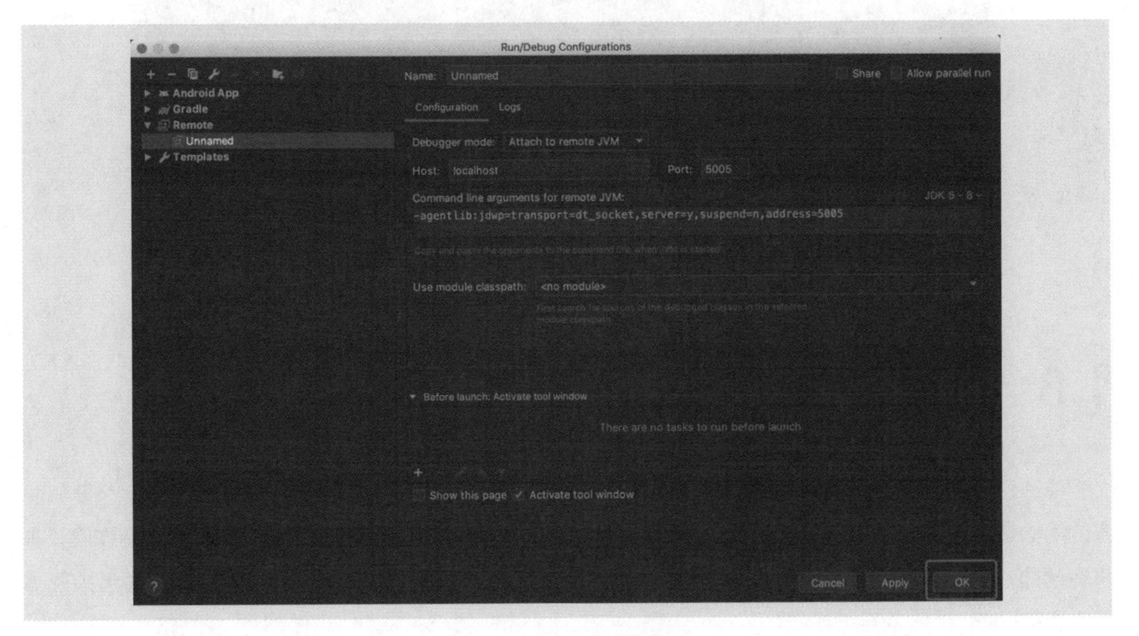

图 1-23　单击 OK 按钮

第三步：执行构建。

在终端执行如下命令，开始构建。

```
./gradlew <任务名> -Dorg.gradle.daemon=false -Dorg.gradle.debug=true
```

注意，操作时，需要把 <任务名> 替换成实际执行的任务。

- -Dorg.gradle.daemon=false：表示不使用守护进程，默认是开启的，也可换成 --no-daemon 选项。
- -Dorg.gradle.debug=true：开始 gradle 进程启动后需要等待调试器连接上才能开始运行。

命令执行之后，可以看到在 Terminal 中整个执行被阻塞了，并输出图 1-24 所示的信息。

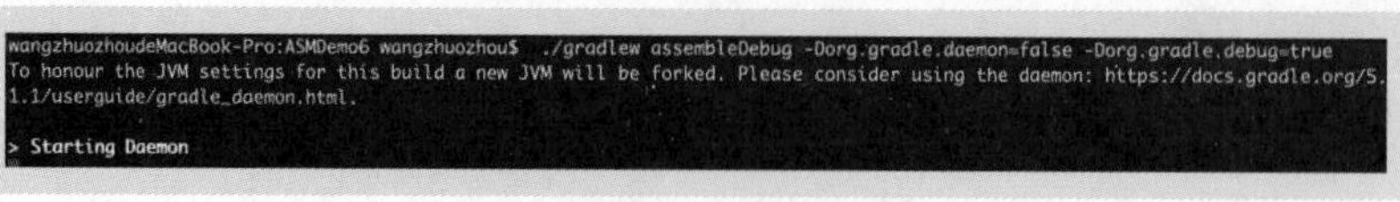

```
wangzhuozhoudeMacBook-Pro:ASMDemo6 wangzhuozhou$  ./gradlew assembleDebug -Dorg.gradle.daemon=false -Dorg.gradle.debug=true
To honour the JVM settings for this build a new JVM will be forked. Please consider using the daemon: https://docs.gradle.org/5.1.1/userguide/gradle_daemon.html.

> Starting Daemon
```

图 1-24　debug 阻塞展示

第四步：启动 Debugger Attach。

选择在第二步新建的 Remote，即 Unnamed，然后单击 debug 按钮，操作如图 1-25 所示。

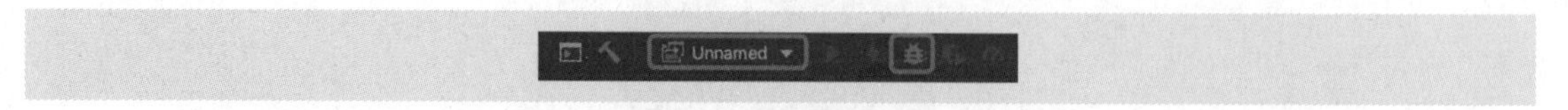

图 1-25　选择 Unnamed 并单击 debug 按钮

第五步：当编译执行到断点处就会停下来，如图 1-26 所示。

图 1-26　debug 进入断点

1.8　小结

本章介绍了 Gradle 基础知识和 Gradle 插件开发中涉及的知识点，读者可能会有疑问：为什么要花很多的篇幅来介绍 Gradle 的基础知识？这是因为 Gradle 的这些知识对后面理解 Android 插件的运作原理非常有帮助，而且整个 Android 应用的构建都依赖 Gradle，所以这些知识对每一个 Android 开发者来说都是需要掌握的。

2. Transform 介绍

Android 使用 Gradle 作为构建工具，为了能构建 Android 应用，对应开发的 Gradle 插件是 Android Gradle Plugin（AGP）。从 AGP 1.5.0-beta1 版本开始，AGP 提供了一个叫 Transform 的 API，该 API 的目标是简化注入自定义类操作，使开发者不必关注任务的处理，并在操作内容上提供更大的灵活性。讲白了就是 Transform API 允许第三方 Gradle 插件能够在编译过程中对 .class 文件进行修改。因为可以修改 .class 文件，自然就可以进行字节码插桩操作。因而市面涌现了大量基于 Transform 的插件，开发出了丰富多彩的功能。

本章将带读者一起了解 Transform 的相关知识，内容包括：

- Android 应用的构建；
- Transform API 介绍；
- Transform 原理介绍。

2.1 Android 应用的构建

在介绍 Transform 之前，先介绍一下 Android 应用的构建流程，以加深读者对 AGP 的认识。

2.1.1 什么是 APK 文件

APK 是 Android Application Package 的缩写，它不是可执行文件，而是一种标准的 ZIP 文件。可以将 APK 文件使用解压缩工具或者拖动到 Android Studio 中打开，打开以后可看到图 2-1 所示的结构。

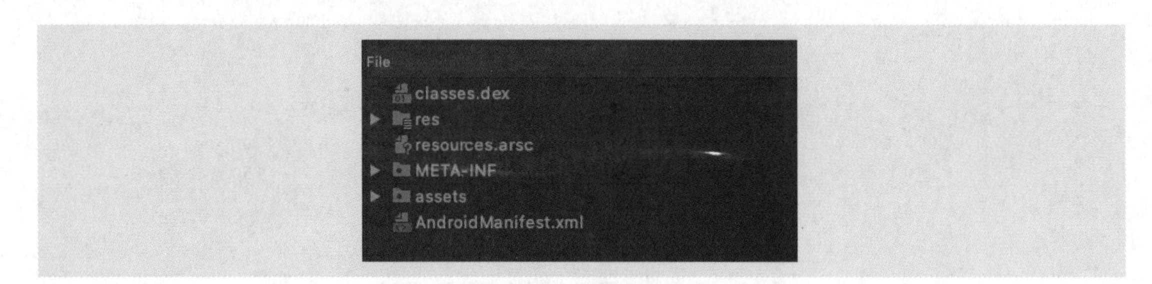

图 2-1　APK 结构

APK 文件由如下几个部分组成。

- **classes.dex：** 该文件是项目中代码编译后最终生成的文件，包含 Android 虚拟机上运行的代码部分。
- **res：** 项目中的资源文件，例如布局、图片等资源。
- **resources.arsc：** 资源文件对应的索引，通过此文件可以定位资源。
- **META-INF：** 此目录中保存了应用的签名和校验信息，用于保证程序的完整性。当生成 APK 时，系统会对目录中的所有内容做一次校验，并将结果保存在这个目录中。手机在安装应用时会对内容再做一次校验并将结果和 META-INF 中的结果进行比较，避免 APK 被恶意篡改。
- **assets：** 项目中使用到的原生资源文件，使用 Asset Mananger 进行管理。

- **AndroidManifest.xml**：用于描述应用的名称、版本、权限、注册的服务等信息。

这些组成部分中，classes.dex 文件比较特殊，2.1.2 小节会介绍。

2.1.2 什么是DEX文件

在介绍 DEX 文件之前先来了解一下 JVM、Dalvik 和 ART 的区别。

- JVM就是Java虚拟机，用来执行Java字节码。
- Dalvik是Google自己设计的用于Android平台的Java虚拟机。
- ART即Android Runtime，是Google为了替代Dalvik而设计的虚拟机，相较Dalvik，它更加高效和省电，并优化了垃圾回收机制，执行的是本地机器码。

这里需要注意一下 Dalvik 和 ART 的区别，在程序运行的过程中，Dalvik 虚拟机会不断地将字节码编译成机器码，而 ART 在程序安装的过程中就已经将所有的字节码重新编译成了机器码，这就是 ART 虚拟机相对 Dalvik 虚拟机来说安装会慢的原因。

DEX（dalvik executable）文件是专为 Dalvik 设计的一种压缩文件，适合内存和处理器速度有限的系统。Dalvik 虚拟机不能直接运行 Java 字节码，需要将编译生成的 .class 文件进行翻译、重构、解释、压缩等处理，这个处理过程由 Google 提供的 DX/D8/R8 等工具完成。通常项目中的所有 .class 字节码文件被处理完以后会生成一个 DEX 文件（可以对比 JAR 包和 .class 文件的关系）。在 DEX 文件中，各个类能够共享公用数据，这在一定程度上减少了冗余，同时文件结构也会更加紧凑。实验表明，DEX 文件比传统的 JAR 包文件小 50% 左右。图 2-2 展示了两种结构的不同。

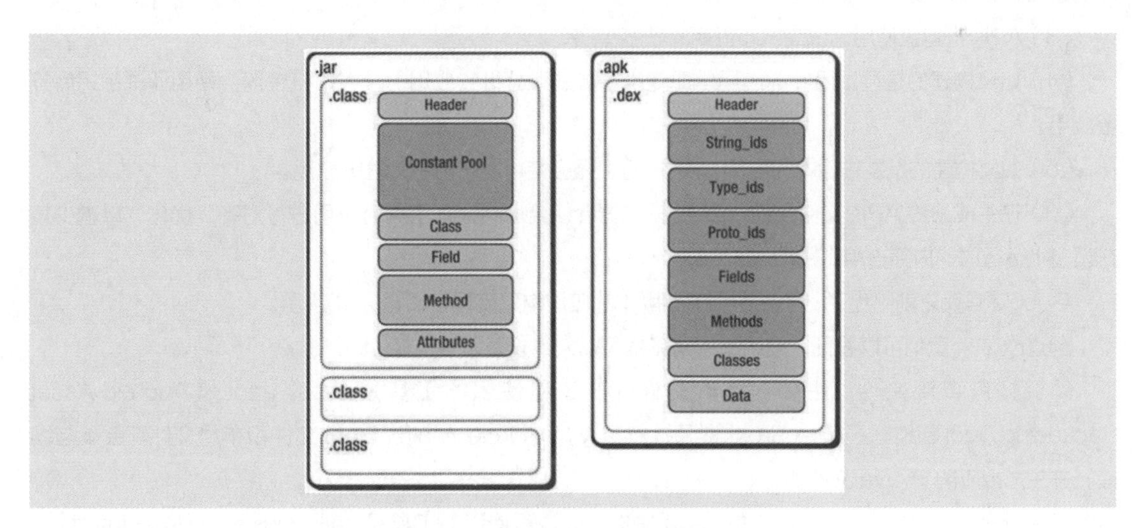

图 2-2 DEX 与 .class 文件结构的对比

了解了什么是 DEX 文件，再来看看典型的 Android 应用的构建流程。

2.1.3 Android应用的构建流程

首先我们看看官方给出的典型 Android 应用的构建流程，如图 2-3 所示。

通过图 2-3 可知，应用的构建流程大概分为如下几步（以下内容摘自 Android 开发者官网）。

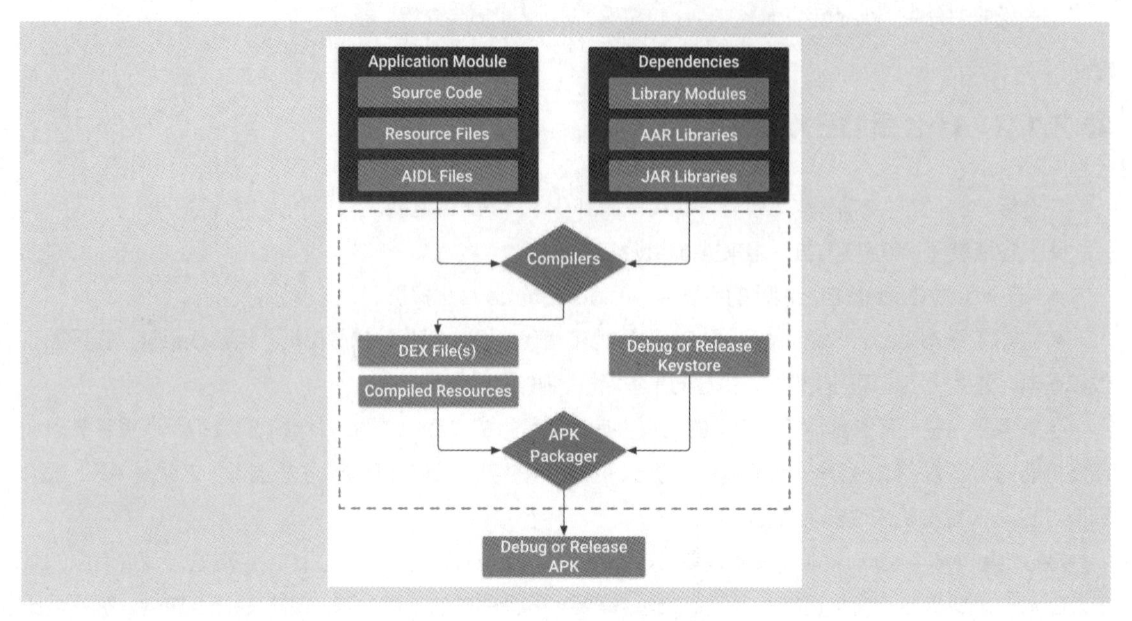

图 2-3　Android 应用的构建流程

（1）编译器将源码转换成 DEX 文件，将其他所有内容转换成已编译资源。

（2）APK 打包器将 DEX 文件和已编译资源合并成单个 APK 文件。不过必须先对 APK 文件进行签名，才能将应用安装并部署到 Android 设备上。

（3）APK 打包器使用调试或发布密钥库签名 APK 文件。

（4）如果构建的是调试版本的应用（即专用于测试和分析的应用），APK 打包器会使用调试密钥库签署应用。

（5）如果构建的是发布版本的应用，APK 打包器会使用发布密钥库签署应用。

（6）在生成最终 APK 文件之前，APK 打包器会使用 zipalign 工具对应用进行对齐、优化，以减少其在设备上运行时的内存占用。

Google 官网之前给出了一个更详细的老版本的应用构建流程，如图 2-4 所示。

通过图 2-4 我们可以看到，应用的构建流程可以大致分为如下步骤。

（1）打包资源文件，生成 R.java 文件：该过程使用的工具是 aapt。aapt 是 Android Asset Packaging Tool 的缩写，即 Android 资源打包工具，AndroidManifest.xml 文件和布局文件都会参与编译，并生成相应的 R.java 文件。

（2）处理 .aidl 文件，生成对应的 .java 文件：该过程使用的工具是 aidl。aidl 是 Android Interface Definition Language 的缩写，即 Android 接口描述语言。aidl 工具会把 .aidl 文件解析成相应的 .java 文件。

（3）编译 .java 文件，生成 .class 文件：该过程使用的工具是 javac。javac 会把项目中的所有 .java 文件（包括 R.java 文件和 .aidl 文件生成的 .java 文件）编译成 .class 文件。

（4）处理 .class 文件，生成 classes.dex 文件：该过程使用的工具是 dx。dx 工具会将所有的 .class 文件（包括第三方库中的 .class 文件 ）转换成 DEX 文件。

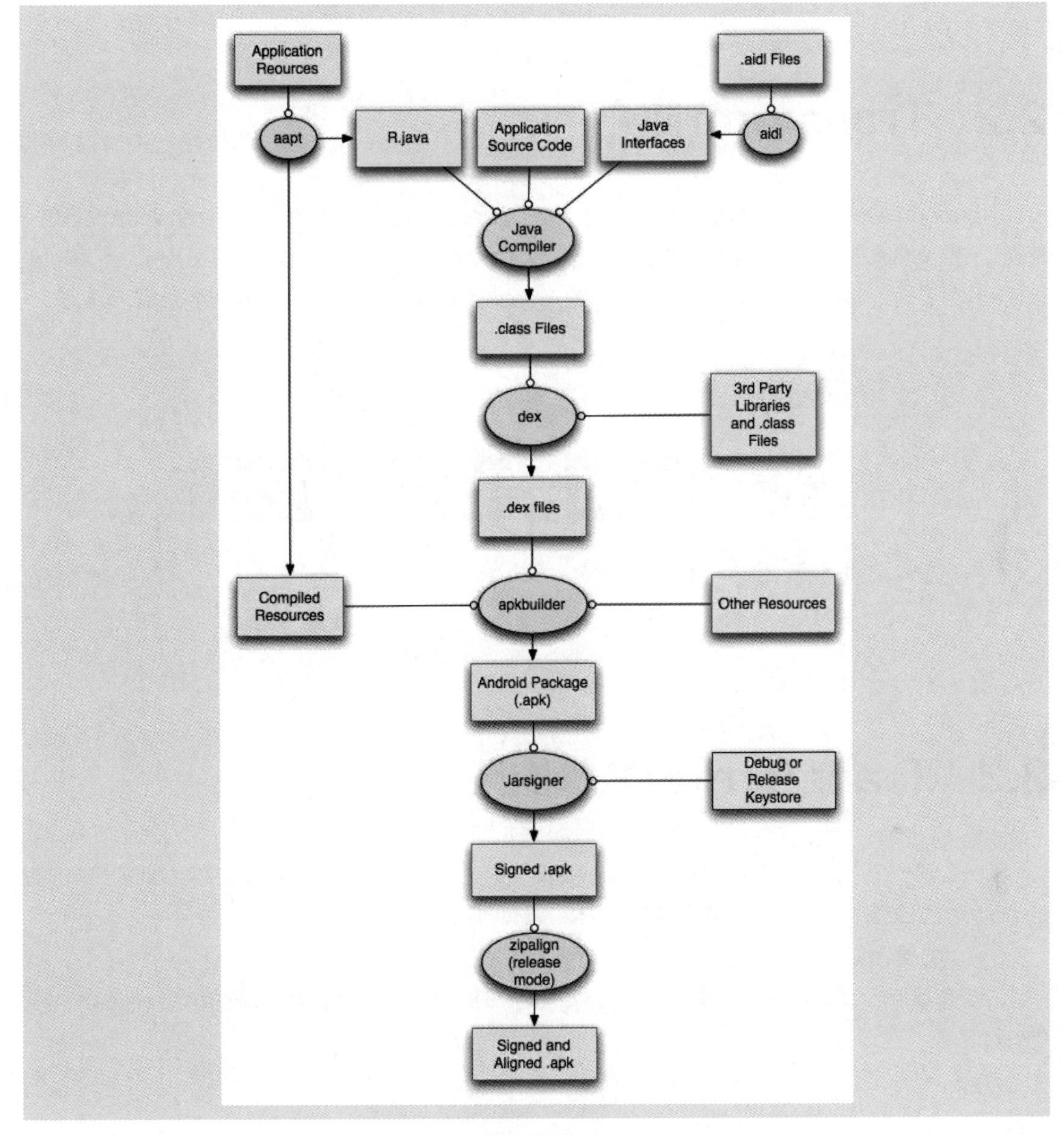

图 2-4 Google 官方老版本的应用构建流程

（5）打包 APK 文件：该过程使用的工具是 apkbuilder。apkbuilder 会将所有的资源（包括未被编译的资源和已被编译的资源）和 .dex 文件打包成 APK 文件。

（6）对 APK 文件进行签名：APK 文件只有被签名了，才能安装到 Android 设备上。

（7）对 APK 文件进行对齐处理：该过程使用的工具是 zipalign。发布正式版的 APK 文件，必须经过字节对齐处理，这样可以减少应用在运行中对内存的使用。

以上是 Android 应用的构建流程，通过以上流程可以发现，AGP 需要处理不同的构建产物，针对不同的输入和输出产物，需要有一个统一的模式来处理，接下来就介绍 AGP 的 Transform。

2.2 Transform简介

Transform 就像 .class 文件的过滤器，内部有一个 Transform Manager 对这些 Transform 进行统一管理。在整个编译的过程中，.class 文件会流入一个 Transform，当前 Transform 加工处理完之后再输出，该输出作为下一个 Transform 的输入，直到所有的 Transform 都处理完成。整个执行顺序如图 2-5 所示。

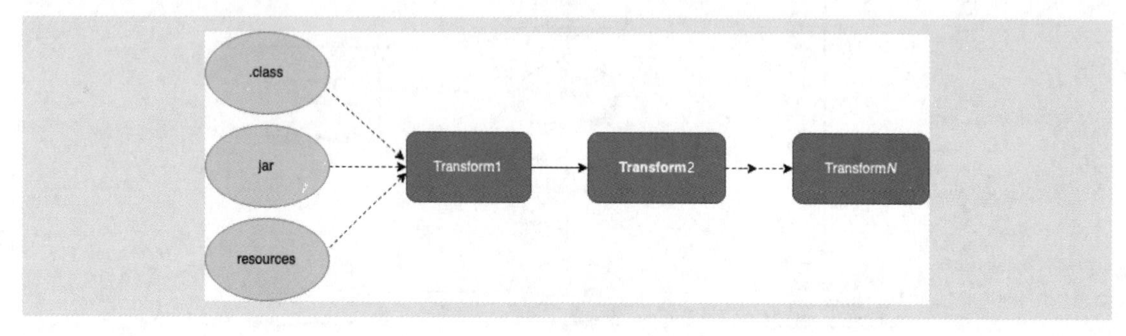

图 2-5　Transform 执行顺序

2.3 Transform的简单应用

首先创建一个 Android 项目，名称是 Chapter2_01，并按照 1.3.2 小节介绍的内容在此项目中创建 buildSrc 类型的插件，之后再对此项目做一些调整，具体调整如下。

（1）为方便后续原理部分的介绍，将 AGP 版本修改为 3.2.0。

（2）在 buildSrc/build.gradle 文件中添加 AGP 的依赖，添加依赖后，就可以使用 Transform API，代码如下。

```
plugins {
      id 'java-library'
}

repositories {
      google()
      mavenCentral()
}

dependencies {
      // 添加 AGP 依赖
      implementation "com.android.tools.build:gradle:3.2.0"
      implementation gradleApi()
}
```

（3）在 buildSrc 的源码部分实现插件内容，这里插件类名定义为 TestTransformPlugin，其代码如下。

```
package cn.sensorsdata.asmbook.buildsrc;
```

```
import com.android.build.gradle.AppExtension;

import org.gradle.api.Plugin;
import org.gradle.api.Project;

public class TestTransformPlugin implements Plugin<Project> {

    @Override
    public void apply(Project project) {
        project.getLogger().warn("Test transform plugin");
        AppExtension appExtension = project.getExtensions().findByType(AppExtension.
class);
        if(appExtension != null){
            appExtension.registerTransform(new MyTransform());
        }
    }
}
```

上述这段代码表示获取 AGP 中提供的 AppExtension，然后调用 appExtension.registerTransform() 方法添加自定义的 MyTransform。继续看 MyTransform 的定义。

（4）MyTransform 的代码如下。

```
package cn.sensorsdata.asmbook.buildsrc;

import com.android.build.api.transform.QualifiedContent;
import com.android.build.api.transform.Transform;
import com.android.build.api.transform.TransformException;
import com.android.build.api.transform.TransformInvocation;
import
    com.android.build.gradle.internal.pipeline.TransformManager;

import java.io.IOException;
import java.util.Set;

public class MyTransform extends Transform {
    @Override
    public String getName() {
        return "chapter2_01";
    }

    @Override
    public Set<QualifiedContent.ContentType> getInputTypes() {
        return TransformManager.CONTENT_CLASS;
    }

    @Override
    public Set<? super QualifiedContent.Scope> getScopes() {
        return TransformManager.SCOPE_FULL_PROJECT;
    }

    @Override
    public boolean isIncremental() {
        return false;
    }

    @Override
```

```
        public void transform(
            TransformInvocation transformInvocation) throws
                            TransformException,
                            InterruptedException,
                            IOException
        {
            System.out.println("=====start transform=====");
            super.transform(transformInvocation);
        }
}
```

MyTransform 继承了 com.android.build.api.transform.Transform 这个类，MyTransform 类中的方法稍后再介绍。

（5）至此插件和 Transform 就定义好了，现在需要把它运行起来。因为使用的是 buildSrc 插件，在 app/build.gradle 工程中直接引入插件。

```
apply plugin:'com.android.application'
apply plugin:cn.sensorsdata.asmbook.buildsrc.TestTransformPlugin
```

然后运行如下命令编译工程。

```
$ ./gradlew clean :app:assembleDebug
```

（6）观察输出结果，首先在控制台中会输出如下内容。

```
$ ./gradlew clean :app:assembleDebug
Starting a Gradle Daemon, 2 incompatible Daemons could not be \
reused, use --status for details

> Configure project :app
Test transform plugin

> Task :app:transformClassesWithChapter2_01ForDebug
=====start transform=====
...

BUILD SUCCESSFUL in 19s
```

在配置阶段输出了“Test transform plugin”，然后执行了一个 Gradle 任务，其名称是“transformClassesWithChapter2_01ForDebug”，通过观察可以发现名称中的“Chapter2_01”就是 MyTransform#getName() 方法返回的值。另外查看 app/build/intermediates/transforms 目录，可以看到有一个文件夹 chapter2_01，这是自定义的 Transform 输出产物的地方，效果如图 2-6 所示。

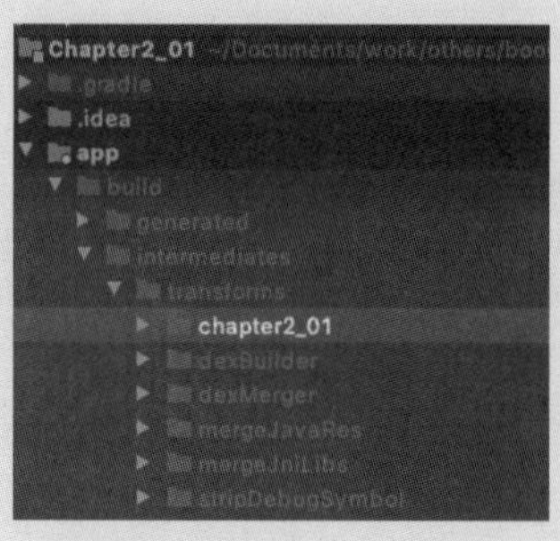

图 2-6 transforms 目录展示

至此简单介绍了 Transform 的应用。本章开头提到 Transform 提供了第三方插件修改 .class 文件的时机，那具体该怎么做呢？接下来详细介绍 Transform 这个类。

2.4 Transform API详细介绍

Transform 是一个抽象类，其定义如下。

```
package com.android.build.api.transform;
...

public abstract class Transform {

      public abstract String getName();

      public abstract Set<ContentType> getInputTypes();

      public Set<ContentType> getOutputTypes() {
          return getInputTypes();
      }

      public abstract Set<? super Scope> getScopes();

      public Set<? super Scope> getReferencedScopes() {
          return ImmutableSet.of();
      }

      public Collection<File> getSecondaryFileInputs() {
          return ImmutableList.of();
      }

      public Collection<SecondaryFile> getSecondaryFiles() {
          return ImmutableList.of();
      }

      public Collection<File> getSecondaryFileOutputs() {
          return ImmutableList.of();
      }

      public Collection<File> getSecondaryDirectoryOutputs() {
          return ImmutableList.of();
      }

      public Map<String, Object> getParameterInputs() {
          return ImmutableMap.of();
      }

      public abstract boolean isIncremental();

      public void transform(
              @NonNull Context context,
              @NonNull Collection<TransformInput> inputs,
```

```
            @NonNull Collection<TransformInput> referencedInputs,
            @Nullable TransformOutputProvider outputProvider,
            boolean isIncremental) throws
                 IOException, TransformException, InterruptedException
    {
    }

    public void transform(@NonNull TransformInvocation transformInvocation)
            throws TransformException, InterruptedException, IOException
    {
        // Just delegate to old method, for code that uses the old API
        //noinspection deprecation
        transform(transformInvocation.getContext(),
                 transformInvocation.getInputs(),
                transformInvocation.getReferencedInputs(),
                transformInvocation.getOutputProvider(),
                transformInvocation.isIncremental());
    }

    public boolean isCacheable() {
        return false;
    }
}
```

下面对 Transform 中的主要方法进行一一介绍。

2.4.1 getName()

该方法的定义如下。

```
/**
   * Returns the unique name of the transform.
   *
   * This is associated with the type of work that the transform does.
   * It does not have to beunique per variant.
   *
   */
@NonNull
public abstract String getName();
```

顾名思义，该方法就是用来指定自定义的 Transform 的名字，需要注意名字的唯一性。在 2.3 节介绍 Transform 的简单使用方法时，已经知道在构建的时候会根据 getName() 的返回值和变体组合成一个任务名字，例如下面这种组合。

```
> Task :app:transformClassesWithChapter2_01ForDebug
```

其实每一个 Transform 都会被转换成 Gradle Task，对应源码中的 TransformTask 这个类，这是 Tranform 最根本的实现原理，关于原理会在 2.7 节中进一步介绍。另外 getName() 的返回值也会作为 Transform 的输出目录的文件夹名，例如图 2-6 所示的结果：app/build/intermediates/transforms/chapter2_01。

2.4.2 getInputTypes()

该方法的定义如下。

```
/**
  * Returns the type(s) of data that is consumed by the Transform.
  * This may be more than one type.
  *
  * This must be of type {@link QualifiedContent.DefaultContentType}
  */
@NonNull
public abstract Set<ContentType> getInputTypes();
```

getInputTypes() 方法返回的是一个 Set 集合，用来指定 Transform 要处理的输入内容的类型。从注释可知，返回值必须是 DefaultContentType 类型的集合。DefaultContentType 的定义如下。

```
/**
  * A content type that is requested through the transform API.
  */
interface ContentType {
      /**
       * Content type name, readable by humans.
       * @return the string content type name
       */
      String name();
      /**
       * A unique value for a content type.
       */
      int getValue();
}
/**
  * The type of of the content.
  */
enum DefaultContentType implements ContentType {
      /**
       * The content is compiled Java code.
       * This can be in a Jar file or in a folder.
       * If in a folder, it is expected to in sub-folders matching package names.
       */
      CLASSES(0x01),
      /** The content is standard Java resources. */
      RESOURCES(0x02);
      private final int value;
      DefaultContentType(int value) {
          this.value = value;
      }
      @Override
      public int getValue() {
          return value;
      }
}
```

DefaultContentType 这个枚举实现了 ContentType 类，其中的枚举值如下。

- CLASSES：表示Transform处理的类型是JAR包或者是包含编译后 .class文件的文件夹。
- RESOURCES：表示标准的Java资源，这里的资源是指Java中除了 .java和 .class以外的文件，一般放在resource文件夹下面。资源文件有很多种，常见的有 .properties文件、.xml文件等。

其实 ContentType 还有一个实现类是 ExtendedContentType，此类中定义了一些系统 Transform 用到的枚举，例如：

- DEX；
- NATIVE_LIBS；
- CLASSES_ENHANCED；
- DATA_BINDING；
- JAVA_SOURCES（已被 @Deprecated标记为过时）；
- DEX_ARCHIVE；
- DATA_BINDING_BASE_CLASS_LOG。

这些资源我们无法使用。另外 TransformManager 已经为我们整合了常用的 ContentType 集合，例如下面的代码部分。

```
public class TransformManager extends FilterableStreamCollection {
        ...
        public static final Set<ContentType> CONTENT_CLASS =
            ImmutableSet.of(CLASSES);
        public static final Set<ContentType> CONTENT_JARS =
            ImmutableSet.of(CLASSES, RESOURCES);
        public static final Set<ContentType> CONTENT_RESOURCES =
            ImmutableSet.of(RESOURCES);
        ...
}
```

通常getInputTypes()方法返回值设置为TransformManager.CONTENT_CLASS即可（可对比2.3节的 MyTransform 代码）。

2.4.3 getScopes()

该方法的定义如下。

```
/**
   * Returns the scope(s) of the Transform.
   * This indicates which scopes the transform consumes.
   */
@NonNull
public abstract Set<? super Scope> getScopes();
```

getInputTypes() 方法定义了 Transform 消费的资源类型，getScopes() 方法则定义了 Transform 消费资源的范围。此方法返回的是 Scope 类型的 Set 集合，Scope 的定义如下。

```
/**
   * Definition of a scope.
```

```
   */
interface ScopeType {
      /**
       * Scope name, readable by humans.
       * @return a scope name.
       */
      String name();
      /**
       * A scope binary flag that will be used to encode directory names.
       * Must be unique.
       * @return a scope binary flag.
       */
      int getValue();
}
/**
   * The scope of the content.
   *
   * <p>
   * This indicates what the content represents,
   * so that Transforms can apply to only part(s)
   * of the classes or resources that the build manipulates.
   */
enum Scope implements ScopeType {
      /** Only the project (module) content */
      PROJECT(0x01),
      /** Only the sub-projects (other modules) */
      SUB_PROJECTS(0x04),
      /** Only the external libraries */
      EXTERNAL_LIBRARIES(0x10),
      /** Code that is being tested by the current variant,
       *including dependencies */
      TESTED_CODE(0x20),
      /** Local or remote dependencies that are provided-only */
      PROVIDED_ONLY(0x40),
      /**
       * Only the project's local dependencies (local jars)
       *
       * @deprecated local dependencies are now processed
       * as {@link #EXTERNAL_LIBRARIES}
       */
      @Deprecated
      PROJECT_LOCAL_DEPS(0x02),
      /**
       * Only the sub-projects's local dependencies (local jars).
       *
       * @deprecated local dependencies are now processed
       * as {@link #EXTERNAL_LIBRARIES}
       */
      @Deprecated
      SUB_PROJECTS_LOCAL_DEPS(0x08);
      private final int value;
      Scope(int value) {
          this.value = value;
      }
```

```
        @Override
        public int getValue() {
            return value;
        }
}
```

可以看到 Scope 也是一个枚举，各个枚举值的解释如表 2-1 所示。

表2-1　Scope类型枚举值说明

序号	Scope类型	说明
1	PROJECT	只作用于当前项目，即使用此插件的模块
2	SUB_PROJECTS	只作用于子项目（模块），PROJECT模块依赖的子模块
3	EXTERNAL_LIBRARIES	只作用于外部的依赖库，PROJECT/SUB_PROJECTS中依赖的JAR、AAR库
4	TESTED_CODE	只作用于测试代码以及测试的依赖项
5	PROVIDED_ONLY	只作用于通过compileOnly方式依赖的库
6	PROJECT_LOCAL_DEPS	只作用于当前项目的本地依赖，例如AAR、JAR（已过期，被EXTERNAL_LIBRARIES替代）
7	SUB_PROJECTS_LOCAL_DEPS	只作用于子项目的本地依赖，例如AAR、JAR（已过期，被EXTERNAL_LIBRARIES替代）

除了以上 7 种类型，还有另外 3 种类型，它们定义在 Internal Scope 枚举中。

- MAIN_SPLIT；
- LOCAL_DEPS；
- FEATURES。

其中，MAIN_SPLIT 是 InstantRun 专用的，LOCAL_DEPS 是 AAR 包专用的。这 3 种类型读者在开发过程中是无法使用的，只能在 AGP 内部使用。类似地，TransformManager 也已经为我们整合了常用的 Scope 集合，如下代码所示。

```
public class TransformManager extends FilterableStreamCollection {

        ...
        public static final Set<Scope> EMPTY_SCOPES = ImmutableSet.of();

        public static final Set<ScopeType> PROJECT_ONLY =
            ImmutableSet.of(Scope.PROJECT);
        public static final Set<Scope> SCOPE_FULL_PROJECT =
            Sets.immutableEnumSet(
                        Scope.PROJECT,
                        Scope.SUB_PROJECTS,
                        Scope.EXTERNAL_LIBRARIES);
        ...
}
```

通常情况下，如何选择类型呢?

假如插件是作用在 app 工程里，就像“com.android.application”这个插件一样，需要处理所有的 .class 文件，此时可以使用 TransformManager.SCOPE_FULL_PROJECT；如果想要在 library 模块中使用，就像“com.android.library”插件只关注当前的 module 工程中的代码，则可以使用 TransformManager.PROJECT_ONLY。具体的使用场景可以根据上面的类型说明自行组合。

2.4.4 transform()

该方法的定义如下。

```
public void transform(@NonNull TransformInvocation transformInvocation)
        throws TransformException, InterruptedException, IOException {
    // Just delegate to old method, for code that uses the old API
    //noinspection deprecation
    transform(transformInvocation.getContext(), transformInvocation.getInputs(),
            transformInvocation.getReferencedInputs(),
            transformInvocation.getOutputProvider(),
            transformInvocation.isIncremental());
}
```

该方法是 Transform 的核心处理逻辑。AGP 会根据重写 Transform 中相关的 Scopes()、Inputs() 配置方法，将需要消费的产物和指定的输出环境封装在 TansformInvocation 对象中，并通过 transform(TransformInvocation transformInvocation) 方法交给用户使用。

TansformInvocation 是一个接口，其方法定义如表 2-2 所示。

表2-2 TransformInvocation接口方法定义

方法	作用
Context getContext()	获取当前 Transform 的上下文信息，通过它可以获取到 LoggingManager 对象、拥有读写权限的临时目录、任务路径、变体名称等
Collection<TransformInput> getInputs()	获取当前 Transform 能够消费的数据，例如获取被消费的 JAR 包或者目录（被消费的数据是由 getScopes() 和 getInputTypes() 结果共同决定的）
Collection<TransformInput> getReferencedInputs()	获取当前 Transform 能够消费的引用类型信息。在 2.4.5 小节介绍 getReferencedScopes() 方法时提到如果只想查看 Transform 消费了哪些内容，可通过此方法获取
Collection<SecondaryInput> getSecondaryInputs()	根据 getSecondaryFiles() 方法返回的内容，获取到额外变动的文件，此方法会在 2.4.8 小节详细介绍
TransformOutputProvider getOutputProvider()	在消费完输入产物后，需要将产物输出到指定的地方，此方法提供了输出的位置信息
boolean isIncremental()	判断此次 Transform 执行是否以增量形式执行

为方便理解，再来看看 TransformInput 这个接口中的定义的方法。

TransformInput 接口方法定义如表 2-3 所示。

表2-3 TransformInput接口方法定义

方法	作用
Collection<JarInput> getJarInputs()	获取包含 JAR 包的数据
Collection<DirectoryInput> getDirectoryInputs()	获取目录中的数据，例如 app 模块工程中的代码，不包括其依赖的 JAR 包和模块。注意，项目中子模块工程中的代码在编译的过程中也会生成 JAR/AAR 包，然后给 app 模块使用

下面通过例子演示 tranform() 方法的使用。该例子的作用是将 Transform 获取到的输入数据不做任何处理并输出到指定的位置。

复制项目 Chapter2_01，将其副本改名为 Chapter2_02。在 buildSrc 那里新建 Transform 的实现类，名称是 CopyFileTransform，然后使用 appExtension.registerTransform() 方法进行注册，具体步骤不再赘述。CopyFileTransform 的代码如下。

```
public class CopyFileTransform extends Transform {
      @Override
      public String getName() {
          return "chapter2_02";
      }

      // 定义消费类型
      @Override
      public Set<QualifiedContent.ContentType> getInputTypes() {
          return TransformManager.CONTENT_CLASS;
      }

      // 定义消费作用域
      @Override
      public Set<? super QualifiedContent.Scope> getScopes() {
          return TransformManager.SCOPE_FULL_PROJECT;
      }

      // 是否支持增量编译
      @Override
      public boolean isIncremental() {
          return false;
      }

      @Override
      public void transform(TransformInvocation transformInvocation)
          throws TransformException, InterruptedException, IOException
      {
          System.out.println("=====start transform=====");
          // 获取需要消费的数据
          Collection<TransformInput> inputCollection =
              transformInvocation.getInputs();
          // 遍历数据
          inputCollection.parallelStream()
              .forEach((TransformInput transformInput) -> {

                  //1. 获取 JAR 包类型的输入
                  Collection<JarInput> jarInputCollection
                      = transformInput.getJarInputs();
                  jarInputCollection.parallelStream().forEach(jarInput -> {
                      // 获取 JAR 包文件
                      File file = jarInput.getFile();
                      // 获取输出的目标文件
                      File outputFile = transformInvocation
                          .getOutputProvider()
```

```
                    .getContentLocation(file.getAbsolutePath(),
                          jarInput.getContentTypes(),
                          jarInput.getScopes(),
                          Format.JAR);
                //将数据复制到指定目录
                try {
                    FileUtils.copyFile(file, outputFile);
                } catch (IOException e) {
                    e.printStackTrace();
                }
            });

            //2. 获取源码编译的文件夹输入
            Collection<DirectoryInput> directoryInputCollection =
                transformInput.getDirectoryInputs();
            directoryInputCollection.parallelStream()
                .forEach(directoryInput -> {
                    //获取源码编译后对应的文件夹
                    File file = directoryInput.getFile();
                    //获取输出的目标
                    File outputDir = transformInvocation
                        .getOutputProvider()
                        .getContentLocation(file.getAbsolutePath(),
                              directoryInput.getContentTypes(),
                              directoryInput.getScopes(),
                              Format.DIRECTORY);
                    //将数据复制到指定目录
                    try {
                        //outputDir不存在，需要创建
                        FileUtils.forceMkdir(outputDir);
                        FileUtils.copyDirectory(file, outputDir);
                    } catch (IOException e) {
                        e.printStackTrace();
                    }
            });
        });
    }
}
```

上述代码依次遍历了 JAR 包和目录中的文件，并通过 transformInvocation 对象获取输出目录，将结果输出到指定的目录中。运行后可以在 build/intermediates/transforms/chapter2_02/ 目录中查看输出的结果，如图 2-7 所示。

观察图 2-7 中的结果，其中 xx.jar 是 jarInput 对应的输出，名字为 0 的文件夹是 directoryInput 对应的输出结果，即 app 模块的项目源码结果。另外还有一个 _content_.json 文件，稍后介绍。先思考如下两个问题。

- 如果不重写 Transform 的 transform() 方法会不会有问题？
- 图 2-7 输出的 JAR 包为什么都是以数字开头的，其对应的真实 JAR 包是什么？

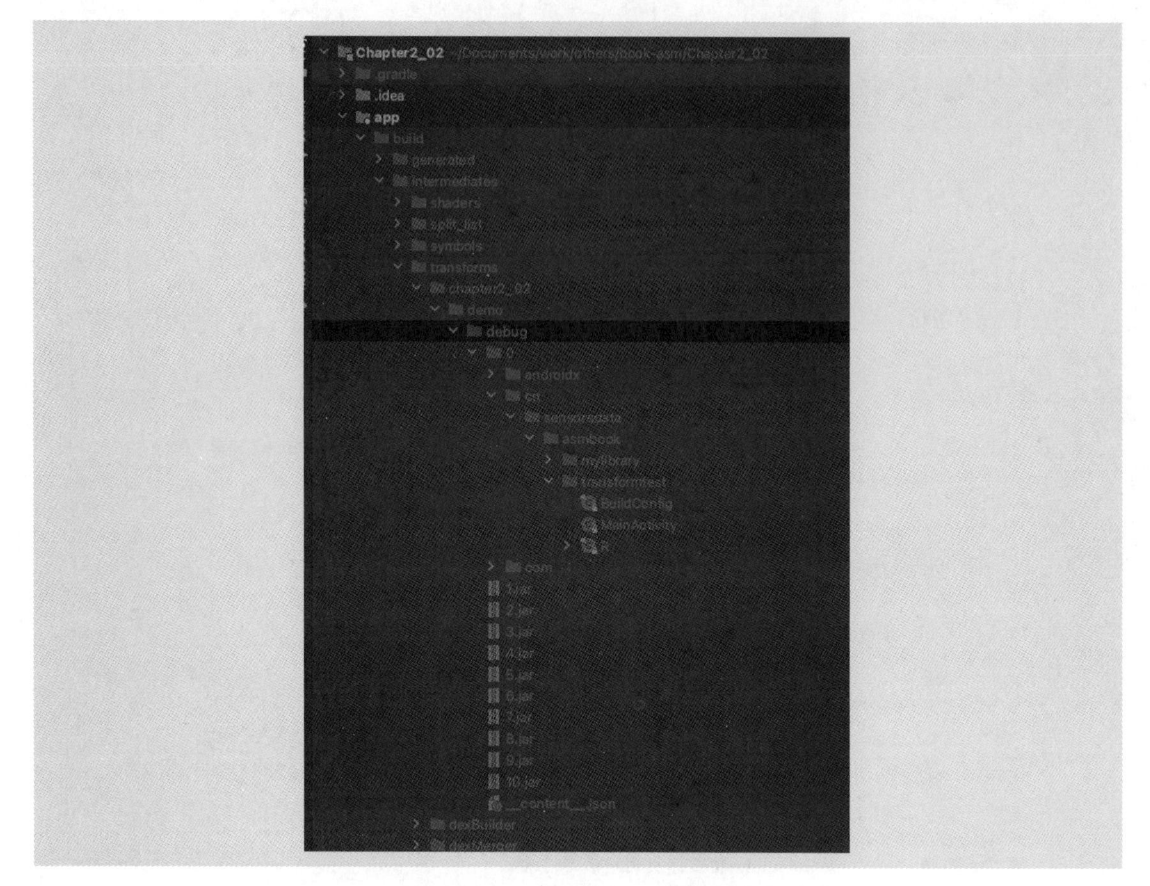

图 2-7　CopyFileTransform 结果展示

（1）如果不重写 transform() 方法会不会有问题？

Transform 是串行运行的，每一个 Transform 会接收上一个 Transform 的输出资源作为其输入资源，假如不重写 transform() 方法，就无法将上一个的输出资源传递给下一个 Transform，所以打包的时候就没有代码，图 2-8 所示为不重写 transform() 方法生成的 APK 反编译的结果。

图 2-8　不重写 transform() 生成的 APK 反编译的结果

因此自定义的 Transform 一定要将资源处理后输出到指定的位置。

（2）为什么输出的 JAR 包都是以数字命名的呢？

输出的文件名是由 transformInvocation.getOutputProvider().getContentLocation() 方法获取的，该方法在类中的定义如下。

```
public interface TransformOutputProvider {

    /**
     * 删除所有的输出内容，该方法在非增量编译的时候非常有用，请大家记住该方法
     */
    void deleteAll() throws IOException;

    /**
     * 根据给定的 Scopes、ContentTypes、Format 返回一个输出目录
     *
     * 如果格式是 Format#DIRECTORY，则返回的结果是一个目录
     * 如果格式是 Format#JAR，则返回结果是一个表示此 JAR 包的文件
     *
     * @param name 为代表文件的唯一值，一般使用文件路径来作为其值
     * @param types 文件类型，参考 CopyFileTransform 中的用法
     * @param scopes 作用域，参考 CopyFileTransform 中的用法
     * @param format 文件格式，参考 CopyFileTransform 中的用法
     * @return the location of the content
     */
    @NonNull
    File getContentLocation(
            @NonNull String name,
            @NonNull Set<QualifiedContent.ContentType> types,
            @NonNull Set<? super QualifiedContent.Scope> scopes,
            @NonNull Format format);
}
```

其中，传给 getContentLocation() 方法的第一个参数 name 需要是唯一的，否则此方法返回重名的文件会导致文件写入失败的异常；types 和 scopes 定义成输入源的 types 和 scopes 即可；format 有两种类型，一种是 JAR，一种是 DIRECTORY，根据输入源获取即可，此方法最终会返回根据输入信息而得到的指定文件路径。至于输出的 JAR 包以数字命名的原因，是 getContentLocation() 方法的实现使用一个数值累加器来当作每个 JAR 包的名字。

如果想知道 0.jar 这个包的真实名称，可以查看 _content_ .json 文件，此文件记录了对应关系，内容如下所示。

```
[
    {
        "name": "/Users/wangzhzh/.gradle/caches/transforms-1/files-1.1/core-\
            runtime-2.0.0.aar/626a3ebec/jars/classes.jar",
        "index": 0,
        "scopes": [
            "EXTERNAL_LIBRARIES"
        ],
        "types": [
            "CLASSES"
        ],
        "format": "JAR",
        "present": true
    },
    {
        "name": "/Users/userHome/.gradle/caches/transforms-1/files-1.1/ \
            drawerlayout-1.0.0.aar/2eb7664c/jars/classes.jar",
        "index": 1,
        "scopes": [
```

```
            "EXTERNAL_LIBRARIES"
        ],
        "types": [
            "CLASSES"
        ],
        "format": "JAR",
        "present": true
    },
    ...
]
```

例如 index 为 0，表示 0.jar 包的名字，name 是原 JAR 包的路径，从路径中可以得出其包名是 core-runtime-2.0.0.aar。

2.4.5 getReferencedScopes()

该方法的定义如下。

```
/**
   * Returns the referenced scope(s) for the Transform.
   * These scopes are not consumed by the Transform. They are provided as inputs,
   * but are still available as inputs for
   * other Transforms to consume.
   *
   * <p>The default implementation returns an empty Set.
   */
@NonNull
public Set<? super Scope> getReferencedScopes() {
      return ImmutableSet.of();
}
```

对比 2.4.3 小节的 getScopes()，getReferencedScopes() 也定义了输入产物的作用域，不过只能查看输入产物，不要求对产物做处理，并且也并不影响后续的 Transform 继续消费输入内容。因此对于 CopyFileTransform 类，可以换一种方式来实现。

首先复制 Chapter2_02，将其副本命名为 Chapter2_03，并将 CopyFileTransform 的内容修改成如下。

```
package cn.sensorsdata.asmbook.buildsrc;
...

public class CopyFileTransform extends Transform {
      @Override
      public String getName() {
          return "chapter2_03";
      }

      @Override
      public Set<QualifiedContent.ContentType> getInputTypes() {
          return TransformManager.CONTENT_CLASS;
      }

      @Override
      public Set<? super QualifiedContent.Scope> getScopes() {
          return ImmutableSet.of(); // 此处返回空的集合
```

```
        }

        @Override
        public Set<? super QualifiedContent.Scope> getReferencedScopes() {
            return TransformManager.SCOPE_FULL_PROJECT;
        }

        @Override
        public void transform(TransformInvocation transformInvocation)
            throws TransformException, InterruptedException, IOException
        {
            System.out.println("=====start transform=====");

            // 测试 getReferencedScopes()
            transformInvocation.getReferencedInputs().stream()
                .forEach(transformInput -> {

                    Collection<DirectoryInput> directoryInputs =
                        transformInput.getDirectoryInputs();
                    Collection<JarInput> jarInputs = transformInput.getJarInputs();
                    System.out.println("directoryInputs size: "
                                     + directoryInputs.size()
                                     + "===jarInputs size: "
                                     + jarInputs.size() + "===");

                    transformInput.getDirectoryInputs().stream()
                        .forEach(directoryInput -> {
                            System.out.println("directoryInput====="
                                         + directoryInput.getFile());
                    });

                    transformInput.getJarInputs().stream().forEach(jarInput -> {
                        System.out.println("jarInput=====" +jarInput.getFile());
                    });
            });
        }

        @Override
        public boolean isIncremental() {
            return false;
        }
    }
```

注意看，上述代码中 getScopes() 方法返回了一个空集合，getReferencedScopes() 方法返回了针对整个项目的作用域，同时在 transform() 方法中通过 TransformInvocation 的 getReferencedInputs() 方法获取了所有的输入数据，并且未对输入数据做任何处理。

2.4.6 isIncremental()

1.2.10 小节介绍了什么是增量编译，简单地说就是如果一个 Task 的输入和输出没有发生变化，那么会跳过该任务（UP-TO-DATE）；如果 Gradle 具有先前任务执行的历史记录，并且自执行以来对任务执行上下文的唯一更改就是输入文件，则 Gradle 能够确定任务需要重新处理哪些输入文件，在这种情况下，将为新增或修改的任何输入文件执行 outOfDate 操作，为所有已删除的输入文件执行 removed 操作。

Transform 并不是一个 Gradle Task，其对应的 TransformTask 是一个 Task，TransformTask 支持增量编译。Transform API 本身提供了 isIncremental() 方法，只有该方法返回结果为 true 的时候，TransformTask 才会将增量构建的信息传递到上下文 TransformInvocation 中，提供给相应的 Transform 使用。

注意 isIncremental() 方法返回 true 的时候只是告诉对应的 TransformTask 支持增量构建，当次构建是否是增量，则需通过 TransformInvocation#isIncremental() 方法来判断。如果当次支持增量构建，只需操作被修改的文件即可。

2.4.7 isCacheable()

isCacheable() 方法的返回值用来告诉 Transform 对应的 TransformTask 是否支持构建缓存，如此可以将 Task 对应的输出缓存起来，当下次再编译的时候如果发现该 Task 已经有对应的构建缓存，则直接复用，从而减少全量构建的时间。

2.4.8 getSecondaryFiles()

该方法的定义如下。

```
/**
   * Returns a list of additional file(s) that this Transform needs to run.
   *
   * <p>Changes to files returned in this list will trigger
   * a new execution of the Transform even if the qualified-content
   * inputs haven't been touched.
   *
   * <p>Each secondary input has the ability to be declared as necessitating
   * a non incremental execution in case of change. This Transform can
   * therefore declare which secondary file changes it supports in incremental mode.
   *
   * <p>The default implementation returns an empty collection.
   */
@NonNull
public Collection<SecondaryFile> getSecondaryFiles() {
      return ImmutableList.of();
}
```

简单地说就是可以将额外的文件作为输入内容添加到 Transform 中，这些文件的变动也会像 .class 文件一样影响 Transform 的执行结果，这些结果同样可以通过 TransformInvocation 获取。

举个例子，复制 Chapter2_02，将其副本命名为 Chapter2_04，并将 CopyFileTransform 的内容修改成如下。

```
public class CopyFileTransform extends Transform {
    @Override
    public String getName() {
        return "chapter2_04";
    }

    //定义消费类型
    @Override
    public Set<QualifiedContent.ContentType> getInputTypes() {
        return TransformManager.CONTENT_CLASS;
    }

    //定义消费作用域
    @Override
    public Set<? super QualifiedContent.Scope> getScopes() {
        return TransformManager.SCOPE_FULL_PROJECT;
    }

    //是否支持增量编译
    @Override
    public boolean isIncremental() {
        return true;
    }

    @Override
    public Collection<SecondaryFile> getSecondaryFiles() {
        File file = new File("/Users/wangzhzh/Documents/work/others/book \
                            -asm/Chapter2_04/log.txt");
        return ImmutableList.of(SecondaryFile.incremental(file));
    }


    @Override
    public void transform(TransformInvocation transformInvocation)
        throws TransformException, InterruptedException, IOException
    {
        System.out.println("=====start transform====="
                          +transformInvocation.isIncremental());
        ...

        Collection<SecondaryInput> secondaryInputs =
            transformInvocation.getSecondaryInputs();
        System.out.println("======secondaryInputs===" + secondaryInputs.size());
        secondaryInputs.stream().forEach(input -> {
            SecondaryFile secondaryFile = input.getSecondaryInput();
            Status status = input.getStatus();
            System.out.println("===transform secondary file===="
                              + secondaryFile.getFile() + "====="
                              + secondaryFile.supportsIncrementalBuild());
            System.out.println("===transform secondary file status===="
                              + status.name());
        });
    }
}
```

上述代码中重写了 getSecondaryFiles() 方法，并且返回了一个本地的文件。清空项目后执行一遍编译，可以发现 secondaryInputs.size() 的值为 0。此时修改 log.txt 文件的内容，再次执行编译，可以发现 secondaryInputs.size() 的值为 1，status.name() 的值变成了 CHANGED。

以上就是对 Transform 中主要 API 的介绍和基本用法，读者可以使用书中的例子测试验证，加深理解。

2.5 Transform 模板

在实际开发中，Transform 的实现方法基本都是一样的。下面给出一个通用做法，读者在使用的时候可以直接按照模板来套用即可。

复制 Chapter2_01，将其副本命名为 Chapter2_05，在 buildSrc 中创建 BoilerplateIncrementalTransform 类，对应的代码如下。

```
package cn.sensorsdata.asmbook.buildsrc;

import com.android.build.api.transform.DirectoryInput;
import com.android.build.api.transform.Format;
import com.android.build.api.transform.JarInput;
import com.android.build.api.transform.QualifiedContent;
import com.android.build.api.transform.Status;
import com.android.build.api.transform.Transform;
import com.android.build.api.transform.TransformException;
import com.android.build.api.transform.TransformInput;
import com.android.build.api.transform.TransformInvocation;
import com.android.build.gradle.internal.pipeline.TransformManager;

import org.apache.commons.codec.digest.DigestUtils;
import org.apache.commons.io.FileUtils;
import org.apache.commons.io.IOUtils;
import org.apache.commons.io.output.ByteArrayOutputStream;

import java.io.File;
import java.io.FileOutputStream;
import java.io.IOException;
import java.io.InputStream;
import java.util.Collection;
import java.util.Enumeration;
import java.util.Map;
import java.util.Set;
import java.util.jar.JarEntry;
import java.util.jar.JarFile;
import java.util.jar.JarOutputStream;

/**
  * 普通写法的样板代码
  */
public class BoilerplateIncrementalTransform extends Transform {

    @Override
    public String getName() {
```

```
    return "boilerplate_incremental";
}

@Override
public Set<QualifiedContent.ContentType> getInputTypes() {
    return TransformManager.CONTENT_CLASS;
}

@Override
public Set<? super QualifiedContent.Scope> getScopes() {
    return TransformManager.SCOPE_FULL_PROJECT;
}

@Override
public boolean isIncremental() {
    return true;
}

@Override
public void transform(TransformInvocation transformInvocation)
    throws TransformException, InterruptedException, IOException {
    System.out.println("=====start transform=====");
    if (!transformInvocation.isIncremental()) {
        transformInvocation.getOutputProvider().deleteAll();
    }
    // 获取需要消费的数据
    Collection<TransformInput> inputCollection
        = transformInvocation.getInputs();
    // 遍历数据
    inputCollection.parallelStream()
    .forEach((TransformInput transformInput) -> {
        //1. 获取 Jar 包类型的输入
        Collection<JarInput> jarInputCollection
            = transformInput.getJarInputs();
        jarInputCollection.parallelStream().forEach(jarInput -> {
            // 处理 Jar 包
            processJarFile(jarInput, transformInvocation);
        });

        //2. 获取源码编译的文件夹输入
        Collection<DirectoryInput> directoryInputCollection
            = transformInput.getDirectoryInputs();
        directoryInputCollection.parallelStream()
            .forEach(directoryInput -> {
                // 处理 directory
                processDirectoryFile(directoryInput, transformInvocation);
            });
    });
}

/**
 * 处理 Jar 包，我们在这个方法中会获取 Jar 包中的文件，
 * 但不对其中的文件做特殊处理，而是直接返回。
 *
 * @param jarInput JarInput
```

```
     * @param transformInvocation TransformInvocation
     */
    private void processJarFile(JarInput jarInput,
                                TransformInvocation transformInvocation)
    {
        // 获取 Jar 包文件
        // 例如: /Users/username/.gradle/caches/transforms-1/files-1.1/
        //appcompat-1.2.0.aar/e80edd062e6d61edb3235af96b64619d/
        //jars/classes.jar
        File file = jarInput.getFile();

        // 获取输出的目标文件
        // 例如: /Users/username/Documents/work/others/book-asm/Chapter2_05/
        //app/build/intermediates/transforms/boilerplate_incremental/
        //debug/40.jar
        File outputFile = transformInvocation.getOutputProvider()
            .getContentLocation(file.getAbsolutePath(),
                jarInput.getContentTypes(),
                jarInput.getScopes(), Format.JAR);

        // 将数据复制到指定目录
        try {
            if (transformInvocation.isIncremental()) {
                switch (jarInput.getStatus()){
                    case REMOVED:
                        FileUtils.forceDelete(outputFile);
                        break;
                    case CHANGED:
                    case ADDED:
                        // 例如修改此 Jar 包文件中的 Class 文件
                        File modifiedJarFile
                            = modifyJar(file,
                                        transformInvocation
                                        .getContext().getTemporaryDir());
                        FileUtils
                            .copyFile(modifiedJarFile != null ?
                                      modifiedJarFile : file, outputFile);
                        break;
                    case NOTCHANGED:
                        break;
                }
            } else {
                // 例如修改此 Jar 包文件中的 Class 文件
                File modifiedJarFile
                    = modifyJar(file,
                                transformInvocation
                                .getContext().getTemporaryDir());
                FileUtils.copyFile(modifiedJarFile != null ?
                                   modifiedJarFile : file, outputFile);
            }
        } catch (IOException e) {
            e.printStackTrace();
        }
    }
```

```
/**
 * 对 Jar 包中的内容进行处理，通常是处理 Class 文件
 *
 * @param file jar 包对应的 File
 * @param tempDir 输出临时文件用的文件夹
 * @return 返回修改过的 Jar 包，如果返回值是 null，表示未成功修改，
 *          在这种情况下直接复制原有文件即可
 */
private File modifyJar(File file, File tempDir) {
    // 这里需要进行判断，可能是一个空 Jar 包，
    // 如果是将会抛出 : ZipException: zip file is empty
    if (file == null || file.length() == 0) {
        return null;
    }
    try {
        // 此处创建 JarFile 对象，注意第二个参数是 false，
        // 表示我们不校验 Jar 包的签名信息
        JarFile jarFile = new JarFile(file, false);
        // 为了防止重名导致覆盖，我们这里取文件 md5 的前 8 位
        String tmpNameHex = DigestUtils
            .md5Hex(file.getAbsolutePath()).substring(0, 8);
        File outputJarFile =
            new File(tempDir, tmpNameHex + file.getName());

        JarOutputStream jarOutputStream =
            new JarOutputStream(new FileOutputStream(outputJarFile));

        // 处理 Jar 包中的内容
        Enumeration<JarEntry> enumeration = jarFile.entries();
        while ((enumeration.hasMoreElements())) {
            JarEntry jarEntry = enumeration.nextElement();
            String entryName = jarEntry.getName();

            // 如果有签名文件，我们忽略签名文件
            if (entryName.endsWith(".DSA") || entryName.endsWith(".SF")) {
                //do nothing 什么都不做
            } else {
                // 创建一个新的 entry，我们会将修改后的内容放在此 entry 中
                JarEntry outputEntry = new JarEntry(entryName);
                // 开始写入一个新的 Jar File entry
                jarOutputStream.putNextEntry(outputEntry);
                // 获取对应 entry 的输入流，我们读取其中的的字节数据，后面需要使用
                try (InputStream inputStream
                     = jarFile.getInputStream(jarEntry))
                {
                    // 获取原 entry 数据
                    byte[] sourceBytes
                        = toByteArrayAndAutoCloseStream(inputStream);
                    byte[] outputBytes = null;
                    // 判断是否是 Class 文件，如果是的话，
                    // 我们就处理对其原始数据进行处理
                    // 比如我们可以使用 ASM、Javassist 对其进行修改
                    if (!jarEntry.isDirectory()
                        && entryName.endsWith(".class"))
                    {
```

```
                outputBytes = handleBytes(sourceBytes);
            }
            jarOutputStream.write(outputBytes == null ?
                                  sourceBytes : outputBytes);
            //结束写入当前的entry，可以开启下一个entry
            jarOutputStream.closeEntry();

        } catch (Exception e) {
            System.err.println(
                "Exception encountered while processing jar: "
                                + file.getAbsolutePath());
            IOUtils.closeQuietly(jarFile);
            IOUtils.closeQuietly(jarOutputStream);
            e.printStackTrace();
            return null;
        }
      }
    }

    IOUtils.closeQuietly(jarFile);
    IOUtils.closeQuietly(jarOutputStream);
    return outputJarFile;
  } catch (IOException e) {
    e.printStackTrace();
  }

  return null;
}

/**
 * 用户可以在这里实现具体的处理原始数据的逻辑，
 * 例如使用ASM、Javassit等工具修改Class文件，然后返回处理后的结果。
 * 此方法直接返回了输入的值。
 *
 * @param data 原始数据
 * @return 修改后的数据
 */
private byte[] handleBytes(byte[] data) {
    //TODO your handle code
    return data;
}

/**
 * 对工程源码中生成的Class文件进行处理
 *
 * @param directoryInput DirectoryInput
 * @param transformInvocation TransformInvocation
 */
private void processDirectoryFile(DirectoryInput directoryInput, TransformInvocation
                                  transformInvocation) {
    //获取源码编译后对应的文件夹
    //例如：/Users/username/Documents/work/others/book-asm/Chapter2_05/app/
    //build/intermediates/javac/debug/compileDebugJavaWithJavac/classes
    File srcDir = directoryInput.getFile();
    System.out.println("src dir====" + srcDir);
```

```
// 获取输出的目标
// 例如：/Users/username/Documents/work/others/book-asm/Chapter2_05/app/
//build/intermediates/transforms/boilerplate_incremental/debug/1
File outputDir = transformInvocation
    .getOutputProvider()
    .getContentLocation(
        srcDir.getAbsolutePath(),
        directoryInput.getContentTypes(),
        directoryInput.getScopes(), Format.DIRECTORY);
// 将数据复制到指定目录
try {
    //outputDir 不存在，需要创建
    FileUtils.forceMkdir(outputDir);
    System.out.println("output dir====" + outputDir);

    // 增量编译
    if (transformInvocation.isIncremental()) {
        System.out.println("======start incremental======");
        Map<File, Status> changedFileMap
            = directoryInput.getChangedFiles();
        changedFileMap.forEach((file, status) -> {
            // 获取变动的文件对应在 output directory 中的位置
            String destFilePath = outputDir.getAbsolutePath()
                + file.getAbsolutePath().replace(
                    srcDir.getAbsolutePath(), "");
            File destFile = new File(destFilePath);
            System.out.println(status +
                                   "===destFilePath====" + destFilePath);
            try {
                switch (status) {
                    case REMOVED:
                        FileUtils.forceDelete(destFile);
                        break;
                    case ADDED:
                    case CHANGED:
                        // 将修改的文件复制到指定位置
                        FileUtils.copyFile(file, destFile);
                        // 获取文件的原始数据
                        byte[] sourceBytes
                            = FileUtils.readFileToByteArray(destFile);
                        // 对原始 Class 数据进行修改
                        byte[] modifiedBytes
                            = handleBytes(sourceBytes);
                        // 如果修改了数据，就将新数据保存到元文件中
                        if (modifiedBytes != null) {
                            FileUtils.writeByteArrayToFile(
                                destFile, modifiedBytes, false);
                        }
                        break;
                    case NOTCHANGED:
                        break;
                }
            } catch (IOException e) {
                e.printStackTrace();
            }
```

```
                });
            }
            // 非增量编译
            else {
                // 将数据复制到输出目录，然后对输出目录中的 Class 文件进行处理
                FileUtils.copyDirectory(srcDir, outputDir);
                // 遍历输出目录中的所有的 Class 文件，包括子目录
                FileUtils.listFiles(outputDir, new String[]{"class"}, true)
                    .parallelStream().forEach(clazzFile -> {
                        try {
                            // 获取文件的原始数据
                            byte[] sourceBytes = FileUtils
                                .readFileToByteArray(clazzFile);
                            // 对原始 Class 数据进行修改
                            byte[] modifiedBytes = handleBytes(sourceBytes);
                            // 如果修改了数据，就将新数据保存到元文件中
                            if (modifiedBytes != null) {
                                FileUtils.writeByteArrayToFile(clazzFile,
                                                               modifiedBytes,
                                                               false);
                            }
                        } catch (IOException e) {
                            e.printStackTrace();
                        }
                });
            }
        } catch (IOException e) {
            e.printStackTrace();
        }
    }

    /** 将输入流转换成 byte 数组 */
    static byte[] toByteArrayAndAutoCloseStream(InputStream input)
        throws Exception
    {
        ByteArrayOutputStream output = null;
        try {
            output = new ByteArrayOutputStream();
            byte[] buffer = new byte[1024 * 4];
            int n = 0;
            while (-1 != (n = input.read(buffer))) {
                output.write(buffer, 0, n);
            }
            output.flush();
            return output.toByteArray();
        } finally {
            IOUtils.closeQuietly(output);
            IOUtils.closeQuietly(input);
        }
    }
}
```

注意上述代码中 isIncremental() 方法返回 true，表示支持增量编译，在 Transform 中需要判断 Tran-sformInvocation#isIncremental() 是否是 true，如果为 true，只处理变动的文件即可。

上面代码展示了如何处理 JAR 包和 Directory 的模板代码，代码中添加了详细的注释，读者可以只关注其中的 handleBytes() 方法，在这个方法中可以使用 ASM、Javassist 等工具处理后返回即可。

另外读者可能对上述代码中对 JAR 包处理的部分会有一些疑问，例如：

- JarEntry到底代表的是什么内容？
- 为什么要忽略DSA和SF文件？

下面为大家简单地介绍一下 JAR 包的结构，方便大家理解相关内容。通常一个 JAR 包的结构如下。

```
example.jar
   |
   +-META-INF
   |  +-MANIFEST.MF
   |  +-x.SF
   |  +-x.DSA
   |  +-versions/
   |
   +-com
   |  +-yourcompany
   |     +-boot
   |        +-project
   |           +-YourClass.class
   |
   +-module-info.class
```

说明如下。

- META-INF是资源配置目录；
- MANIFEST.MF定义了与包相关的打包工具、程序入口、扩展等信息；
- x.SF和x.DSA定义了此JAR包的签名信息，可以验证签名信息，防止包被他人修改；
- com.yourcompary.boot.project.YourClass.class对应.class文件；
- module-info.class是Java 9中的模块化信息。

当然 JAR 包其实就是包含特殊信息的 ZIP 文件，里面可以添加很多信息，例如 README、LICENSE 等。现在来回答关于 JAR 包的两个问题。

● JarEntry到底代表的是什么内容？JarEntry对应着JAR包中的文件，不仅是class文件，还包括.MF/.SF/.DSA/module-info.class，以及其他所有文件。

● 为什么要忽略DSA和SF文件？因为程序很可能会修改JAR包中的内容，内容发生变化后，签名和加密信息就不正确了，所以选择忽略。

读者理解了这些内容，再去看前文的代码就很好理解了。另外需要注意处理 JAR 包和 Directory 的不同，前文代码中在处理 Directory 的时候是将文件直接复制到 output 目录中再处理；而 JAR 包则是先在临时文件夹中处理好以后再复制到 output 目录中，效果是一样的。

2.6 并发编译

为了加快构建的速度，可以在 Transform 中使用并发的特性。其实 2.5 节介绍的模板代码中已经使用了并发编译，使用的是 Java 8 中的 Stream API 的 parallelStream()。当然，读者也可以使用其他的

ForkJoin 框架来处理。

2.7 Transform 原理介绍

至此我们对 Transform 的用法已经比较熟悉了，那么其原理是什么呢？

前文有提到，Transform 最终会对应到 TransformTask 这个 Gradle 任务上，那么就来一探究竟。首先从 AGP 的入口开始。

1. BaseExtension#registerTransform()

```
private final List<Transform> transforms = Lists.newArrayList();
private final List<List<Object>> transformDependencies = Lists.newArrayList();
public void registerTransform(@NonNull Transform transform, Object... dependencies) {
      transforms.add(transform);
      transformDependencies.add(Arrays.asList(dependencies));
}
public List<Transform> getTransforms() {
      return ImmutableList.copyOf(transforms);
}
```

可以看到所有的自定义 Transform 通过 registerTransform() 方法添加到 transforms 列表中，并提供 getTransforms() 方法用于获取所有的 Transform，getTransforms() 被 TaskManager#createPostCompilationTasks() 方法所使用。

2. TaskManager#createPostCompilationTasks()

此方法是在 TaskManager#createCompileTask() 方法中调用，先来看看 createCompileTask() 方法。

```
protected void createCompileTask(@NonNull VariantScope variantScope) {
      JavaCompile javacTask = createJavacTask(variantScope);
      addJavacClassesStream(variantScope);
      setJavaCompilerTask(javacTask, variantScope);
      createPostCompilationTasks(variantScope);
}
```

从代码可以看出，此方法是创建编译任务，先将 JavaCompile 这个编译任务添加到 Gradle 中，再调用 createPostCompilationTasks() 方法创建编译后的任务，该方法代码如下。

```
public void createPostCompilationTasks(
          @NonNull final VariantScope variantScope) {
      checkNotNull(variantScope.getTaskContainer().getJavacTask());
      final BaseVariantData variantData = variantScope.getVariantData();
      final GradleVariantConfiguration config = variantData.getVariantConfiguration();
      TransformManager transformManager = variantScope.getTransformManager();
      // ---- Code Coverage first -----
      boolean isTestCoverageEnabled =
              config.getBuildType().isTestCoverageEnabled()
                      && !config.getType().isForTesting()
                      && !variantScope.getInstantRunBuildContext().isInInstantRunMode();
      if (isTestCoverageEnabled) {
          createJacocoTransform(variantScope);
      }
      maybeCreateDesugarTask(variantScope, config.getMinSdkVersion(), transformManager);
```

```
        AndroidConfig extension = variantScope.getGlobalScope().getExtension();
        // 创建并添加 Merge Java Resources Transform
        createMergeJavaResTransform(variantScope);
        // ----- External Transforms -----
        // apply all the external transforms
        // 获取所有的 Transform，并且遍历
        List<Transform> customTransforms = extension.getTransforms();
        List<List<Object>> customTransformsDependencies =
            extension.getTransformsDependencies();
        // 遍历 Transform
        for (int i = 0, count = customTransforms.size(); i < count; i++) {
            Transform transform = customTransforms.get(i);
            List<Object> deps = customTransformsDependencies.get(i);
            transformManager
                        // 调用 transformManager 的 addTransform() 方法
                    .addTransform(taskFactory, variantScope, transform)
                    .ifPresent(
                            t -> {
                                if (!deps.isEmpty()) {
                                    t.dependsOn(deps);
                                }
                                // if the task is a no-op then we make assemble task
                                //depend on it
                                if (transform.getScopes().isEmpty()) {
                                    variantScope.getTaskContainer()
                                        .getAssembleTask().dependsOn(t);
                                }
                            });
        }
        // 其余代码是添加系统自带的一些 Transform，例如 MultiDex 和 InstantRun 类型的 Transform
        ...
}
```

其中createJacocoTransform()、maybeCreateDesugarTask()、createMergeJavaResTransform()等方法会创建并添加一些预先执行的 Transform，然后通过 extension.getTransforms() 获取用户自定义的 Transform。这里遍历并处理所有的 Transform，最后调用 TransformManager#addTransform()方法。

3. TransformManager#addTransform()

我们再来看看 addTransform() 方法中的逻辑。

```
public <T extends Transform> Optional<TransformTask> addTransform(
        @NonNull TaskFactory taskFactory,
        @NonNull TransformVariantScope scope,
        @NonNull T transform,
        @Nullable TransformTask.ConfigActionCallback<T> callback) {
    // 校验 Transfrom input/output types、scopes、names 是否符合要求
    if (!validateTransform(transform)) {
        // validate either throws an exception, or records the problem during sync
        // so it's safe to just return null here
        return Optional.empty();
    }
    List<TransformStream> inputStreams = Lists.newArrayList();
    // 获取此 Transform 对应的 Task 名称，此方法对应我们在控制台中看到的 ××× 任务的名字
```

```
        // 例如：transformClassesWithBoilerplate_incrementalForDemoDebug
        String taskName = scope.getTaskName(getTaskNamePrefix(transform));
          // 获取引用类型的对应的输入和输出，特别是 TransformStream、Transform 的输入和输出都是通过
          // 此类来获取的
        // get referenced-only streams
        List<TransformStream> referencedStreams = grabReferencedStreams(transform);
       // 根据 Transform 的 getInputTypes 和 getScopes 的返回值获取匹配的 TransformStream 作为
       // 构建输入流，并构建输出
       // 流作为下一个 Transform 的输入流的来源之一，注意输入流也可能有多个。其实这就是我们在第 1 章介
       // 绍的通过输入和输出来添加
            // Gradle Task 依赖方式的内容
        // find input streams, and compute output streams for the transform
        IntermediateStream outputStream = findTransformStreams(
                transform,
                scope,
                inputStreams,
                taskName,
                scope.getGlobalScope().getBuildDir());
        ...
        transforms.add(transform);
        // create the task...
        // 从这里可以最终看出根据输入和输出以及 Transform 的一些其他相关信息，我们构建了一个 TransformTask 类
        // 这个类继承自 DefaultTask，同时被添加到了 TaskContainer 中
        TransformTask task =
                taskFactory.create(
                        new TransformTask.ConfigAction<>(
                                scope.getFullVariantName(),
                                taskName,
                                transform,
                                inputStreams,
                                referencedStreams,
                                outputStream,
                                recorder,
                                callback));
        return Optional.ofNullable(task);
}
```

说明如下。

- validateTransform()会校验通过Transform#getInputTypes()、getScopes()获取的值是否正确，例如判断返回的是否是Transform中的定义值；是否使用了不能使用的值（PROVIDED_ONLY、TESTED_CODE）以及是否使用了过期的、不推荐使用的值。
- getTaskNamePrefix()方法根据Transform的getName()和getInputTypes()的值拼装字符串，其规则是$inputType1 + And + $inputType2 + … + And + $inputTypeN + With + $transformName + For，以BoilerplateIncrementalTransform的实现得出的结果是transformClassesWithBoilerplate_incrementalFor。scope.getTaskName()方法会根据当前的构建变体获取对应的字符串，规则是productFlavors + buildTypes，以Chapter2_05这个Demo为例，最终的输出结果是transformClassesWithBoilerplate_incrementalForDemoDebug。
- findTransformStreams()方法会根据Transform的getInputTypes()和getScopes()的值以及TransformManager中的List<TransformStream> streams的值来确认输入源为List<TransformStream>

inputStreams，并构建OutputStream，然后添加到TransformManager#streams()中，添加以后可以作为下一个Transform的输入。其实这就是1.2.10小节介绍的通过输入和输出来添加Gradle Task依赖关系的内容。

- taskFactory.create()方法根据相关信息最终构建了一个TransformTask类，TransformTask继承自DefaultTask，这就证明了我们之前说的Transform对应Gradle Task。

接着再来查看 TransformTask 的 transform() 方法。

4. TransformTask#transform()

```
@TaskAction
void transform(final IncrementalTaskInputs incrementalTaskInputs)
        throws IOException, TransformException, InterruptedException {
    ...
    GradleTransformExecution executionInfo = preExecutionInfo.toBuilder()
                .setIsIncremental(isIncremental.getValue()).build();
    recorder.record(
            ExecutionType.TASK_TRANSFORM,
            executionInfo,
            getProject().getPath(),
            getVariantName(),
            new Recorder.Block<Void>() {
                @Override
                public Void call() throws Exception {
                    // 调用 transform() 方法
                    transform.transform(
                            new TransformInvocationBuilder(TransformTask.this)
                                    .addInputs(consumedInputs.getValue())
                                            .addReferencedInputs(referencedInputs.
getValue())
                                    .addSecondaryInputs(changedSecondaryInputs.getValue())
                                    .addOutputProvider(
                                            outputStream != null
                                                    ? outputStream.asOutput(
                                                            isIncremental.
getValue())
                                                    : null)
                                    .setIncrementalMode(isIncremental.getValue())
                                    .build());
                    if (outputStream != null) {
                        outputStream.save();
                    }
                    return null;
                }
            });
}
```

可以看到 TransformTask#transform() 方法使用 @TaskAction 注解，Task 在执行的时候会调用，另外此方法中调用了 Transform#transform() 方法。

以上就是 Transform 原理的基本介绍，读者可以按照该思路进一步的探索。另外从 AGP 7.0 开始，Transform 已经被标记为 Deprecated（废弃）。读者可能觉得刚学完就废弃了，那岂不是很亏。其实不是这样的，AGP 7.0 虽然将 Transform 标记为废弃，但是并没有给出 Transform 的替代方案，关于这部分知识会在 9.6 节详细介绍。另外，一个好的插件也是要兼容新老版本的，Transform 也是绕不过去的知识点。

2.8 小结

本章详细地介绍了 Transform 的使用和原理，Transform 最终对应到一个 Gradle Task，这部分内容归根结底还是 Gradle 相关的知识，只不过通过 AGP 的 Transform，读者能以更加简单的方式去处理 .class 文件。在模板代码中提供了处理 .class 文件对应的字节数组的入口 [(handleBytes() 方法]，其中提到可以使用 ASM 和 Javassist 等工具来修改 .class 文件，而在此之前必须对 .class 文件的结构有所了解，才能利用好 ASM 和 Javassist 这些工具。下一章将介绍字节码相关的知识。

另外从 AGP 8.0 开始，Transform API 将被移除，关于如何适配 AGP 8.0 将在 9.6 节详细介绍。

3. 字节码基础

根据前两章介绍的内容，读者应该知道如何在 Android 中定义 Gradle 插件，不过只知道这些内容还不够，还缺少一个很重要的能力，就是操作 .class 文件。操作 .class 文件的框架有很多种，比较出名的有 ASM 和 Javassist，它们的简单对比如表 3-1 所示。

表3-1 ASM与Javassist对比

特点	ASM	Javassist
核心包体积	366 KB（v9.2）	772KB（v3.28）
性能	高、更灵活	低于ASM
API封装程度	低	高
要求	需要精通 .class文件格式和JVM指令	了解 .class文件结构和JVM指令
学习难度	大	中等
文档	烦琐、不容易理解、教材少	资料丰富、简单易懂

从表 3-1 可以看出，如果你是初学者或者急于实现功能，建议使用 Javassist，毕竟上手快；如果需要考虑性能和灵活性，则建议选择 ASM。但是 ASM 的学习曲线是非常陡峭的，对很多初学者来说是不友好的，也正因为如此，本书才有存在的价值。

无论使用上述哪种字节码框架，都需要对 .class 文件结构和字节码指令的知识有所了解才行，本章就先介绍 .class 文件结构。

3.1 Java虚拟机

我们知道 Java 的口号是“Write once, run anywhere”，翻译为中文就是一次编译，到处运行。这里说的是 Java 语言的跨平台性，可以在 Windows、Linux、macOS 平台上运行，但是运行 Java 语言的虚拟机并不是跨平台的。在计算机科学领域有一句名言：“计算机科学领域的任何问题都可以通过增加一个间接的中间层来解决”，这个用于描述 JVM（Java Virtual Machine，Java 虚拟机）再合适不过了。为了让 Java 语言具有跨平台的特性，Java 虚拟机需要对不同的平台特性进行适配，屏蔽底层的系统差异。也就是 Java 虚拟机这个“中间层”使 Java 语言具有了跨平台的特性。其实 Java 虚拟机真正能执行的也不是 Java 语言，而是其编译后的 .class 文件中的内容。因此任何编程语言被编译成 .class 文件，都可以运行在 Java 虚拟机上，类似 Java 语言的有 Kotlin、Groovy、Scala 等。

Java 虚拟机就是一个能运行 .class 文件内容的程序，只要一个程序能正确地处理和执行 .class 文件中的内容，那么它就是一个“虚拟机程序”，比较出名的虚拟机有 Oracle 的 HotSpot VM、IBM 的 J9 虚拟机以及 Google 的 Dalvik/ART Android 虚拟机。Google 的 Android 虚拟机并不是严格意义上的 JVM，是其为了在移动设备上运行程序的一个 JVM 变种。

3.2 javap工具介绍

.class 文件就是一个普通的二进制文件，只不过其内容是按照一定规则生成的。既然是有规则的，那就可以借助工具展示其中的内容，这里要介绍的工具是 JDK 提供的 javap。javap 能够对 .class 文件中的字节码进行反编译，通过它，可以比较直观地查看 .class 的内部细节，这对于分析和排查问题非常有帮助。

其使用方式如下。

```
javap [options] classes
```

其中参数 classes 表示 .class 文件，options 为可选选项，可选选项包括：

```
-help  --help  -?          // 输出此用法信息
-version                   // 版本信息
-v  -verbose               // 输出附加信息
-l                         // 输出行号和本地变量表
-public                    // 仅显示公共类和成员
-protected                 // 显示受保护的 / 公共类和成员
-package                   // 显示程序包 / 受保护的 / 公共类和成员（默认）
-p  -private               // 显示所有类和成员
-c                         // 对代码进行反汇编
-s                         // 输出内部类型签名
-sysinfo                   // 显示正在处理的类的系统信息（路径、大小、日期、MD5 散列）
-constants                 // 显示最终常量
-classpath <path>          // 指定查找用户类文件的位置
-cp <path>                 // 指定查找用户类文件的位置
-bootclasspath <path>      // 覆盖引导类文件的位置
```

以下面的 Java 代码为例来介绍 javap 各选项，在此之前先使用 javac 工具将其编译成 HelloJava.class 文件。

```
// 源码
package cn.sensorsdata;

public class HelloJava {
      private int privateField = 10;
      public int publicField = 11;
      int defaultField = 12;

      public int add(int a, int b){
          int result = a + b;
          return result;
      }

      void run(){
          new Thread(()->{

          });
      }

      private void foo(){
      }
}
```

1. javap 不添加任何选项

```
$ javap HelloJava

// 输出结果
警告：二进制文件 HelloJava 包含 cn.sensorsdata.HelloJava
Compiled from "HelloJava.java"
```

```
public class cn.sensorsdata.HelloJava {
    private int privateField;
    public int publicField;
    int defaultField;
    public cn.sensorsdata.HelloJava();
    public int add(int, int);
    void run();
    private void foo();
    private static void lambda$run$0();
}
```

从输出结果可以看到 private 类型的成员（privateField 字段、foo() 方法）没有显示出来，默认情况下 javap 只会显示 public、protected 和默认修饰符的成员，如果想要显示 private 成员需要添加 -p 选项。

注意，上述演示的 javap 指令后面的 Hellojava.class 表示 HelloJava.java 的 .class 文件，也可以不写后面的 .class。

2. -p 显示所有成员

```
$ javap -p HelloJava

// 输出结果
警告： 二进制文件 HelloJava 包含 cn.sensorsdata.HelloJava
Compiled from "HelloJava.java"
public class cn.sensorsdata.HelloJava {
    private int privateField;
    public int publicField;
    int defaultField;
    public cn.sensorsdata.HelloJava();
    public int add(int, int);
    void run();
    private void foo();
    private static void lambda$run$0();
}
```

从输出结果中可以看到显示了 private 成员，并且能够发现多了一个 lambdarun0() 方法，而此方法在源码中是不存在的。其实 Java 源码是给开发者看的，而编译后的字节码是给机器看的，因此 Java 源码和编译后的结果不同是很正常的，这里的 lambdarun0() 是编译器生成的方法。

3. -s 输出方法描述符信息

```
$ javap -s HelloJava

// 输出结果
警告： 二进制文件 HelloJava 包含 cn.sensorsdata.HelloJava
Compiled from "HelloJava.java"
public class cn.sensorsdata.HelloJava {
    public int publicField;
      descriptor: I
    int defaultField;
      descriptor: I
    public cn.sensorsdata.HelloJava();
      descriptor: ()V

    public int add(int, int);
      descriptor: (II)I
```

```
    void run();
      descriptor: ()V
}
```

方法描述符又叫方法签名，是 .class 文件中标识方法的参数和返回值的一种字符串，例如 HelloJava 中的 add() 方法，其参数是 int 类型的，我们使用大写的“I”来表示，返回值也是 int 类型的，所以也用“I”来表示，将其整合在一起就是“(II)I”，圆括号外面的“I”是方法的返回值类型，这就是 add() 方法的签名信息。不同的参数类型有不同的描述符，3.3 节进一步介绍方法签名。

4. -c 对代码进行反汇编

```
$ javap -c HelloJava


// 输出结果
警告：二进制文件 HelloJava 包含 cn.sensorsdata.HelloJava
Compiled from "HelloJava.java"
public class cn.sensorsdata.HelloJava {
    public cn.sensorsdata.HelloJava();
      Code:
         0: aload_0
         1: invokespecial #1                  // Method java/lang/Object."<init>":()V
         4: return

    public int add(int, int);
      Code:
         0: iload_1
         1: iload_2
         2: iadd
         3: istore_3
         4: iload_3
         5: ireturn

    ...
}
```

以输出的 add() 方法为例，其中 iload_1、iload_2、ireturn 这些符号就是 Java 虚拟机中的字节码指令，它们代表了源码中 add() 方法体中的代码逻辑。.class 文件就是一个二进制文件，以 iload_1 指令为例，它对应的二进制是 00011011，十六进制是 0x1B，这显然是不方便记忆的。因此为方便开发者理解和记忆，就形成了助记符，iload_1 就是 0x1B 的助记符。关于字节码指令，会在第 4 章中详细介绍。

5. -v 显示详细信息

```
$ javap -v HelloJava

// 输出结果
警告：二进制文件 HelloJava 包含 cn.sensorsdata.HelloJava
Classfile /Users/wangzhzh/Documents/.../HelloJava.class
    Last modified 2022-3-22; size 1103 bytes
    MD5 checksum 5841c8a99dd4adc2165797023cd15e61
    Compiled from "HelloJava.java"
public class cn.sensorsdata.HelloJava
    minor version: 0
    major version: 52
    flags: ACC_PUBLIC, ACC_SUPER
Constant pool:
```

```
    #1 = Methodref #9.#25  // java/lang/Object."<init>":()V
    #2 = Fieldref  #8.#26  // cn/sensorsdata/HelloJava.privateField:I
    #3 = Fieldref  #8.#27  // cn/sensorsdata/HelloJava.publicField:I
    #4 = Fieldref  #8.#28  // cn/sensorsdata/HelloJava.defaultField:I
    #5 = Class     #29     // java/lang/Thread
    #6 = InvokeDynamic      #0:#34          // #0:run:()Ljava/lang/Runnable;
    #7 = Methodref  #5.#35 // java/lang/Thread."<init>":(Ljava/lang/Runnable;)V
    #8 = Class              #36             // cn/sensorsdata/HelloJava
    #9 = Class              #37             // java/lang/Object
   #10 = Utf8               privateField
   ...
   #51 = Class              #53             // java/lang/invoke/MethodHandles
   #52 = Utf8               java/lang/invoke/MethodHandles$Lookup
   #53 = Utf8               java/lang/invoke/MethodHandles
{
   public int publicField;
     descriptor: I
     flags: ACC_PUBLIC

   int defaultField;
     descriptor: I
     flags:

  ...

   public int add(int, int);
     descriptor: (II)I
     flags: ACC_PUBLIC
     Code:
       stack=2, locals=4, args_size=3
          0: iload_1
          1: iload_2
          2: iadd
          3: istore_3
          4: iload_3
          5: ireturn
       LineNumberTable:
         line 9: 0
         line 10: 4

  ...
}
SourceFile: "HelloJava.java"
InnerClasses:
      public static final #48= #47 of #51;
BootstrapMethods:
   0: #31 invokestatic java/lang/invoke/LambdaMetafactory.metafactory: \
          (Ljava/lang/invoke/MethodHandles$Lookup;Ljava/lang/String; \
          Ljava/lang/invoke/MethodType;Ljava/lang/invoke/MethodType; \
          Ljava/lang/invoke/MethodHandle;Ljava/lang/invoke/MethodType;) \
          Ljava/lang/invoke/CallSite;
     Method arguments
       #32 ()V
       #33 invokestatic cn/sensorsdata/HelloJava.lambda$run$0:()V
       #32 ()V
```

上述结果为方便阅读省略了很多输出，从中可以看到，添加了 -v 选项输出了很多的信息，包括版本信息、修改时间、源码信息、访问标识信息、常量池信息、方法描述符信息、字节码指令、调试信息等。-v 也是实际工作中使用最多的一个选项。

6. 最佳组合

通常使用 -v 选项已经可以获得比较详细的信息，不过只使用 -v 无法显示 private 信息，所以在使用 javap 指令的时候都会带上 -p 选项，这样就能将 private 成员以及编译器生成的方法显示出来，指令如下。

```
$ javap -v -p <classes>
```

javap 工具在分析字节码和学习 .class 文件结构时的作用非常重大，在实际的工作中也会经常用到。与字节码相关的问题通常比较难以排查，往往需要从 .class 文件的字节码开始分析，后续章节会频繁地使用到这个工具。

3.3 特定名称介绍

3.3.1 字段描述符、方法描述符

描述符（descriptor）是一个描述字段和方法的类型的字符串，它分为字段描述符（field descriptor）和方法描述符（method descriptor）。

字段描述符用来描述类、实例或局部变量。例如在 .class 文件中对于基本类型，使用 I 来表示 int 型，C 表示 char 类型，J 表示 long 类型。对于引用类型，使用“L 类的全限定名 ;”来表示，例如 java.lang.Object 类型字段的描述符是“Ljava/lang/Object;”，以英文分号结尾。对于数组类型，使用“[基本类型或者引用类型描述符”来表示，例如 String[] 数组的表示方式是“[Ljava/lang/String;”。表 3-2 列出了所有的字段描述符。

表3-2 字段描述符

字段	类型	描述
B	byte	用于表示基本类型
C	char	用于表示字符类型
D	double	用于表示双精度类型
F	float	用于表示单精度浮点型
I	int	用于表示整型
S	short	用于表示有符号短整型
J	long	用于表示长整型
Z	boolean	用于表示布尔型
L类的全限定名;	reference	用于表示类、接口
[	reference	用于表示一维数组
[[	reference	表示二维数组就用两个方括号 [[，*N*维数组就用*N*个方括号

方法描述符用来描述方法的参数和返回值，其形式是“（0 个或者多个字段描述符）返回值的字段描述

符”。例如，使用 javap 显示 HelloJava.class 中的方法描述符。

```
$ javap -s -p HelloJava

警告：二进制文件 HelloJava 包含 cn.sensorsdata.HelloJava
Compiled from "HelloJava.java"
public class cn.sensorsdata.HelloJava {
    public cn.sensorsdata.HelloJava();
      descriptor: ()V

    public int add(int, int);
      descriptor: (II)I

    void run();
      descriptor: ()V

    private void foo();
      descriptor: ()V

    private static void lambda$run$0();
      descriptor: ()V
}
```

其中构造方法的描述符是 ()V，它表示构造方法无参数且返回值为空，这里大写的 V 表示方法的返回值为 void；add() 方法的描述符是 (II)I，它表示方法有两个 int 类型参数，且返回值也是 int 类型；run() 方法的描述符是 ()V，它表示无参且返回值是 void。

再比如下面这个方法。

```
String foo(int i, double d, Object object){ ... }
```

它的参数部分由基本类型 int（描述符是 *I*）+ 基本类型 double（描述符是 D）+ 引用类型 java.lang.Object（描述符是 Ljava/lang/Object;）组成，参数部分描述符组合起来是“IDLjava/lang/Object;”，返回值是引用类型 java.lang.String，对应的字段描述符是“Ljava/lang/String;”，最终的结果是“(IDLjava/lang/Object)Ljava/lang/String;”。

3.3.2 全限定名

3.3.1 小节中提到了一个概念——“全限定名”，什么是“全限定名”呢？在介绍之前，先来看看 java.lang.Class 类中的如下几个方法。

```
class Class{
        ...
        public String getName()
        public String getSimpleName()
        public String getCanonicalName()
        public String getTypeName()
}
```

对于这几个方法的返回值，相信大部分读者如果没有专门研究过，应该都是分不清楚的，接下来简单介绍这几个方法。

（1）getName() 以字符串形式返回由该类对象表示的实体（类、接口、数组类、基元类型或 void）的名称。因此根据这里实体的不同，它的值也是不一样的，具体规则如下。

- 当是引用类型时，返回的是该类或接口的包名 + 类名，例如String.class.getName()，其返回值为“java.lang.String”。
- 当是基本类型或者void类型时，其返回值类型名，例如int.class.getName()，其返回值为“int”。
- 当是数组类型时，其返回值由数组的维度 + 数组的组成类型决定，例如String[].class.getName()，其返回值为“java.lang.String[]”；如果是int[].class.getName()，其返回值为“int[]”。

关于getName()还有两点需要注意，一个是对于内部类的值，例如内部类Map.Entry，内部类也是引用类型，Map.Entry.class.getName()的返回值为“java.util.Map$Entry”，即内部类和外部类使用$链接。另一个是对于Class.forName(String className)方法，这里的className参数的取值对应着getName()的输出。

（2）getSimpleName()以字符串形式返回类在源码中的名称。如果是匿名类，就会返回空字符串。表3-3列举了几个例子。

表3-3　getSimpleName()例子

例子	结果
String.class.getSimpleName()	"String"
Map.Entry.class.getSimpleName()	"Entry"
new Object(){}.getClass().getSimpleName()	""

从表3-3可以看出，getSimpleName()返回的结果中不包含包名等路径信息。

（3）getCanonicalName()获取Java语言规范中定义的基础类（underlying class）的规范名称，如果一个基础类是局部的或匿名的，或数组的组件类型没有规范名称，则返回null。可以从Java源码的角度来理解这个方法，getCanonicalName()返回的是Class对象在源码中的使用方式。表3-4列举了几个例子。

表3-4　getCanonicalName()例子

例子	结果
String.class.getCanonicalName()	"java.lang.String"
Map.Entry.class.getCanonicalName()	"java.util.Map.Entry"
new Object(){}.getClass().getCanonicalName()	null
String[].class.getCanonicalName()	"java.lang.String[]"
int[].class.getCanonicalName()	"int[]"

从表3-4中可以看出，getCanonicalName()方法返回的值包含包路径，而且其定义的写法与Java源码中是一致的，需要注意的是匿名类返回的结果是null。其实匿名类在编译时还是会被映射到一个具体的实现类中，很多框架会使用getCanonicalName()的结果作为某种标识使用，这种场景就需要考虑null值的情况。

（4）getTypeName()返回该Class对象的信息字符串（informative string）。该结果与getCanonicalName()的结果很相似，不过getTypeName()返回的类路径使用$来表示内部类或者匿名内部类。表3-5列举了几个例子。

表3-5 getTypeName()例子

例子	结果
String.class.getTypeName()	"java.lang.String"
Map.Entry.class.getTypeName()	"java.util.Map$Entry"
new Object(){}.getClass().getTypeName()	"cn.sensorsdata.asmbook.Main$3"
String[].class.getTypeName()	"java.lang.String[]"
int[].class.getTypeName()	"int[]"

可以看到匿名类也有对应的结果，这里的结果为什么是“cn.sensorsdata.asmbook.Main$3”呢？简单地说，匿名类在编译的时候会根据所在的类也生成一个 .class 文件。

上面介绍了 class 中几个与 name 相关的方法，那它们又与“全限定名”有什么关系呢？这里有两个目的。

（1）总结一下 class 中这几个与 name 相关的方法的不同点；

（2）“全限定名”实际上与 Class.getName() 方法的结果很相似，只不过是将 getName() 结果中的“.”改成“/”、使用字段描述符替换了基本类型。

“全限定名”是 .class 文件中使用的、用于表示类的信息的字符串，也称为内部名（internal name）。.class 文件结构中出现的类或接口的名称，都是通过全限定形式来表示的，例如对于 java.lang.Object 这个类，在 .class 文件中的表现形式是 java/lang/Object。下面通过 javap 展示 .class 文件中显示的全限定名。

```
Constant pool:
     ...
     #5 = Class              #26           // cn/sensorsdata/HelloJava
     #6 = Class              #27           // java/lang/Object
    ...
    #26 = Utf8               cn/sensorsdata/HelloJava
    #27 = Utf8               java/lang/Object
```

可以从上面索引 #26、#27 看到常量池类型为 CONSTANT_Utf8_info 结构所表示的 HelloJava 类名和 Object 类的全限定名。

3.3.3 <init> 和 <cinit>

在 class 内部使用 <init> 来表示构造方法的方法名，使用 <cinit> 来表示静态初始化代码块的方法名，例如下面这个例子。

```
public class CInit {
       static { //<cinit>
           int a = 10;
       }
       public static void main(String[] args) {
           new CInit();
       }
}
```

上述代码中定义了一个静态代码块，并在 main() 方法中调用了 CInit 类的构造方法，使用 javac 编译后再用 javap 命令来查看 main() 方法的字节码信息。

```
public static void main(java.lang.String[]);
     descriptor: ([Ljava/lang/String;)V
     flags: ACC_PUBLIC, ACC_STATIC
     Code:
       stack=2, locals=1, args_size=1
          0: new           #2                  // class cn/sensorsdata/CInit
          3: dup
          4: invokespecial #3                  // Method "<init>":()V
          7: pop
          8: return
       LineNumberTable:
         line 5: 0
         line 6: 8
static {};
     descriptor: ()V
     flags: ACC_STATIC
     Code:
       stack=1, locals=1, args_size=0
          0: bipush        10
          2: istore_0
          3: return
```

在 main() 方法的字节码中可以看到 invokespecial 执行调用的方法是 <init>，这就是构造方法的名字，而不是 CInit。对于 <cinit>，因其在此无法直观地展示名称，读者现在只需要知道它是指静态代码块即可。

3.4 .class 文件结构

3.4.1 初识 .class 文件

以下面这段代码为例，首先将其编译成 .class 文件后，然后使用十六进制工具打开并查看其内容。

```
//源码
package cn.sensorsdata;

public class HelloJava {
      public int add(int a, int b){
          int result = a + b;
          return result;
      }
}
```

这里使用 010 Editor 这个工具打开上面代码生成的 .class 文件，效果如图 3-1 所示。

其中图 3-1 的上半部分中阴影部分是 .class 内容的十六进制的显示，图 3-1 的下半部分中阴影部分是 010 Editor 这个工具为方便查看 .class 文件提供的便捷工具，它显示了整个 .class 文件的结构信息，可以单击其中的各个项快速定位到红框部分对应的字节码，这对于学习和研究 .class 文件结构很有帮助。

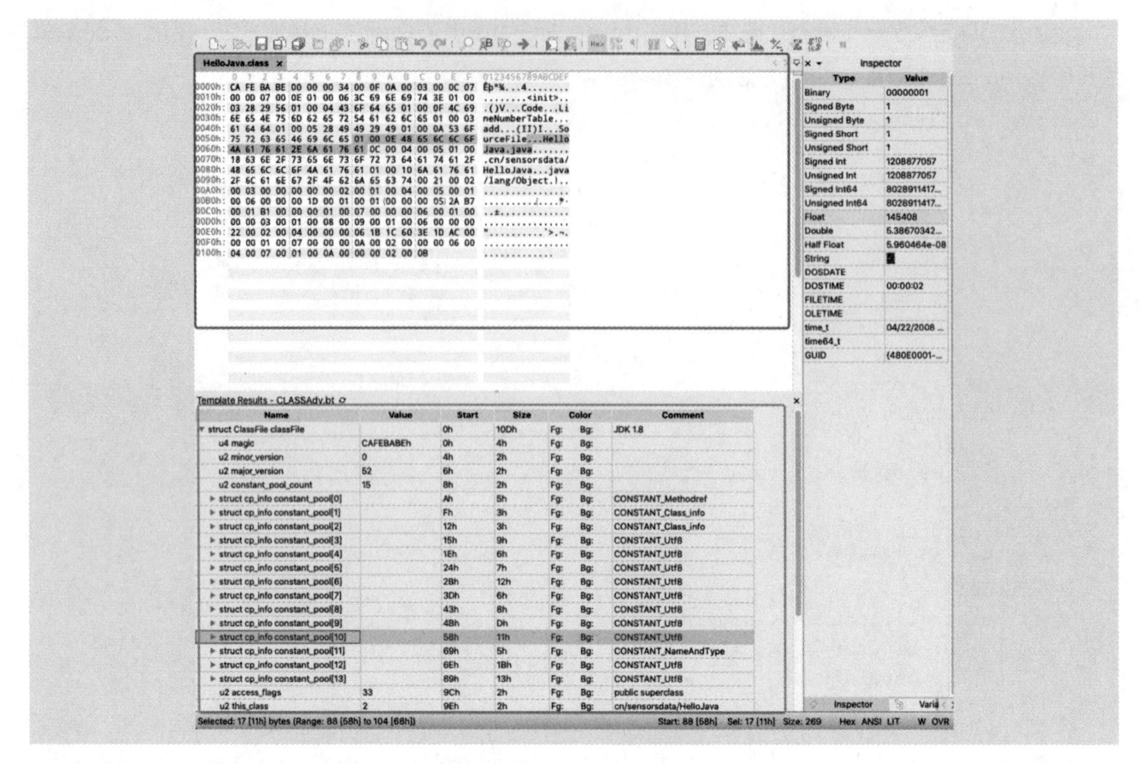

图 3-1　010 Editor 效果展示

3.4.2　.class 文件的组成

Java 自诞生至今一直保持着良好的向后兼容性，这与 .class 文件结构的稳定密不可分。.class 文件由一组以二进制位为基础单位的字节流组成（Java 虚拟机读取的不一定是 .class 文件，也可以是对应的数据流）。《Java 虚拟机规范》中说明，.class 文件结构采用类似于 C 语言结构体的伪结构来描述，另外还定义了一组专用的数据类型来表示 .class 文件的内容，包括 u1、u2、u4，分别代表 1 个字节、2 个字节和 4 个字节的无符号整数。

.class 的文件结构如下。

```
ClassFile{
      u4 magic;
      u2 minor_version;
      u2 major_version;
      u2 constant_pool_count;
      cp_info constant_pool[constant_pool_count - 1];
      u2 access_flags;
      u2 this_class;
      u2 super_class;
      u2 interfaces_count;
      u2 interfaces[interfaces_count];
      u2 fields_count;
      field_info fields[fields_count]
      u2 methods_count;
      method_info methods[methods_count];
```

```
        u2 attributes_count;
        attribute_info attributes[attributes_count];
}
```

可以看到 .class 文件由如下 10 个部分组成。

- 魔数（magic）;
- 版本号（minor_version + major_version）;
- 常量池（constant_pool）;
- 访问标识（access_flags）;
- 类索引（this_class）;
- 父类索引（super_class）;
- 接口索引（interfaces）;
- 字段表（fields）;
- 方法表（methods）;
- 属性表（attributes）。

上面这些内容都很重要，接下来会对这 10 个部分进行一一介绍。在此之前约定使用 ClassFile 来表示 .class 文件的结构，原因是 .class 是一种文件扩展名，并非文件的结构，使用 .class 来描述并不准确。而且从《Java 虚拟机规范》中的使用以及 javap 显示的结果用的都是 ClassFile，后续内容将使用 ClassFile 代替 .class。

1. 魔数

魔数通常用在文件开头的几个字节中，用于辨识特定的文件格式，例如 .jpg 文件以 0xFFD8FF 开头、.png 文件以 0x89504E47 开头。制定文件的人可以自由选择魔数，只要不引起混淆即可，最好是有一定的特殊意义。

使用十六进制工具打开 HelloJava.class 文件，可以看到开头的 4 个字节是 0xCAFEBABE，如图 3-2 所示。

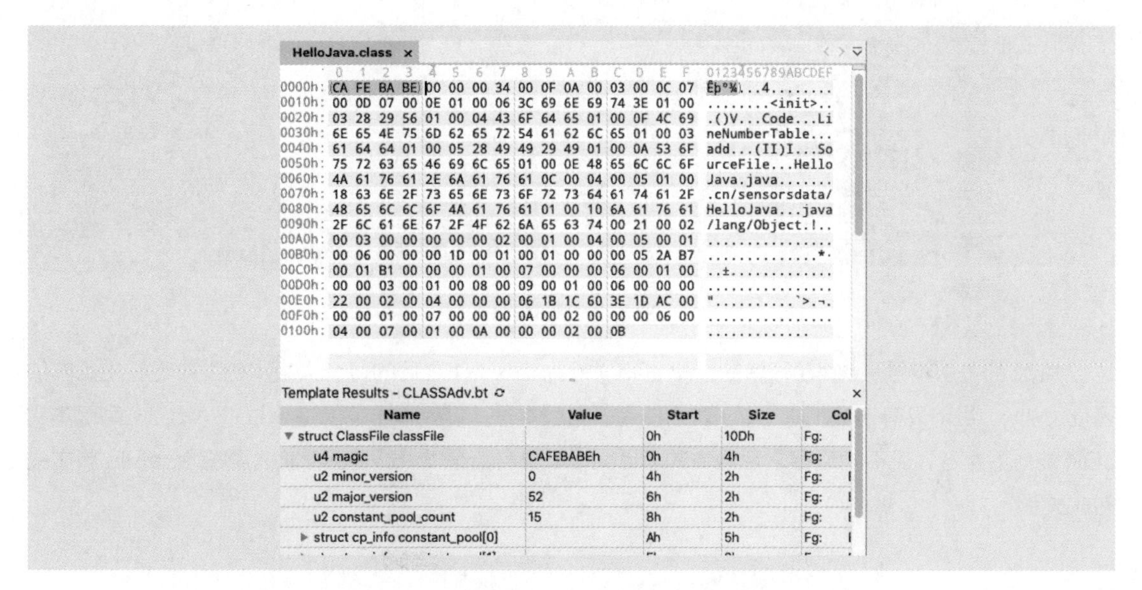

图 3-2 使用 010 Editor 展示 ClassFile 魔数

“Java 之父”James Gosling 曾说过 Java 魔数的由来，大致意思是 Gosling 经常跟朋友去一个叫 St Michael’s Alley 的地方吃午餐。根据当地传说，在那个时代，GradleFul Dead 摇滚乐队经常在这里演出，此餐馆因此出名了。当乐队中的吉他手 Jerry 去世后，人们在此地对其进行了纪念，因此 Gosling 将这个地方称为 Cafe Dead。可以将 CAFEDEAD 看成一个十六进制数，刚好此时 Gosling 正在整理一些文件的格式代码，需要两个魔数，一个用于持久化对象，一个用于 .class 文件。Gosling 将 CAFEDEAD 用在了持久化对象上，然后用 CAFEDEAD 的前 4 个字符 CAFE，灵机一动又选择了一个 BABE，将其组合成 CAFEBABE 用于 .class 文件中。Gosling 选择这两个魔数的时候也没考虑太多，最终 CAFEBABE 成为 .class 文件的魔数，CAFEDEAD 成为持久化对象的魔数，但是持久化对象技术没多久真的“DEAD”了。魔数的唯一作用就是确定这个文件是否为一个被虚拟机所接受的 .class 文件，Java 虚拟机加载 .class 文件的时候会做很多的格式校验，校验魔数只是其中基本的一个，如果校验魔数不正确，会抛出 java.lang.ClassFormatError。

2. 版本号

紧跟着魔数后面的 4 个字节分别表示副版本号（minor_version）和主版本号（major_version），它们共同构成了 ClassFile 文件的格式版本号，如图 3-3 所示。

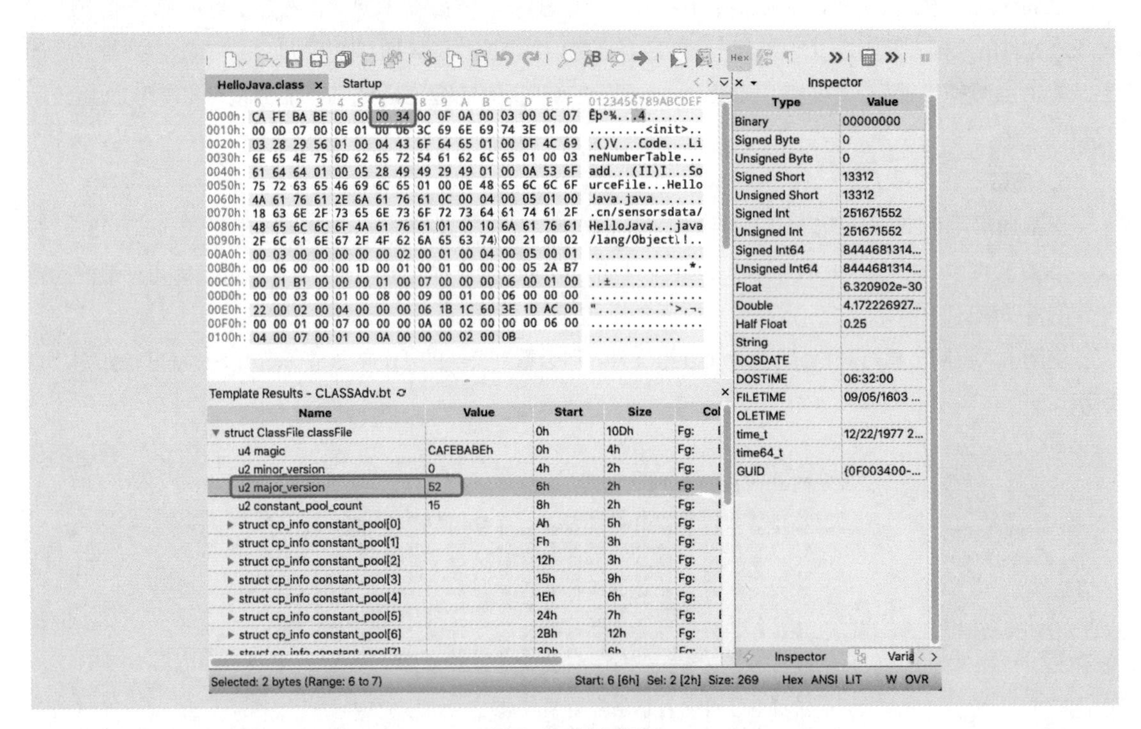

图 3-3　版本信息

Java 的主版本号是从 45 开始的，即 JDK 1.0。JDK 1.1 之后的每个大版本的发布是在主版本的基础上加 1。目前只有 JDK1.0 ~ JDK1.1 使用了副版本号，其版本号范围是 45.0 ~ 45.3。高版本的 JDK 能向下兼容，但不能运行之后版本的 ClassFile 文件。表 3-6 列出了 JDK 和 ClassFile 版本的对应关系。

表3-6　JDK和ClassFile版本的对应关系

Java SE Version	Major Version	发布时间
1.0.2	45(0x2D)	1996.5
1.1	45(0x2D)	1997.2
1.2	46(0x2E)	1998.12
1.3	47(0x2F)	2000.5
1.4	48(0x30)	2002.2
1.5	49(0x31)	2004.9
6	50(0x32)	2006.12
7	51(0x33)	2011.7
8	52(0x34)	2014.3
9	53(0x35)	2017.9
10	54(0x36)	2018.3
11	55(0x37)	2018.9
12	56(0x38)	2019.3
13	57(0x39)	2019.9
14	58(0x3A)	2020.3
15	59(0x3B)	2020.9
16	60(0x3C)	2021.3

Java 从 2017 年开始施行每半年发布一个版本。

3. 常量池

在版本之后是常量池数据，常量池在整个 ClassFile 中是比较重要的一块内容，ClassFile 中的其他结构都可能引用常量池的数据，可以将常量池看成 ClassFile 里的资源仓库，例如方法名、描述符、字符串、long 类型的值等都存储常量池中。

常量池的结构定义如下。

```
{
        u2 constant_pool_count;
        cp_info constant_pool[constant_pool_count - 1];
}
```

其中，两个字节的 constant_pool_count 表示常量池的大小，也就是 cp_info（常量池表）的大小。从上面的结构可以看出 cp_info 的实际长度为 constant_pool_count -1，这是因为常量池表的索引是从 1 开始的，而不是从 0 开始的。如图 3-4 所示，constant_pool_count 的十进制值是 15，表示常量池中有 14 个项，其对应的索引是 1 ~ 14。设计者将索引 0 的位置空出来是有特殊考虑的，例如某些结构中指向常量池的索引值的数据，在特定情况下需要表达“不引用任何一个常量池中的项”的时候，可以将其值设置为0，这种情况在后面分析字节码的时候会经常遇到。读者需要注意的是，在ClassFile结构中的其他部分，只有常量池表中的索引是这么规定的，其他组成部分的结构都是从 0 开始的。

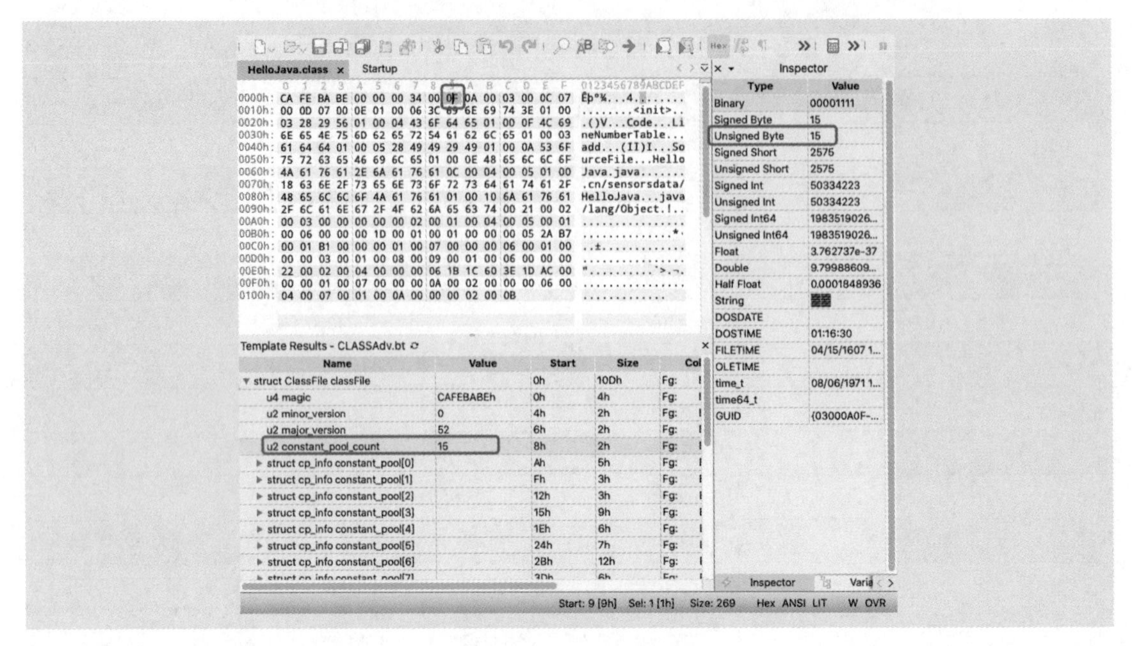

图 3-4　常量池大小展示

为更加直观地理解，使用 javap 来看 HelloJava.class 文件中的常量池信息。

```
$ javap -v -p HelloJava

Constant pool:
    #1 = Methodref          #3.#12         // java/lang/Object."<init>":()V
    #2 = Class              #13            // cn/sensorsdata/HelloJava
    #3 = Class              #14            // java/lang/Object
    #4 = Utf8               <init>
    #5 = Utf8               ()V
    #6 = Utf8               Code
    #7 = Utf8               LineNumberTable
    #8 = Utf8               add
    #9 = Utf8               (II)I
   #10 = Utf8               SourceFile
   #11 = Utf8               HelloJava.java
   #12 = NameAndType        #4:#5          // "<init>":()V
   #13 = Utf8               cn/sensorsdata/HelloJava
   #14 = Utf8               java/lang/Object
```

从上面的结果中可以看到，常量池的索引是从 #1 ~ #14，#0 并没有显示出来。

接下来，再看看 cp_info 的结构。在《Java 虚拟机规范》中将此种结构称为表（table），可以简单地理解为常量池中的具体项的结构，其通用结构如下。

```
cp_info{
      u1 tag;
      //剩余部分的具体结构由 tag 的值决定
}
```

一个字节的 tag 决定了其余部分的结构，截至 JDK 13，共有 17 个类型。tag 类型都形如 CONSTANT_{TYPE}_info，即以 CONSTANT 开头和以 info 结尾，具体的值如表 3-7 所示。

表3-7　常量池类型

类型	tag 值	简述
CONSTANT_Utf8_info	1	UTF-8编码的字符串
CONSTANT_Integer_info	3	整型字面量
CONSTANT_Float_info	4	浮点型字面量
CONSTANT_Long_info	5	长整型字面量
CONSTANT_Double_info	6	双精度型字面量
CONSTANT_Class_info	7	类或者接口的符号引用
CONSTANT_String_info	8	字符串类型字面量
CONSTANT_Fieldref_info	9	字段的符号引用
CONSTANT_Methodref_info	10	类方法的符号引用
CONSTANT_InterfaceMethodref_info	11	接口方法的符号引用
CONSTANT_NameAndType_info	12	字符或方法的部分符号引用
CONSTANT_MethodHandle_info	15	方法句柄
CONSTANT_MethodType_info	16	方法类型
CONSTANT_Dynamic_info	17	表示一个动态计算常量
CONSTANT_InvokeDynamic_info	18	表示一个方法的动态调用点
CONSTANT_Module_info	19	表示一个模块
CONSTANT_Package_info	20	表示一个模块中开发或者导出的包

其中 CONSTANT_MethodHandle_info、CONSTANT_MethodType_info、CONSTANT_InvokeDynamic_info 是 Java 7 新增的类型，用于支持动态方法，例如 Java 8 中的 Lambda 表达式就是基于此来实现的。Java 9 为了支持模块化功能又加入了 CONSTANT_Module_info 和 CONSTANT_Package_info 这两种类型。而 CONSTANT_Dynamic_info 是 JDK 11 新增的用于改善 invokedynamic 的指令。因篇幅限制以及从使用角度来说，本书将只介绍 Java 9 之前的常量池类型。

接下来重点介绍表 3-7 中的几个重要类型（未按表格中的顺序介绍）。

（1）CONSTANT_Integer_info 和 CONSTANT_Float_info 型常量的结构。

CONSTANT_Integer_info 和 CONSTANT_Float_info 结构表示 4 个字节（int 和 float）的数字常量，它们的结构相似。

```
CONSTANT_Integer_info{
      u1 tag;
      u4 bytes;
}
CONSTANT_Float_info{
      u1 tag;
      u4 bytes;
}
```

其中 CONSTANT_Integer_info 结构中 tag 的值是 3，CONSTANT_Float_info 中的是 4。查看下面这段代码对应的常量池结果。

```
public class IntAndFloat {
      public final int a = 999999;
      public final float b = 100;
}
```

我们使用 javap 指令查看常量池结果。

```
Constant pool:
    #1 = Methodref          #7.#18         // java/lang/Object."<init>":()V
    #2 = Integer            999999
    #3 = Fieldref           #6.#19         // cn/sensorsdata/IntAndFloat.a:I
    #4 = Float              100.0f
    #5 = Fieldref           #6.#20         // cn/sensorsdata/IntAndFloat.b:F
    ...
```

可以看到常量池中索引 #2 的类型是 Integer，值是 999999，索引 #4 的类型是 Float，值是 100.0f。也可以通过工具获得更加直观的结果，如图 3-5 所示。

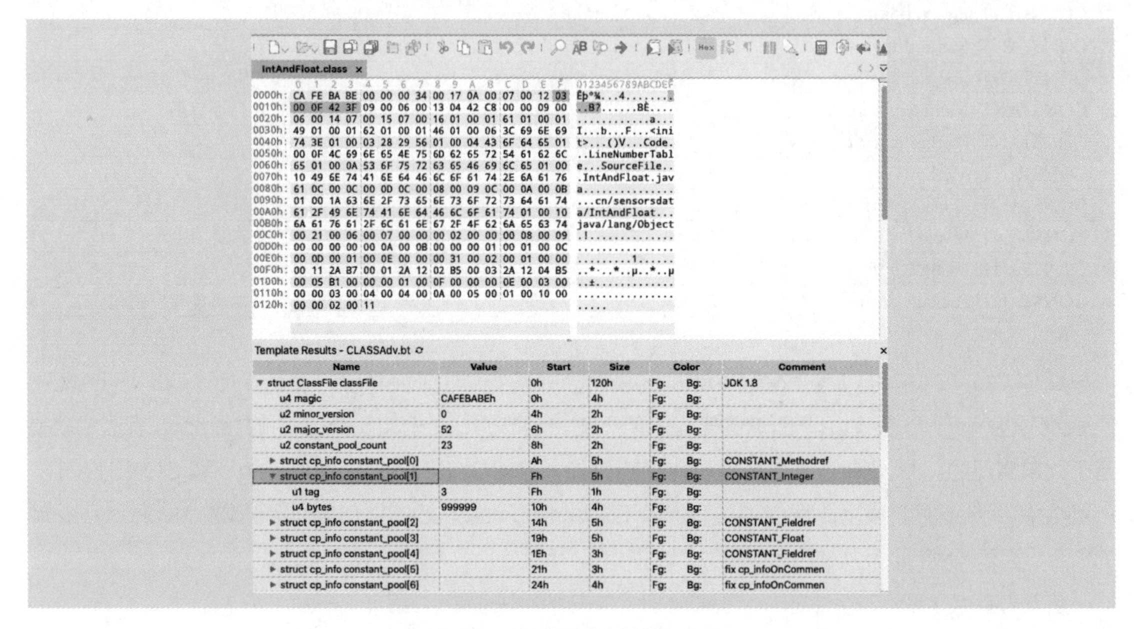

图 3-5　常量池中值展示

注意，Java 语言中的 byte、char、short、boolean 类型的常量在常量池中都会被当作 int 类型来处理。

（2）CONSTANT_Long_info 和 CONSTANT_Double_info 型常量的结构。

CONSTANT_Long_info 和 CONSTANT_Double_info 结构表示 8 个字节（long 和 double）的数字常量，它们的结构也很相似。

```
CONSTANT_Long_info{
    u1 tag;
    u4 hight_bytes;
    u4 low_bytes;
}
CONSTANT_Double_info{
    u1 tag;
    u4 hight_bytes;
    u4 low_bytes;
}
```

其中 CONSTANT_Long_info 结构中 tag 的值是 5，CONSTANT_Double_info 中的是 6。在 ClassFile 的常量池表中，所有的 8 字节常量均占两个表成员空间。用下面这个例子来说明。

```
public class LongAndDouble {
      public final long LONG_VALUE = 199;
}
```

使用 javap 命令查看常量池信息如下。

```
Constant pool:
     #1 = Methodref         #6.#16        // java/lang/Object."<init>":()V
     #2 = Long              199l
     #4 = Fieldref          #5.#17        // cn/sensorsdata/LongAndDouble.LONG_VALUE:J
     ...
```

上述结果中并没有看到索引 #3，实际上是 #2 和 #3 代表了 CONSTANT_Long_info 结构，下一项的索引是从 #4 开始的。

（3）CONSTANT_Utf8_info 型常量的结构。

CONSTANT_Utf8_info 结构用于表示字符常量的值，其结构如下。

```
CONSTANT_Utf8_info{
      u1 tag;
      u2 length;
      u1 bytes[length]
}
```

其中 CONSTANT_Utf8_info 结构的 tag 的值是 1，length 的值是 bytes 数组的长度，bytes 存储的是经过修改过的 UTF-8 编码的字节数组，用于表示一个字符串。在常量池中凡是涉及字符串的结构，最终都会指向一个 CONSTANT_Utf8_info 结构。例如下面这个例子。

```
public class StringResult {
      public final String info = "Hello World";
}
```

使用 javap 查看常量池信息如下。

```
Constant pool:
     #1 = Methodref   #5.#15   // java/lang/Object."<init>":()V
     #2 = String      #16      // Hello World
     #3 = Fieldref    #4.#17   // cn/sensorsdata/StringResult.info:Ljava/lang/String;
     #4 = Class       #18      // cn/sensorsdata/StringResult
     #5 = Class           #19          // java/lang/Object
     #6 = Utf8            info
     #7 = Utf8            Ljava/lang/String;
     #8 = Utf8            ConstantValue
     #9 = Utf8            <init>
    #10 = Utf8            ()V
    #11 = Utf8            Code
    #12 = Utf8            LineNumberTable
    #13 = Utf8            SourceFile
    #14 = Utf8            StringResult.java
    #15 = NameAndType     #9:#10       // "<init>":()V
    #16 = Utf8            Hello World
    #17 = NameAndType     #6:#7        // info:Ljava/lang/String;
    #18 = Utf8            cn/sensorsdata/StringResult
    #19 = Utf8            ava/lang/Object
```

可以看到索引 #16 代表的是 CONSTANT_utf8_info 结构，其值为“Hello World”，还可以看到 #2 最终指向了 #16 的值，#2 其实是一个 CONSTANT_String_info 结构，这就是字符串在 ClassFile 中的

表述形式。接下来再来看看 CONSTANT_String_info 结构。

（4）CONSTANT_String_info 型常量的结构。

Constant_String_info 结构用于表示 string 类型的常量对象，其结构如下。

```
CONSTANT_String_info{
      u1 tag;
      u2 string_index;
}
```

其中 tag 的值为 8，string_index 为指向 CONSTANT_Utf8_info 结构的有效索引。所以可以看到 CONSTANT_Utf8_info 是真正存字符串的地方，CONSTANT_String_info 会指向它。使用工具查看 StringResult.class，结果如图 3-6 所示。

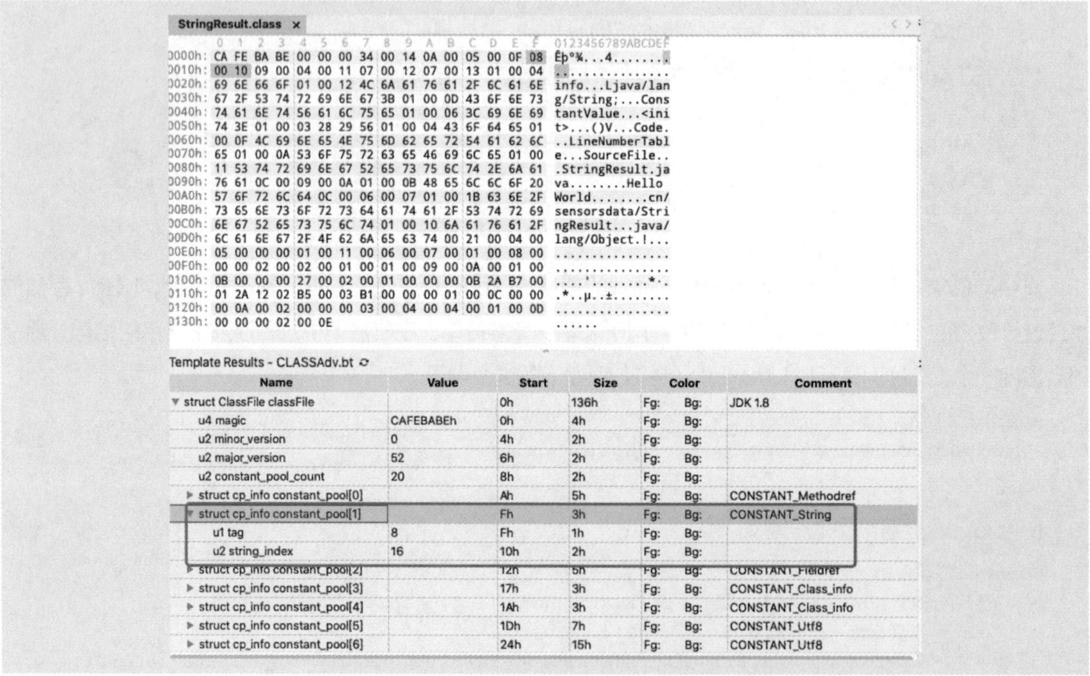

StringResult.class ×

```
       0  1  2  3  4  5  6  7  8  9  A  B  C  D  E  F   0123456789ABCDEF
0000h: CA FE BA BE 00 00 00 34 00 14 0A 00 05 00 0F 08  Êþº¾...4........
0010h: 00 10 09 00 04 00 11 07 00 12 07 00 13 01 00 04  ................
0020h: 69 6E 66 6F 01 00 12 4C 6A 61 76 61 2F 6C 61 6E  info...Ljava/lan
0030h: 67 2F 53 74 72 69 6E 67 3B 01 00 0D 43 6F 6E 73  g/String;...Cons
0040h: 74 61 6E 74 56 61 6C 75 65 01 00 06 3C 69 6E 69  tantValue...<ini
0050h: 74 3E 01 00 03 28 29 56 01 00 04 43 6F 64 65 01  t>...()V...Code.
0060h: 00 0F 4C 69 6E 65 4E 75 6D 62 65 72 54 61 62 6C  ..LineNumberTabl
0070h: 65 01 00 0A 53 6F 75 72 63 65 46 69 6C 65 01 00  e...SourceFile..
0080h: 11 53 74 72 69 6E 67 52 65 73 75 6C 74 2E 6A 61  .StringResult.ja
0090h: 76 61 0C 00 09 00 0A 01 00 0B 48 65 6C 6C 6F 20  va........Hello 
00A0h: 57 6F 72 6C 64 0C 00 06 00 07 01 00 1B 63 6E 2F  World........cn/
00B0h: 73 65 6E 73 6F 72 73 64 61 74 61 2F 53 74 72 69  sensorsdata/Stri
00C0h: 6E 67 52 65 73 75 6C 74 01 00 10 6A 61 76 61 2F  ngResult...java/
00D0h: 6C 61 6E 67 2F 4F 62 6A 65 63 74 00 21 00 04 00  lang/Object.!...
00E0h: 05 00 00 00 01 00 11 00 06 00 07 00 01 00 08 00  ................
00F0h: 00 00 02 00 02 00 01 00 01 00 09 00 0A 00 01 00  ................
0100h: 0B 00 00 00 27 00 02 00 01 00 00 00 0B 2A B7 00  ....'........*·.
0110h: 01 2A 12 02 B5 00 03 B1 00 00 00 01 00 0C 00 00  .*..µ..±........
0120h: 00 0A 00 02 00 00 00 03 00 04 00 04 00 01 00 0D  ................
0130h: 00 00 00 02 00 0E                                ......
```

Template Results - CLASSAdv.bt

Name	Value	Start	Size	Color		Comment
▼ struct ClassFile classFile		0h	136h	Fg:	Bg:	JDK 1.8
u4 magic	CAFEBABEh	0h	4h	Fg:	Bg:	
u2 minor_version	0	4h	2h	Fg:	Bg:	
u2 major_version	52	6h	2h	Fg:	Bg:	
u2 constant_pool_count	20	8h	2h	Fg:	Bg:	
▶ struct cp_info constant_pool[0]		Ah	5h	Fg:	Bg:	CONSTANT_Methodref
▼ struct cp_info constant_pool[1]		Fh	3h	Fg:	Bg:	CONSTANT_String
u1 tag	8	Fh	1h	Fg:	Bg:	
u2 string_index	16	10h	2h	Fg:	Bg:	
▶ struct cp_info constant_pool[2]		12h	5h	Fg:	Bg:	CONSTANT_Fieldref
▶ struct cp_info constant_pool[3]		17h	3h	Fg:	Bg:	CONSTANT_Class_info
▶ struct cp_info constant_pool[4]		1Ah	3h	Fg:	Bg:	CONSTANT_Class_info
▶ struct cp_info constant_pool[5]		1Dh	7h	Fg:	Bg:	CONSTANT_Utf8
▶ struct cp_info constant_pool[6]		24h	15h	Fg:	Bg:	CONSTANT_Utf8

图 3-6　CONSTANT_String_info 结果展示

从图 3-6 中可以看到 CONSTANT_String_info 结构中的 string_index 值为 16，指向索引 #16 的位置，正好与介绍 CONSTANT_Utf8_info 结构时使用 javap 查看到的结果相对应。

（5）CONSTANT_Class_info 型结构。

CONSTANT_Class_info 结构用于表示类或接口，其格式如下。

```
CONSTANT_Class_info{
      u1 tag;
      u2 name_index;
}
```

其中第一个字节的 tag 的值为 7，紧接着两个字节的 name_index 为指向 CONSTANT_Utf8_info 结构的有效索引，指向的 CONSTANT_Utf8_info 的值为当前类或者接口的全限定名。

（6）CONSTANT_NameAndType_info 型结构。

CONSTANT_NameAndType_info 结构用于表示字段或方法，其格式如下。

```
CONSTANT_NameAndType_info{
      u1 tag;
      u2 name_index;
      u2 descriptor_index;
}
```

这个结构包括 3 个部分：

- 一个字节的tag的值固定为12；
- 紧跟着tag的是两个字节长度的name_index，它为指向CONSTANT_Utf8_info结构的有效索引，用于表示字段或方法的名字；
- 在name_index后面的两个字节长度的descriptor_index为指向CONSTANT_Utf8_info结构的有效索引，用于表示字段或方法的描述符。

例如下面这个常量池所示，索引 #13 表示的构造方法名和构造方法的描述符，分别指向常量池索引 #5 和 #6 的位置，索引 #5 为构造方法名，#6 为构造方法描述。

```
Constant pool:
     #1 = Methodref          #4.#13  // java/lang/Object."<init>":()V
     #2 = Class              #14     // cn/sensorsdata/CInit
     #3 = Methodref          #2.#13  // cn/sensorsdata/CInit."<init>":()V
     #4 = Class              #15     // java/lang/Object
     #5 = Utf8               <init>
     #6 = Utf8               ()V
     #7 = Utf8               Code
     #8 = Utf8               LineNumberTable
     #9 = Utf8               main
    #10 = Utf8               ([Ljava/lang/String;)V
    #11 = Utf8               SourceFile
    #12 = Utf8               CInit.java
    #13 = NameAndType        #5:#6          // "<init>":()V
    #14 = Utf8               cn/sensorsdata/CInit
    #15 = Utf8               java/lang/Object
```

（7）CONSTANT_Fieldref_info、CONSTANT_Methodref_info、CONSTANT_InterfaceMethodref_info 结构。

这 3 个常量类型结构比较类似，分别代表字段、方法和接口方法，他们的结构如下。

```
// 字段
CONSTANT_Fieldref_info{
      u1 tag;
      u2 class_index;
      u2 name_and_type_index;
}

// 方法
CONSTANT_Methodref_info{
      u1 tag;
      u2 class_index;
      u2 name_and_type_index;
}

// 接口方法
CONSTANT_InterfaceMethodref_info{
      u1 tag;
```

```
        u2 class_index;
        u2 name_and_type_index;
}
```

结构中包括 3 个部分。

- CONSTANT_Fieldref_info的tag值为9，CONSTANT_Methodref_info的tag值为10，CONSTANT_InterfaceMethodref_info的tag值为11；
- class_index为指向CONSTANT_Class_info结构的有效索引，用于表示类或接口；
- name_and_type_index为指向CONSTANT_NameAndType_info结构的有效索引，用于描述字段或方法的名字和描述符。

可以看到这 3 种结构也是用于描述方法和字段的，但是与 CONSTANT_NameAndType_info 不同的是，有一个 class_index 索引，用于表示具体是哪一个类的方法和字段。例如下面这个例子对应的常量池数据。

```
public class MethodRefMain {
      public int add(int a, int b){
          return a + b;
      }

      public static void main(String[] args) {
          new MethodRefMain().add(1, 2);
      }
}

// 常量池信息
Constant pool:
     #1 = Methodref      #5.#16    // java/lang/Object."<init>":()V
     #2 = Class          #17       // cn/sensorsdata/MethodRefMain
     #3 = Methodref      #2.#16    // cn/sensorsdata/MethodRefMain."<init>":()V
     #4 = Methodref      #2.#18    // cn/sensorsdata/MethodRefMain.add:(II)I
     #5 = Class          #19       // java/lang/Object
     #6 = Utf8           <init>
     #7 = Utf8           ()V
     #8 = Utf8           Code
     #9 = Utf8           LineNumberTable
    #10 = Utf8           add
    #11 = Utf8           (II)I
    #12 = Utf8           main
    #13 = Utf8           ([Ljava/lang/String;)V
    #14 = Utf8           SourceFile
    #15 = Utf8           MethodRefMain.java
    #16 = NameAndType    #6:#7            // "<init>":()V
    #17 = Utf8           cn/sensorsdata/MethodRefMain
    #18 = NameAndType    #10:#11          // add:(II)I
    #19 = Utf8           java/lang/Object
```

其中索引 #4 所代表的是 CONSTANT_Methodref_info 结构，对应着 MethodRefMain 类中的 add() 方法；该结构中的 class_index 为 #2，表示类的信息，即 MethodRefMain，name_and_type_index 为 #18，表示 add() 方法的名字和描述符信息。

（8）CONSTANT_MethodHandle_info、CONSTANT_MethodType_info、CONSTANT_InvokeDynamic_info 结构。

这3个结构是Java 7为了支持动态语言调用而引入的3个常量池类型，在Java 8中配合invokedynamic字节码指令实现了Lambda表达式的语法功能。理解这3个结构对理解Lambda表达式的实现机制非常有帮助，在8.4节中有专门对Lambda的设计思想和Hook方式的介绍。在此之前先来认识这3种常量池结构。

首先是CONSTANT_MethodHandle_info，其结构如下。

```
CONSTANT_MethodHandle_info{
      u1 tag;
      u1 reference_kind;
      u2 reference_index;
}
```

此结构用于表示方法句柄（方法句柄包含一个方法的基本信息，例如方法名称、方法描述符和访问修饰符的结构），各项说明如下。

- tag的值为15。
- reference_kind表示方法句柄的类型，其值的范围是1~9,reference_kind值用于表明在方法句柄中具体执行什么样的操作，其值具体如表3-8所示。
- reference_index的值对应常量池的有效索引，具体指向何种类型的常量池类型由reference_kind的值决定，总结如下。

（a）当reference_kind的值是1、2、3、4时，reference_index所指向的索引处的常量池类型必须是CONSTANT_Fieldref_info结构。

（b）当reference_kind的值是5、8时，reference_index所指向的索引处的常量池类型必须是CONSTANT_Methodref_info结构。

（c）当reference_kind的值是6、7时，reference_index所指向的索引处的常量池类型与.class文件的版本号有关，即当小于52.0（JDK 8）时，reference_index所指向的结构必须是CONSTANT_Methodref_info结构；当不小于52.0（JDK 8）时，reference_index所指向的结构是CONSTANT_Methodref_info或CONSTANT_InterfaceMethodref_info结构。

（d）当reference_kind的值是9时，reference_index所指向的索引处的常量池类型必须是CONSTANT_InterfaceMethodref_info结构。

（e）当reference_kind的值是5、6、7、9时，reference_index所指向的结构所表示的方法名称不能是<init>或<clinit>。

（f）当reference_kind的值是8时，reference_index所指向的CONSTANT_Methodref_info结构的方法名称必须是<init>。

表3-8所示为粗略对应关系，严格标准需要参照上面描述的reference_kind和reference_index规则。再来看CONSTANT_MethodType_info结构。

```
CONSTANT_MethodType_info{
      u1 tag;
      u2 descriptor_index;
}
```

表3-8 reference_kind和reference_index的对应关系

reference_kind值	描述	reference_index
1	REF_getField	CONSTANT_Fieldref_info
2	REF_getStatic	CONSTANT_Fieldref_info
3	REF_putField	CONSTANT_Fieldref_info
4	REF_putStatic	CONSTANT_Fieldref_info
5	REF_invokeVirtual	CONSTANT_Methodref_info
6	REF_invokeStatic	CONSTANT_Methodref_info
7	REF_invokeSpecial	CONSTANT_Methodref_info CONSTANT_InterfaceMethodref_info
8	REF_newInvokeSpecial	CONSTANT_Methodref_info
9	REF_invokeInterface	CONSTANT_InterfaceMethodref_info

此结构比较好理解，表示一个方法的描述符信息，包括参数、返回值。其中tag的值为16，descriptor_index必须指向CONSTANT_Utf8_info结构。

最后来看看CONSTANT_InvokeDynamic_info结构。

```
CONSTANT_InvokeDynamic_info{
      u1 tag;
      u2 bootstrap_method_attr_index;
      u2 name_and_type_index;
}
```

此结构用于表示invokedynamic指令所用到的引导方法（bootstrap method）以及引导方法需要动态调用的名称、参数和返回类型，并可以给引导方法传入一系列成为静态参数的常量。其各部分说明如下。

- tag的值为18。
- bootstrap_method_attr_index的值必须是对.class文件中BooststrapMethods属性的bootstrap_methods表的有效引用。
- name_and_type_index的值必须是常量池中类型是CONSTANT_NameAndType_info结构的有效引用。

对于这3种结构，读者可以编译下面这段代码，然后使用javap查看常量池结果，其中涉及的名词、Lambda原理等知识，会在介绍Lambda表达式的Hook方式时，对这一部分内容再进行详细介绍，感兴趣的读者可以直接查看8.4节。

```
public class LambdaMain {
      public void foo(){
          Consumer<String> consumer = o -> {

          };
      }
}
```

本节介绍了ClassFile中的17种常量池类型，内容有些多，希望大家能够自己动手操作，借助javap或者十六进制工具的结果来加深理解。下面继续介绍ClassFile结构中的类访问标识符。

4. 访问标识

跟在常量池结构后面的是两个字节的类访问标识（access_flags），用于标识类或接口的访问信息，完整的类访问标识信息如表 3-9 所示。

表3-9　类访问标识

标识名称	十六进制	描述
ACC_PUBLIC	0x0001	标识是否是public
ACC_FINAL	0x0010	标识是否是final，只有类才能设置此值，接口不能设置
ACC_SUPER	0x0020	已不再使用，自JDK 1.0.2之后，都会带上此标识
ACC_INTERFACE	0x0200	标识ClassFile是一个接口
ACC_ABSTRACT	0x0400	标识是否是abstract
ACC_SYNTHETIC	0x1000	标识是否由编译器生成，非源码生成
ACC_ANNOTATION	0x2000	标识是否是注解类
ACC_ENUM	0x4000	标识是否是枚举类
ACC_MODULE	0x8000	标识是否是一个模块

从表 3-9 可以看出类访问标识的标识值对应两个字节（16 位）中的每一位，所以总共可以定义 16 个标识，目前已经使用了 9 个。不同的标识可以相互组合在一起使用，例如下面代码中定义的 MyClass 类。

```
public final class MyClass{
}
```

其标识为 0x0001(ACC_PUBLIC) | 0x0010(ACC_FINAL) | 0x0020(ACC_SUPER) = 0x0031（十进制等于 49），从 JDK 1.0.2 之后都会添加上 ACC_SUPER 标识，使用 javap 工具查看 MyClass 的结果如下。

```
$ javap -v -p MyClass

Classfile /Users/wangzhzh/.../MyClass.class
public final class cn.sensorsdata.MyClass
    minor version: 0$
    major version: 52
    flags: ACC_PUBLIC, ACC_FINAL, ACC_SUPER
```

上述结果的 flags 部分展示了 MyClass 的访问标识信息。也可以通过工具更加直观地显示，如图 3-7 所示。

需要注意的是，各个标识之间也并非全都能互相组合，例如 ACC_INTERFACE 和 ACC_FINAL 就不能同时使用，因为这违反了 Java 的语法规定。

5. 类索引、父类索引、接口索引

类索引（this_class）、父类索引（super_class）以及接口索引（interfaces_count、interfaces[interfaces_count]）这 3 个结构用来表示一个类或接口的继承关系。this_class 和 super_class 都是 u2 类型的数据，它们的值指向常量池中的结果。this_class 的值为当前类的全限定名。super_class 的值为父类的全限定名，所有的类都有父类，如果没有显示声明，那么其父类是 java.lang.Object。interfaces_count 表示实现的接口数目，也是一个 u2 类型的数据，interfaces[interfaces_count]

中的值表示当前类实现的接口，其中具体的项也是引用常量池中的结果，常量池中的值为接口的全限定名。

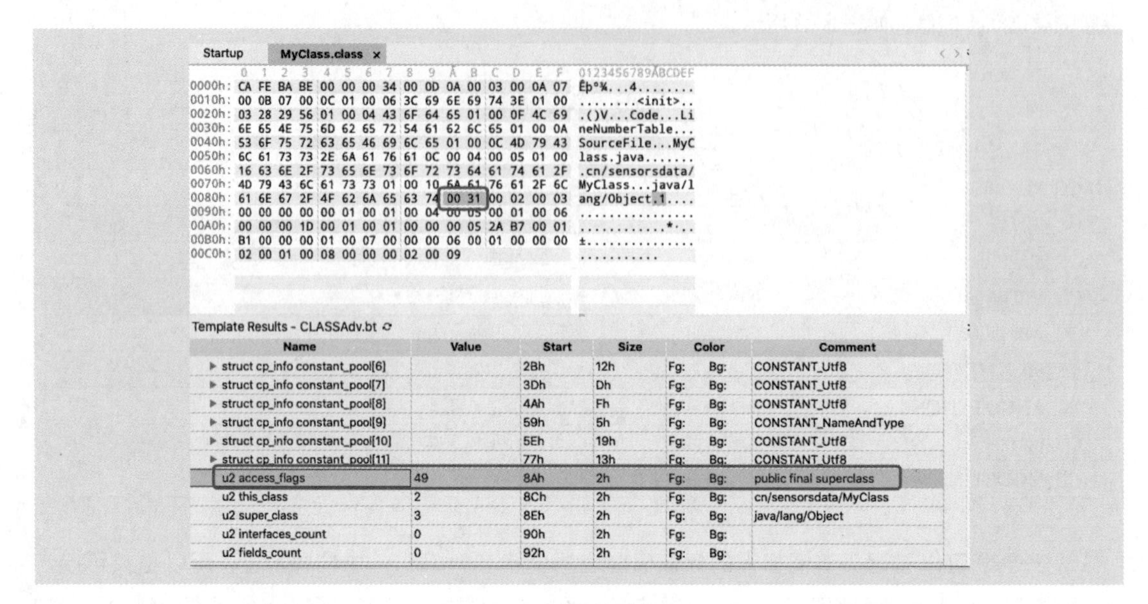

图 3-7　直观展示访问标识

以如下代码为例。

```
public final class ThisSuperInterfaceMain implements Runnable {
    @Override
    public void run() {
    }
}
```

ThisSuperInterfaceMain 实现了 Runnable 接口，将其编译后使用 javap 查看其常量池信息，结果如下。

```
Constant pool:
    #1 = Methodref      #3.#12      // java/lang/Object."<init>":()V
    #2 = Class          #13         // cn/sensorsdata/ThisSuperInterfaceMain
    #3 = Class          #14         // java/lang/Object
    #4 = Class          #15         // java/lang/Runnable
    #5 = Utf8           <init>
    #6 = Utf8           ()V
    #7 = Utf8           Code
    #8 = Utf8           LineNumberTable
    #9 = Utf8           run
   #10 = Utf8           SourceFile
   #11 = Utf8           ThisSuperInterfaceMain.java
   #12 = NameAndType    #5:#6           // "<init>":()V
   #13 = Utf8           cn/sensorsdata/ThisSuperInterfaceMain
   #14 = Utf8           java/lang/Object
   #15 = Utf8           java/lang/Runnable
```

相应的类索引、父类索引和接口索引信息如图 3-8 所示。

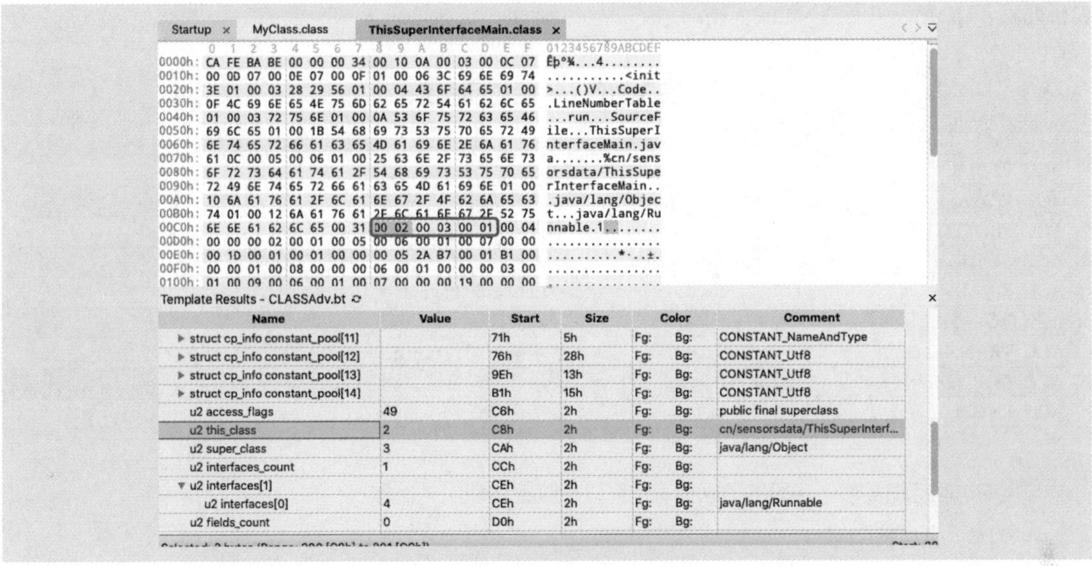

Startup × MyClass.class ThisSuperInterfaceMain.class ×

```
       0  1  2  3  4  5  6  7  8  9  A  B  C  D  E  F  0123456789ABCDEF
0000h: CA FE BA BE 00 00 00 34 00 10 0A 00 03 00 0C 07  Êþº¾...4........
0010h: 00 0D 07 00 0E 07 00 0F 01 00 06 3C 69 6E 69 74  ...........<init
0020h: 3E 01 00 03 28 29 56 01 00 04 43 6F 64 65 01 00  >...()V...Code..
0030h: 0F 4C 69 6E 65 4E 75 6D 62 65 72 54 61 62 6C 65  .LineNumberTable
0040h: 01 00 03 72 75 6E 01 00 0A 53 6F 75 72 63 65 46  ...run...SourceF
0050h: 69 6C 65 01 00 1B 54 68 69 73 53 75 70 65 72 49  ile...ThisSuperI
0060h: 6E 74 65 72 66 61 63 65 4D 61 69 6E 2E 6A 61 76  nterfaceMain.jav
0070h: 61 0C 00 05 00 06 01 00 25 63 6E 2F 73 65 6E 73  a.......%cn/sens
0080h: 6F 72 73 64 61 74 61 2F 54 68 69 73 53 75 70 65  orsdata/ThisSupe
0090h: 72 49 6E 74 65 72 66 61 63 65 4D 61 69 6E 01 00  rInterfaceMain..
00A0h: 10 6A 61 76 61 2F 6C 61 6E 67 2F 4F 62 6A 65 63  .java/lang/Objec
00B0h: 74 01 00 12 6A 61 76 61 2F 6C 61 6E 67 2F 52 75  t...java/lang/Ru
00C0h: 6E 6E 61 62 6C 65 00 31 00 02 00 03 00 01 00 04  nnable.1........
00D0h: 00 00 00 02 00 01 00 05 00 06 00 01 00 07 00 00  ................
00E0h: 00 1D 00 01 00 01 00 00 00 05 2A B7 00 01 B1 00  ..........*·..±.
00F0h: 00 00 01 00 08 00 00 00 06 00 01 00 00 00 03 00  ................
0100h: 01 00 09 00 06 00 01 00 07 00 00 00 19 00 00 00  ................
```

Template Results - CLASSAdv.bt

Name	Value	Start	Size	Color	Comment
▶ struct cp_info constant_pool[11]		71h	5h	Fg: Bg:	CONSTANT_NameAndType
▶ struct cp_info constant_pool[12]		76h	28h	Fg: Bg:	CONSTANT_Utf8
▶ struct cp_info constant_pool[13]		9Eh	13h	Fg: Bg:	CONSTANT_Utf8
▶ struct cp_info constant_pool[14]		B1h	15h	Fg: Bg:	CONSTANT_Utf8
u2 access_flags	49	C6h	2h	Fg: Bg:	public final superclass
u2 this_class	2	C8h	2h	Fg: Bg:	cn/sensorsdata/ThisSuperInterf...
u2 super_class	3	CAh	2h	Fg: Bg:	java/lang/Object
u2 interfaces_count	1	CCh	2h	Fg: Bg:	
▼ u2 interfaces[1]		CEh	2h	Fg: Bg:	
u2 interfaces[0]	4	CEh	2h	Fg: Bg:	java/lang/Runnable
u2 fields_count	0	D0h	2h	Fg: Bg:	

图 3-8 类索引、父类索引和接口索引信息

结合常量池信息和图 3-8 所示的 this_class、super_class、interfaces_count 和 interfaces[1] 可以看到，this_class 的值为 0x0002、super_class 的值为 0x0003，它们指向常量池索引 #2 和 #3 的位置，对比常量池中的数据，可以看到当前类和父类的全限定名分别是 cn/sensorsdata/ThisSuperInterfaceMain 和 java/lang/Object。interfaces_count 的值为 0x0001，表示实现了一个接口，通过工具也可以看到 u2 interfaces[0] 的值为 0x0004，指向常量池索引为 #4 的位置，对比可发现其值为 java/lang/Runnable。

6. 字段表

在接口索引之后的是字段表（fields），字段表用于描述类或者接口中定义的字段，这里的字段不包括方法体内的局部变量，特指类变量（static 修饰的变量）和实例变量。在 ClassFile 中与字段表相关的结构如下。

```
u2 fields_count;
field_info fields[fields_count]
```

两个字节的 fields_count 表示字段的数目（不包括父类中的字段），其中每个项使用 field_info 结构来表示。field_info 的结构如下。

```
field_info{
      u2 access_flags;
      u2 name_index;
      u2 descriptor_index;
      u2 attribute_count;
      attribute_info attributes[attributes_count]
}
```

上述结构分 3 个部分，各部分的说明如下。

（1）access_flags。

类似 3.4.6 小节的类访问标识，字段也有访问标识，也是两个字节，最多可以定义 16 个标识，详细的

介绍如表 3-10 所示。

表3-10　字段访问标识

标识名称	十六进制	描述
ACC_PUBLIC	0x0001	字段被声明为 public
ACC_PRIVATE	0x0002	字段被声明为 private
ACC_PROTECTED	0x0004	字段被声明为 protected
ACC_STATIC	0x0008	字段被声明为 static
ACC_FINAL	0x0010	字段被声明为 final
ACC_VOLATILE	0x0040	字段被声明为 volatile
ACC_TRANSIENT	0x0080	字段被声明为 transient，该字段不会被持久化
ACC_SYNTHETIC	0x1000	表示非源码生成，而是由编译器生成
ACC_ENUM	0x4000	表示这是一个枚举类

与类的访问标识类似，字段的访问标识也可以相互组合使用，例如：

```
public static final float PI = 3.1415f;
```

其访问标识就是 0x0001(ACC_PUBLIC)|0x0008(ACC_STATIC) | 0x0010(ACC_FINAL) = 0x0019。同样，字段的访问标识也不能随意组合，例如 ACC_PUBLIC 与 ACC_PRIVATE 这个组合便是违反 Java 语法规则的。

（2）name_index 和 descriptor_index。

name_index 的值为指向常量池的有效索引，其结构是 CONSTANT_Utf8_info，用来表示当前字段的名称。descriptor_index 也是指向常量池的有效索引，其结构也是 CONSTANT_Utf8_info，用来描述当前字段的类型，即字段描述符。

（3）attribute。

attribute 表示字段的属性表。Java 的泛型是从 JDK 1.5 开始支持的，可是通过 javap 查看字段和方法的描述符信息可以看到，其中并不包括泛型信息（与 C# 语言相比，Java 的泛型又被称为假泛型），那么泛型内容存储在何处呢？其实字段的泛型（signature）信息就存储在字段属性表中。属性将会在 3.4.10 小节详细介绍。

7. 方法表

在 ClassFile 中，方法表的结构如下。

```
u2 methods_count;
method_info methods[methods_count]
```

两个字节的 methods_count 表示方法的数目，不包括父类中的方法。每个方法都用 method_info 结构来表示，method_info 的结构如下。

```
method_info{
      u2 access_flags;
      u2 name_index;
      u2 descriptor_index;
      u2 attribute_count;
      attribute_info attributes[attributes_count]
}
```

从中可以看到方法表和字段表的结构有很多相似的地方，同样分 3 个部分，每个部分的说明如下。

（1）access_flags。

access_flags 是方法的访问标识符，其各项描述信息如表 3-11 所示。

表 3-11　方法访问标识

标识名称	十六进制	描述
ACC_PUBLIC	0x0001	方法被声明为 public
ACC_PRIVATE	0x0002	方法被声明为 private
ACC_PROTECTED	0x0004	方法被声明为 protected
ACC_STATIC	0x0008	方法被声明为 static
ACC_FINAL	0x0010	方法被声明为 final
ACC_SYNCHRONIZED	0x0020	方法被声明为 synchronized
ACC_BRIDGE	0x0040	方法被声明为 bridge，由编译器产生
ACC_VARARGS	0x0080	表示方法带有可变参数
ACC_NATIVE	0x0100	方法被声明为 native
ACC_ABSTRACT	0x0400	方法被声明为 abstract
ACC_STRICT	0x0800	方法被声明为 strictfp，使用 FP-strict 浮点模式
ACC_SYNTHETIC	0x1000	表示非源码生成，而是由编译器生成

类似于类和字段的访问标识符，方法的标识符之间也是可以相互组合使用的，例如 ACC_PUBLIC | ACC_STATIC，但是 ACC_ABSTRACT 与 ACC_FINAL 就不能组合使用，这违反 Java 语法规则。另外，观察可以发现在类、方法、字段中同时出现的标识符具有相同的值，例如 ACC_PUBLIC 的值都是 0x0001，这也是为什么在 ASM 框架中标识符不是按照类、字段、方法进行分类的一个原因，需要读者理解这些标识符，这样在 ASM 中使用才不会出错。

（2）name_index 和 descriptor_index。

name_index 的值为指向常量池的有效索引，其结构是 CONSTANT_Utf8_info，用来表示当前方法的名称，方法名可能是 <init> 或 <clinit>。descriptor_index 也是指向常量池的有效索引，其结构也是 CONSTANT_Utf8_info，用来描述当前方法的类型，即方法描述符。

（3）attribute。

行文至此你可能会疑问，方法的结构中包含访问标识符、名称、方法描述符信息，那么方法中的代码在哪里呢？与字段的泛型信息存放在字段属性表中类似，方法的代码是存放在方法属性表中一个名为 Code 的属性中，属性表是 ClassFile 中最有扩展性的结构，下面就来介绍属性表。

8. 属性表

通过前面介绍的知识我们知道字段中有属性表、方法中有属性表，还有 ClassFile 结构的最后一个结构也是属性表。不像 ClassFile 文件的结构有严格的顺序要求，属性表的要求相对宽松很多，只要属性名不重复即可，甚至读者也可以实现自己的属性，自定义的属性只要 Java 虚拟机能够识别即可。如果 Java 虚拟机遇到无法识别的属性，应当忽略掉。

属性表的通用结构如下。

```
attribute_info{
      u2 attribute_name_index;
      u4 attribute_length;
      u1 info[attribute_length]
}
```

其中 attribute_name_index 为指向 CONSTANT_Utf8_info 结构的索引，表示属性的名称；attribute_ length 和 info 表示 info 数组的长度和具体的数组内容。

表 3-12 列出的属性是按照在 JDK 版本中出现的顺序排列的，表中同时列出了属性被使用的位置以及 ClassFile 版本号。

表 3-12 列出了 JDK 1 到 JDK 8 中的所有属性，因篇幅限制，接下来只对其中的部分属性进行介绍，更多详细的介绍读者可以参考官方文档。

表3-12　ClassFile属性列表

属性名	使用位置	JDK 版本	ClassFile 版本号
ConstantValue	field_info	1.0.2	45.3
Code	method_info	1.0.2	45.3
Exceptions	method_info	1.0.2	45.3
SourceFile	ClassFile	1.0.2	45.3
LineNumberTable	Code	1.0.2	45.3
LocalVariableTable	Code	1.0.2	45.3
InnerClasses	ClassFile	1.1	45.3
Synthetic	ClassFile,field_info,method_info	1.1	45.3
Deprecated	ClassFile,field_info,method_info	1.1	45.3
EnclosingMethod	ClassFile	5.0	49.0
Signature	ClassFile,field_info,method_info	5.0	49.0
SourceDebugExtension	ClassFile	5.0	49.0
LocalVariableTypeTable	Code	5.0	49.0
RuntimeVisibleAnnotations	ClassFile,field_info,method_info	5.0	49.0
RuntimeInvisibleAnnotations			49.0
RuntimeVisibleParameterAnnotations	method_info	5.0	49.0
RuntimeInvisibleParameterAnnotations		5.0	49.0
AnnotationDefault	method_info	5.0	49.0
StackMapTable	Code	6	50.0
BootstrapMethods	ClassFile	7	51.0
RuntimeVisibleTypeAnnotations	ClassFile,field_info,method_info,Code	8	52.0
RuntimeInvisibleTypeAnnotations		8	52.0
MethodParameters	method_info	8	52.0

（1）ConstantValue 属性。

在 Java 虚拟机中，不同类型的变量的初始化顺序是不一样的，例如对于 int x = 123 和 static int x = 123，虚拟机对这两种变量的赋值方式是不同的。

对于非 static 变量，其初始赋值是在实例构造方法（即 <init>）中；对于类变量（即 static 修饰的变量）是在类构造器（即 <clinit>）中赋值；而对于 final + static 修饰的变量，并且变量的类型是基本类型或 string 类型，那么其值是用 ConstantValue 属性记录的。ConstantValue 属性位于 field_info 结构的属性中，被用来存放常量的初始值。

ConstantValue 属性的结构如下。

```
ConstantValue_attribute{
      u2 attribute_name_index;
      u4 attribute_length;
      u2 constantvalue_index;
}
```

其中 attribute_name_index 为指向 CONSTANT_Utf8_info 结构的索引，用以表示属性名的值，此处为“ConstantValue”；attribute_length 是一个为 2 的固定值；constantvalue_index 的值是对常量池的一个有效索引，该索引处的值就是 ConstantValue 的值，并且索引处的结构必须是表 3-13 中的一个，即基本类型加上 string 类型。

表3-13　ConstantValue 支持的类型

字段类型	常量池结构
int、short、char、byte、boolean	CONSTANT_Integer_info
float	CONSTANT_Float_info
double	CONSTANT_Double_info
string	CONSTANT_String_info

以下面这个类为例。

```
public class ConstantValueMain {
      private static final int A = 10;
      private static int B = 11;
      private int C = 12;

      public ConstantValueMain(){
      }
}
```

类中定义了 3 个成员字段 A、B、C，其中 A 是常量，B 是静态的成员变量，C 是普通的成员变量。ConstantValueMain 编译以后使用 javap 工具查看字节码信息，观察这 3 个变量的初始化位置。

```
$ javap -v -p ConstantValueMain.class

// 输出结果
public class cn.sensorsdata.ConstantValueMain

Constant pool:
     #1 = Methodref       #5.#19      // java/lang/Object."<init>":()V
     #2 = Fieldref        #4.#20      // cn/sensorsdata/ConstantValueMain.C:I
     #3 = Fieldref        #4.#21      // cn/sensorsdata/ConstantValueMain.B:I
     #4 = Class           #22         // cn/sensorsdata/ConstantValueMain
     #5 = Class           #23         // java/lang/Object
     #6 = Utf8            A
     #7 = Utf8            I
     #8 = Utf8            ConstantValue
     #9 = Integer         10
    #10 = Utf8            B
    #11 = Utf8            C
    #12 = Utf8            <init>
    #13 = Utf8            ()V
```

```
    #14 = Utf8               Code
    #15 = Utf8               LineNumberTable
    #16 = Utf8               <clinit>
    #17 = Utf8               SourceFile
    #18 = Utf8               ConstantValueMain.java
    #19 = NameAndType        #12:#13        // "<init>":()V
    #20 = NameAndType        #11:#7         // C:I
    #21 = NameAndType        #10:#7         // B:I
    #22 = Utf8               cn/sensorsdata/ConstantValueMain
    #23 = Utf8               java/lang/Object
{
    private static final int A; //(1)
      descriptor: I
      flags: ACC_PRIVATE, ACC_STATIC, ACC_FINAL
      ConstantValue: int 10

    private static int B; //(2)
      descriptor: I
      flags: ACC_PRIVATE, ACC_STATIC

    private int C; //(3)
      descriptor: I
      flags: ACC_PRIVATE

    public cn.sensorsdata.ConstantValueMain();//(6)
      descriptor: ()V
      flags: ACC_PUBLIC
      Code:
        stack=2, locals=1, args_size=1
           0: aload_0
           1: invokespecial #1     // Method java/lang/Object."<init>":()V
           4: aload_0
           5: bipush        12
           7: putfield      #2     // Field C:I //(7)
          10: return

    static {};//(4)
      descriptor: ()V
      flags: ACC_STATIC
      Code:
        stack=1, locals=0, args_size=0
           0: bipush        11
           2: putstatic     #3     // Field B:I //(5)
           5: return
}
SourceFile: "ConstantValueMain.java"
```

首先观察位置（1）处的变量 A 的定义，字节码结果中显示了变量 A 的字段描述符（descriptor）、访问标识符（flags）以及 ConstantValue 属性的值，而位置（2）处的静态变量 B 和位置（3）处的变量 C 就没有 ConstantValue 属性，这是变量 A 与它们的区别。

再借助工具查看变量 A 的结构，如图 3-9 所示。

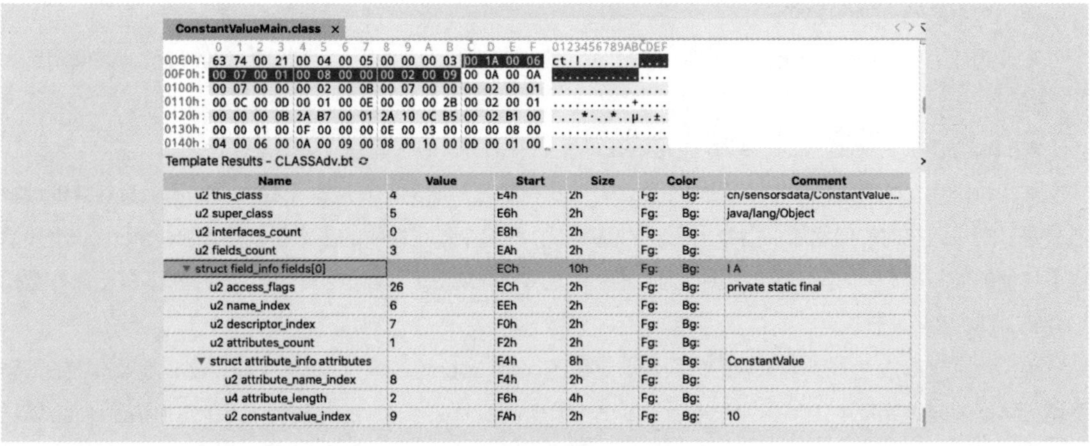

图 3-9　ConstantValue 属性

图 3-9 展示了变量 A 的结构，可以从 A 的结构中看到 ConstantValue 属性，constantvalue_index 指向索引 #9，其值可以从工具的 Comment 一栏中看到结果是 10，也可以从 javap 的结果中查看索引 #9 的位置结果是 10。

而静态变量 B 的初始化方式在位置（4）处的静态代码块中（即类构造器 <cinit>() 方法中），位置（5）处的字节码（字节码在会在第 4 章介绍）表示给变量 B 赋值。类似变量 B 这样的静态变量会按照其在类中声明的顺序依次在类构造器方法中初始化。

变量 C 的初始化是在位置（6）处的构造方法中（即实例构造器 <init>() 方法中）初始化，位置（7）处的字节码表示给变量 C 赋值。类似 C 这样的变量也会按照其声明的顺序依次在实例构造器中初始化。

关于这部分的内容的介绍，读者可以结合 4.9 小节内容一起理解。

（2）Code 属性。

Code 属性是 ClassFile 中非常重要的一部分，该属性定义在 method_info 结构中，用于记录方法中的代码。不过并非所有的方法都存储在 Code 属性，例如 abstract() 和 native() 方法就不允许存储在 Code 属性。

Code 属性的结构如下。

```
Code_attribute{
      u2 attribute_name_index;
      u4 attribute_length;
      u2 max_stack;
      u2 max_locals;
      u4 code_length;
      u1 code[code_length];
      u2 exception_table_length;
      {
          u2 start_pc;
          u2 end_pc;
          u2 handler_pc;
          u2 catch_type;
      } exception_table[exception_table_length];
      u2 attribute_count;
      attribute_info attributes[attribute_count];
}
```

该结构中各部分的介绍如下。

- attribute_name_index为指向常量池CONSTANT_Utf8_info结构的有效索引，该值固定为“Code”。
- attribute_length为Code属性内容的长度。
- max_stack为操作数栈的最大深度。方法执行的任意时刻的深度都不会大于此值，其具体的计算方式是：指令入栈时的值增加，指令出栈时的值减少，在这个过程中栈的最大值就是max_stack。一般情况下栈值的加减是1，但是对long和double相关的指令的加减是2，void相关的指令则是0。关于操作数栈的介绍，请参考4.4节。
- max_locals为局部变量表的最大值，这里的max_locals的单位是槽。对于Java虚拟机中的数据类型（详细内容请参考4.10节）如byte、char、short、int、boolean、float、reference、returnAddress等长度不超过32位的数据类型，每个局部变量占一个槽；对于long和double此种长度为64位的数据类型，每个局部变量占用两个槽。方法的参数以及方法中使用的局部变量的存储都需要局部变量表，但也不是代码中有多少个局部变量，局部变量表就有多少个值，局部变量表中的槽可以被复用，当然最大值越小越好，这就表示占用的内存就更少。很多优化字节码的工具，其中优化的一项就是局部变量表。想象一下，一个方法有两个参数，但是方法中并没有使用这两个参数，那么这两个参数所占用的槽就可以被方法中其他局部变量拿来使用，以此来减少内存的开销。关于局部变量的介绍，请参考4.3节。
- code_length和code用来存储字节码指令信息，其中code_length表示字节码指令数组的长度，code是u1类型，表示所有的指令都用一个字节表示。
- exception_table_length和exception_table用于表示方法中相关的异常信息。其中start_pc、end_pc、handler_pc表示指向字节码指令中偏移的位置，catch_type表需要捕获的异常类型的索引，指向常量池中的CONSTANT_Class_info结构。可以这么理解这几个值：当start_pc和end_pc之间（包括start_pc，不包括end_pc）的字节码指令发生了异常的时候，就判断异常的类型是否与catch_type一致，如果一致，虚拟机就跳到handler_pc指向的位置继续执行。如果catch_type的值为0，则表示可以处理任意异常，且都要到handler_pc的位置继续执行，可以用此来实现finally语法。读者可以参考4.7.4节来详细了解try-catch-finally实现的原理。
- attribute_count和attribute_info表示Code属性中相关的附加属性，例如LineNumberTable、LocalVariableTable、LocalVariableTypeTable、StackMapTable、RuntimeVisibleTypeAnnotations、RuntimeInvisibleTypeAnnotations等属性。以LineNumberTable为例，此属性用来存储源码中行号和字节码偏移量的对应关系，可以作为调试使用，它不是必需的，就算没有也不会有问题，只不过当发生异常的时候无法匹配到源码中的对应行。StackMapTable也是一个比较重要的属性，我们在后文会介绍。

下面通过一个具体的例子来了解 Code 属性。

```
public class CodeAttribute {

    void foo(int x) {
        try {
            x = 1;
        } catch (RuntimeException exception) {
            x = 2;
        } finally {
```

```
                x = 3;
            }
        }
}
```

编译 CodeAttribute 以后使用 javap 工具查看 foo() 方法中的字节码信息，结果如下。

```
void foo(int);
      descriptor: (I)V
      flags:
      Code:
        stack=1, locals=4, args_size=2 //(1)
           0: iconst_1
           1: istore_1
           2: iconst_3
           3: istore_1
           4: goto          20
           7: astore_2
           8: iconst_2
           9: istore_1
          10: iconst_3
          11: istore_1
          12: goto          20
          15: astore_3
          16: iconst_3
          17: istore_1
          18: aload_3
          19: athrow
          20: return
        Exception table://(2)
            from    to  target type
               0     2     7   Class java/lang/RuntimeException
               0     2    15   any
               7    10    15   any
        LineNumberTable://(3)
          line 7: 0
          line 11: 2
          line 12: 4
          line 8: 7
          line 9: 8
          line 11: 10
          line 12: 12
          line 11: 15
          line 12: 18
          line 13: 20
        StackMapTable: number_of_entries = 3 //(4)
          frame_type = 71 /* same_locals_1_stack_item */
            stack = [ class java/lang/RuntimeException ]
          frame_type = 71 /* same_locals_1_stack_item */
            stack = [ class java/lang/Throwable ]
          frame_type = 4 /* same */
```

通过上述结果可以看到，位置(1)处显示了max_stack和max_local的值，注意这里的args_size = 2，表示方法有两个参数。读者可能会有疑问，foo() 不是只有一个参数吗？这是因为对于实例方法，默认其第一个参数是 this。位置（2）处显示了方法的异常信息表，其中 from 和 to 分别对应 start_pc 和 end_pc；target 对应 handler_pc，表示当出现异常时继续执行字节码的位置；type 对应 catch-type，表示异常的类型。位置（3）处表示源码中行号和字节码偏移对照。位置（4）处则是 StackMapTable 属性的值，接下来介绍该属性。

（3）StackMapTable 属性。

Java 虚拟机中，一个类从被加载到虚拟机内存中开始，到卸载出内存为止，它的生命周期有 7 个阶段：加载、验证、准备、解析、初始化、使用、卸载。在验证阶段会对 ClassFile 的格式（魔数是不是 CAFEBABE）、元数据（是否继承了 final 类）、字节码、符号引用等数据进行验证，防止虚拟机加载了有害的数据。

字节码验证可以保证指令的格式正确，即跳转指令能够指向合法的位置、所有的指令操作的类型是正确的。问题是字节码本身没有明确的类型信息，例如 iconst 指令创建一个 integer 值，然后将其存储在局部变量表中的 1 这个槽中，现在这个槽是一个 int 类型，如果又将一个 float 类型存在此处并且还是将其按照 int 类型对待，那么就会出错。

DK 1.6 之前按照一定规则来推断所有的类型，这可能需要不止遍历一遍字节码，并且效率也不高。为了解决这个问题，Oracle 添加了一个新的验证器，新的验证器速度更快并且只需遍历一遍字节码指令。因为方法中的字节码不能改变并且不能存储新验证器需要的元数据，所以从 Java 6 开始，新增了 StackMapTable 属性用于保存元数据。为了使元数据信息更小、更高效，因此只记录跳转指令的跳转目标位置的信息，在跳转目标之间所有的代码都是线性的，即不存在其他跳转指令。每一个这样的位置我们用一个栈映射帧（stack map frame）来表示，为了减少数据量，栈映射帧只保存与前一个帧不同的地方，即当前帧由上一个帧计算得出。另外每一个方法都有一个隐式的帧，这个隐式的帧由方法的描述信息中获得，这个隐式的帧又称为初始帧并且不会在 StackMapTable 属性中显示出来。

下面对 StackMapTable 属性进行详细的介绍，此部分内容需要读者对字节码有所了解，读者可以先学习第 4 章知识，再来学习此部分内容。

StackMapTable 属性的结构如下。

```
StackMapTable_attribute{
      u2 attribute_name_index;
      u4 attribute_length;
      u2 number_of_entries;
      stack_map_frame entries[number_of_entries]
}
```

其中 attribute_name_index 为指向常量池 CONSTANT_Utf8_info 结构的有效索引，其值固定为“StackMapTable”。attribute_length 表示 StackMapTable 属性的长度，不包括 attribute_length 和 attribute_name_index 这 6 个字节。number_of_entries 表示 stack_map_frame 结构体的长度。stack_map_frame 即栈映射帧，一个 StackMapTable 包含 0 个或多个栈映射帧。每一个栈映射帧显式或者隐式地指定了字节码的偏移量，并且指定了此偏移量所在位置的局部变量和操作数栈中项所需要的校验类型（verification type）。

stack_map_frame 的结构如下。

```
union stack_map_frame {
      same_frame;
      same_locals_1_stack_item_frame;
      same_locals_1_stack_item_frame_extended;
      chop_frame;
      same_frame_extended;
```

```
        append_frame;
        full_frame;
}
```

上面的联合体表示 stack_map_frame 中的类型只可能是其中一个，而不是同时存在多个，这些类型有一个通用的结构如下。

```
u1 frame_type;
u2 offset_delta
```

即一个字节的帧类型和两个字节的字节码偏移量，如果帧中没有显式指明 offset_delta，那么 offset_delta 这个值是隐式的。至于为什么使用偏移量的增量，而不是字节码实际的偏移量呢？因为通过这种方式可以确保栈映射帧是顺序的，并且可以通过 offset_delta + 1 来根据每个显式帧计算出下一个帧的偏移量。

下面就来看看 stack_map_frame 中每种类型的定义。

（1）帧类型（frame type）为 same_frame 的取值范围是 [0, 63]。如果帧类型是 same_frame，则表示该帧对比前一个帧有相同的局部变量，并且操作数栈为空，同时 offset_delta 的值等于 frame_type 的值。same_frame 的结构如下。

```
same_frame {
        u1 frame_type = SAME; /* 0-63 */
}
```

（2）帧类型为 same_locals_1_stack_item_frame 的取值范围是 [64, 127]。如果帧类型是 same_locals_1_stack_item_frame，则表示此帧对比前一个帧有相同的局部变量，并且操作数栈中有一项，同时 offset_delta 的值等于 frame_type - 64，并且结构中有一个 verification_type_info，用于表示操作数栈中的类型，其结构如下。

```
same_locals_1_stack_item_frame {
        u1 frame_type = SAME_LOCALS_1_STACK_ITEM; /* 64-127 */
        verification_type_info stack[1];
}
```

verification_type_info 结构用于表示操作数栈中的成员类型，其结构在此步进行声明。

（3）frame_type 中范围在 [128, 246] 之间的值暂未使用，留作未来扩展。

（4）帧类型为 same_locals_1_stack_item_frame_extended 的取值是 247。如果帧类型是 same_locals_1_stack_item_frame_extended，则表示此帧对比前一个帧有相同的局部变量，并且操作数栈中有一项。与 same_locals_1_stack_item_frame 不同的是，该类型必须指明 offset_delta 的值。其结构如下。

```
same_locals_1_stack_item_frame_extended {
        u1 frame_type = SAME_LOCALS_1_STACK_ITEM_EXTENDED; /* 247 */
        u2 offset_delta;
        verification_type_info stack[1];
}
```

（5）帧类型为 chop_frame 的取值范围是 [248, 250]。如果是此类型，表示操作数栈为空，并且与上一个帧相比，局部变量中缺少了 *k* 个项，其中 *k* 的值为 251 - frame_type。其结构如下。

```
chop_frame {
        u1 frame_type = CHOP; /* 248-250 */
        u2 offset_delta;
}
```

（6）帧类型为 same_frame_extended 的取值是 251。如果是此类型，表示与前一帧有相同的局部变量并且操作数栈为空。与 same_frame 不同的是，此类型需要指明 offset_delta 的值。其结构如下。

```
same_frame_extended {
      u1 frame_type = SAME_FRAME_EXTENDED; /* 251 */
      u2 offset_delta;
}
```

（7）帧类型为 append_frame 的取值范围是 [248, 250]。如果是此类型，表示操作数栈为空，并且与前一帧相比，局部变量表中除了新增的 *k* 个项外都与前一帧的局部变量相同，*k* 的值为 frame_type－251。其结构如下。

```
append_frame {
      u1 frame_type = APPEND; /* 252-254 */
      u2 offset_delta;
      verification_type_info locals[frame_type - 251];
}
```

（8）帧类型为 full_frame 的取值是 255。如果是此类型，那么会给出局部变量和操作数栈中详细的数目和对应的 verification_type_info 信息。其结构如下。

```
full_frame {
      u1 frame_type = FULL_FRAME; /* 255 */
      u2 offset_delta;
      u2 number_of_locals;
      verification_type_info locals[number_of_locals];
      u2 number_of_stack_items;
      verification_type_info stack[number_of_stack_items];
}
```

以上就是 StackMapTable 相关的基本信息，下面用一个具体的例子看看其表现形式。

```
void foo() {
      int i = 10;
      if (i > 11) {
          int j = 16;
      } else {
          int m = 15;
      }
      int k = 29;
}
```

上面这个方法对应的字节码信息如下。

```
void foo();
      descriptor: ()V
      flags:
      Code:
        stack=2, locals=3, args_size=1
           0: bipush        10
           2: istore_1
           3: iload_1
           4: bipush        11
           6: if_icmple     15
           9: bipush        16
          11: istore_2
          12: goto          18
```

```
        15: bipush        15
        17: istore_2
        18: bipush        29
        20: istore_2
        21: return
      StackMapTable: number_of_entries = 2 //(1)
        frame_type = 252 /* append */ //(2)
          offset_delta = 15
          locals = [ int ]
        frame_type = 2 /* same */ //(3)
```

从位置（1）处的 number_of_entries = 2 可知，StackMapTable 中包含两个栈映射帧。首先根据方法的签名得出隐式初始帧的局部变量是 [this]，操作数栈为空。第一个栈映射帧的类型是 append [如位置（2）所示]，局部变量中多了一个 [int]，我们从字节码偏移 6: if_cmple 15 的位置可以看出，对应字节码偏移位置 15 的局部变量是 [this, int]，操作数栈也为空，比初始帧的局部变量多了 [int]，其 offset_delta = 15。第二个栈映射帧的类型是 same [如位置（3）所示]，说明局部变量和前一帧是相同的并且操作数栈为空，根据字节码偏移 12: goto 18 可以得出其局部变量也是 [this, int]，说明是 same，same 类型的 offset_delta 等于 frame_type 的值，此处的字节码偏移的位置为前一帧的 offset_delta 15 加上 frame_type 加 1，即 18。

在 Android 虚拟机中对 ClassFile 的 StackMapTable 属性没有做要求，可以不需要这个结构。不过从 AGP 4.2.0 开始，Android D8 会开始对字节码进行校验，如果缺少或者对 StackMapTable 校验失败都会给出警告，在使用 ASM 操作 ClassFile 的时候要保留 StackMapTable 属性。如果不希望虚拟机校验，可以给虚拟机配置如下参数。

```
org.gradle.jvmargs= -Xverify:none
```

（4）Signature 属性。

Signature 是 JDK 5 引入的属性，如果类中用到了泛型，泛型信息记录在这个属性中。该属性的使用范围是 ClassFile、method_info、field_info，也就是类、方法、字段中都可以使用泛型。在 JDK 5 之前 Java 是不支持泛型的，为了支持泛型同时保持向后兼容（既保持 ClassFile 文件结构的稳定），所以只能通过新增 Signature 属性，并将泛型信息保存在属性中，也就是说编译后的字节码中不包含泛型信息，我们称为泛型擦除，Java 语言的这种泛型方式称为伪泛型。即 Java 泛型无法像 C# 那样将泛型类型和普通类型一样对待。

关于泛型擦除，下面举个例子来说明。

```
public class Node<T> {
      public T data;

      public Node(T data) { this.data = data; }

      public void setData(T data) {
          System.out.println("Node.setData");
          this.data = data;
      }
}
```

在上面这个例子中定义了泛型 <T>，T 没有指定任何边界（bound）。这里的边界是指能够确定 T 类型的值，例如对于 class Node<T extends Thread>，这里 T 的边界就是 java.lang.Thread，如果没有指定边界，那么其边界类型默认是 java.lang.Object。所以对这个例子来说，Java 编译器在编译时会将所有的 T 使用 java.lang.Object 替换，即类似：

```
public class Node {
      public Object data;

      public Node(Object data) { this.data = data; }

      public void setData(Object data) {
          System.out.println("Node.setData");
          this.data = data;
      }
}
```

对于有边界的情况，则使用边界值来替换，在此不多做介绍。

关于泛型擦除有一种情况需要注意一下，例如下面这个例子的运行结果。

```
public class MyNode extends Node<Integer> {

      public MyNode(Integer data) {
          super(data);
      }

      public void setData(Integer data) {
          System.out.println("MyNode.setData");
          super.setData(data);
      }

      public static void main(String[] args) {
          MyNode mn = new MyNode(5);
          Node n = mn; // 注意这里的赋值
          n.setData("Hello");      //(1)
          Integer x = mn.data;
      }
}
```

上面的例子当运行 main() 方法的时候，会在位置（1）处抛出如下异常。

```
Exception in thread "main" java.lang.ClassCastException: java.lang.String cannot be cast to java.
lang.Integer
```

首先对于上面的语法规则，编译器是可以正常编译通过的，但是一旦运行就会存在很大的问题，因为 mn 所接收的参数类型已经由 integer 转变成了 string，这显然是不符合预期逻辑的，这就是泛型擦除带来的问题。为了解决这个问题以及保持 Java 语言的多态特性，编译器会生成一个桥接方法来对数据进行强制转换，具体可以使用 javap 来看看 MyNode.class 背后的本质原因（注意，Java 中的很多语法规则都是使用桥接的方式来实现的，例如 Lambda 表达式的实现）。

```
//MyNode.class

//setData() 方法 1
public void setData(java.lang.Integer);
    descriptor: (Ljava/lang/Integer;)V
    flags: ACC_PUBLIC
    Code:
      stack=2, locals=2, args_size=2
         0: getstatic     #2  // Field java/lang/System.out:Ljava/io/PrintStream;
```

```
        3: ldc           #3  // String MyNode.setData
          5: invokevirtual #4    // Method java/io/PrintStream.println:(Ljava/lang/
String;)V
        8: aload_0
        9: aload_1
         10: invokespecial #5    // Method cn/sensorsdata/Node.setData:(Ljava/lang/
Object;)V
       13: return

//setData() 方法 2
public void setData(java.lang.Object);
    descriptor: (Ljava/lang/Object;)V
    flags: ACC_PUBLIC, ACC_BRIDGE, ACC_SYNTHETIC
    Code:
      stack=2, locals=2, args_size=2
        0: aload_0
        1: aload_1
        2: checkcast     #11   // class java/lang/Integer //(1)
        5: invokevirtual #12   // Method setData:(Ljava/lang/Integer;)V
        8: return
```

从上面的结果可以看到有两个 setData() 方法，其中第二个 setData() 方法的访问标识是 ACC_PUBLIC、ACC_BRIDGE、ACC_SYNTHETIC，表示由编译器生成的桥接方法。通过其中的字节码可以看出，MyNode 的实现中会将 Object 类型强制转换成 integer 类型 [如位置（1）所示]，这就是上面抛出 ClassCastException 的本质原因，知道这一点有助于我们更好地理解 Java 的泛型。

下面来看看 Signature 属性的结构。

```
Signature_attribute {
      u2 attribute_name_index;
      u4 attribute_length;
      u2 signature_index;
}
```

其中 attribute_name_index 为指向常量池 CONSTANT_Utf8_info 结构的有效索引，其值固定为“Signature”。attribute_length 为值为 2 的固定项。signature_index 也为指向常量池 CONSTANT_Utf8_info 结构的有效索引，用于表示类、方法、字段的泛型签名信息。Oracle 的 HotSpot 虚拟机在加载类或者链接类的时候并不检查 Signature 属性，Java SE 中提供的一些用于查询泛型信息的 API，其核心逻辑就是查询 Signature 属性中的数据，例如 Field.getGenericType()、Class.getGenericSuperclass() 等方法。

Signature 中 signature_index 存储的泛型信息是一个按照一定规则生成的字符串，这个规则相对来说有一些复杂。Java 语言的类型签名分为引用类型签名（reference type signature）和基本类型签名（primitive type signature），其结构如下。

```
JavaTypeSignature: ReferenceTypeSignature BaseType
```

其中 BaseType 与 3.3 节中的内容类似，必须是 B、C、D、F、I、J、S、Z 中的一个；ReferenceTypeSignature（引用类型签名）表示 Java 语言中的引用，例如类、接口、类型变量（type variable）以及数组类型。

引用类型又包括 ClassTypeSignature、TypeVariableSignature 和 ArrayTypeSignature。其中类的类型签名（class type signature）表示一个类（可能已经被参数化了，例如 List<String> 是 List<T>

的参数化类型）或接口类型；类型变量签名（type variable signature）可以理解为 List<T> 中 T 的签名信息；数组类型签名（array type signature）表示数组类型的维度。

详细的 ReferenceTypeSignature 的结构如下。

```
ReferenceTypeSignature: ClassTypeSignature TypeVariableSignature ArrayTypeSignature
ClassTypeSignature: L [PackageSpecifier] SimpleClassTypeSignature {ClassTypeSignatureSuffix}
PackageSpecifier: Identifier / {PackageSpecifier}
SimpleClassTypeSignature: Identifier [TypeArguments]
TypeArguments: < TypeArgument {TypeArgument} >
TypeArgument: [WildcardIndicator] ReferenceTypeSignature *
WildcardIndicator: + -
ClassTypeSignatureSuffix: . SimpleClassTypeSignature
TypeVariableSignature: T Identifier
ArrayTypeSignature: JavaTypeSignature
```

以上是引用类型签名的规则。现在知道 Java 类型签名包括 BaseType 和 ReferenceTypeSignature 这两种，那么 Java 语言中的类签名（class signature）、方法签名（method signature）和字段签名（field signature）的结构又是怎样的呢?

下面列出了这 3 种签名的结构。

（1）类签名规则。

```
ClassSignature: [TypeParameters] SuperclassSignature {SuperinterfaceSignature}
TypeParameters: <TypeParameter {TypeParameter} >
TypeParameter: Identifier ClassBound {InterfaceBound}
ClassBound: [ReferenceTypeSignature]
InterfaceBound: ReferenceTypeSignature
SuperclassSignature: ClassTypeSignature
SuperinterfaceSignature: ClassTypeSignature
```

注意上述结构要结合 JavaTypeSignature 对比着看。

接下来通过一个例子来分析类的签名信息，示例如下。

```
public class SignatureMain<E, V extends Number>
        extends ArrayList<E> implements Comparable<V>
{
    @Override
    public boolean add(E e) {
        return super.add(e);
    }

    @Override
    public int compareTo(V o) {

        List<String> list = new ArrayList<>();

        return 0;
    }
}
```

类 SignatureMain 对应的签名如下。

```
<E:Ljava/lang/Object;V:Ljava/lang/Number;>Ljava/util/ArrayList<TE;>;Ljava/lang/
Comparable<TV;>;
```

对照类签名中的结构:

- <E:Ljava/lang/Object;V:Ljava/lang/Number;> 是对 SignatureMain<E, V extends Number> 的描述，其规则是 TypeParameters；
- Ljava/util/ArrayList<TE;>; 是对 SuperclassSignature 规则的描述；
- Ljava/lang/Comparable<TV;>; 是对 SuperinterfaceSignature 规则的描述。

可能大家还是对此有点迷糊，图 3-10 所示为对 <E:Ljava/lang/Object;V:Ljava/lang/Number;> 部分按照类签名的规则进行拆解展示，其中标明了每一部分对应的类型。

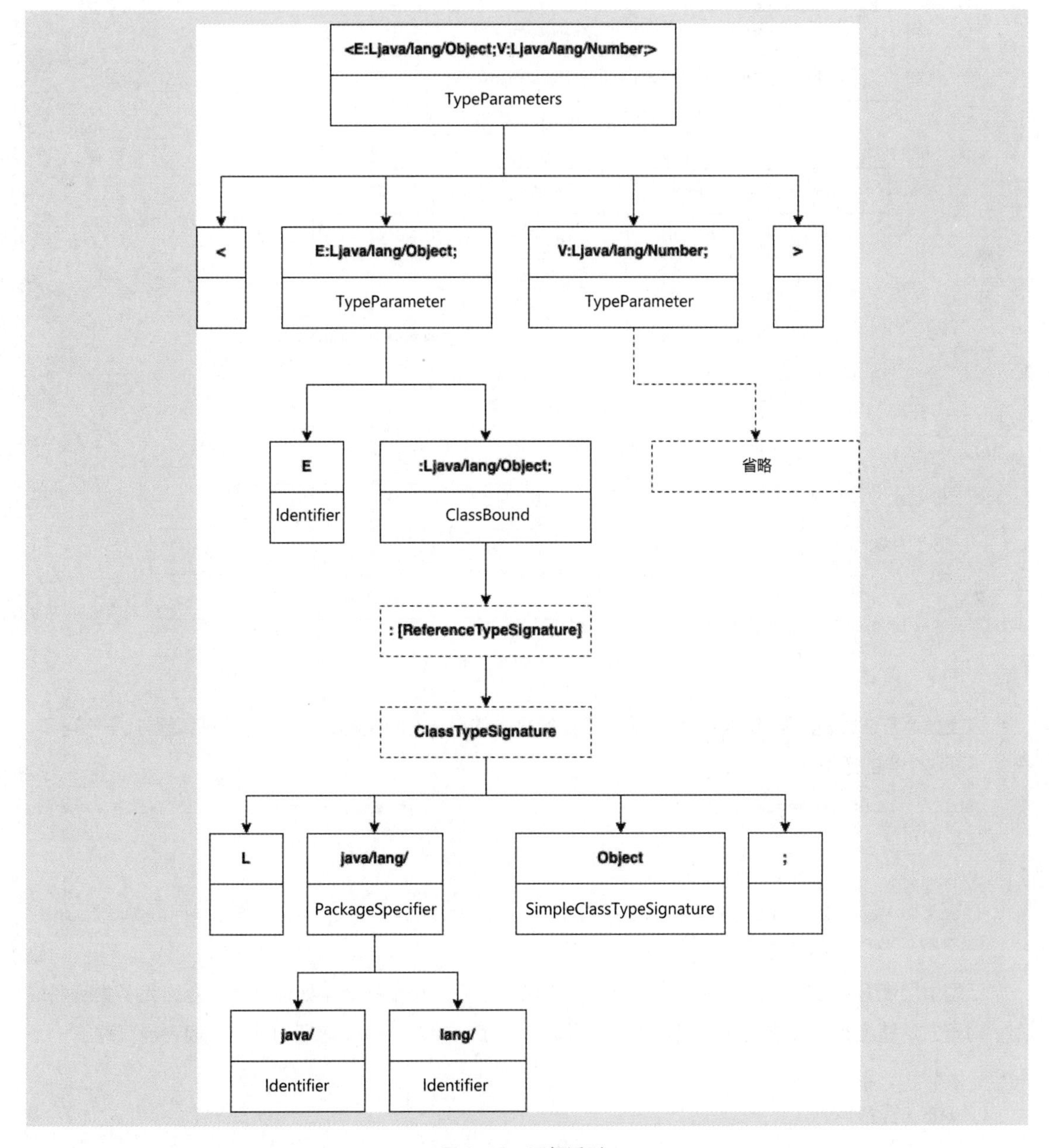

图 3-10　图解泛型 1

按照上面的拆解方式，再来分析 Ljava/util/ArrayList<TE;>;，对应的结果如图 3-11 所示。

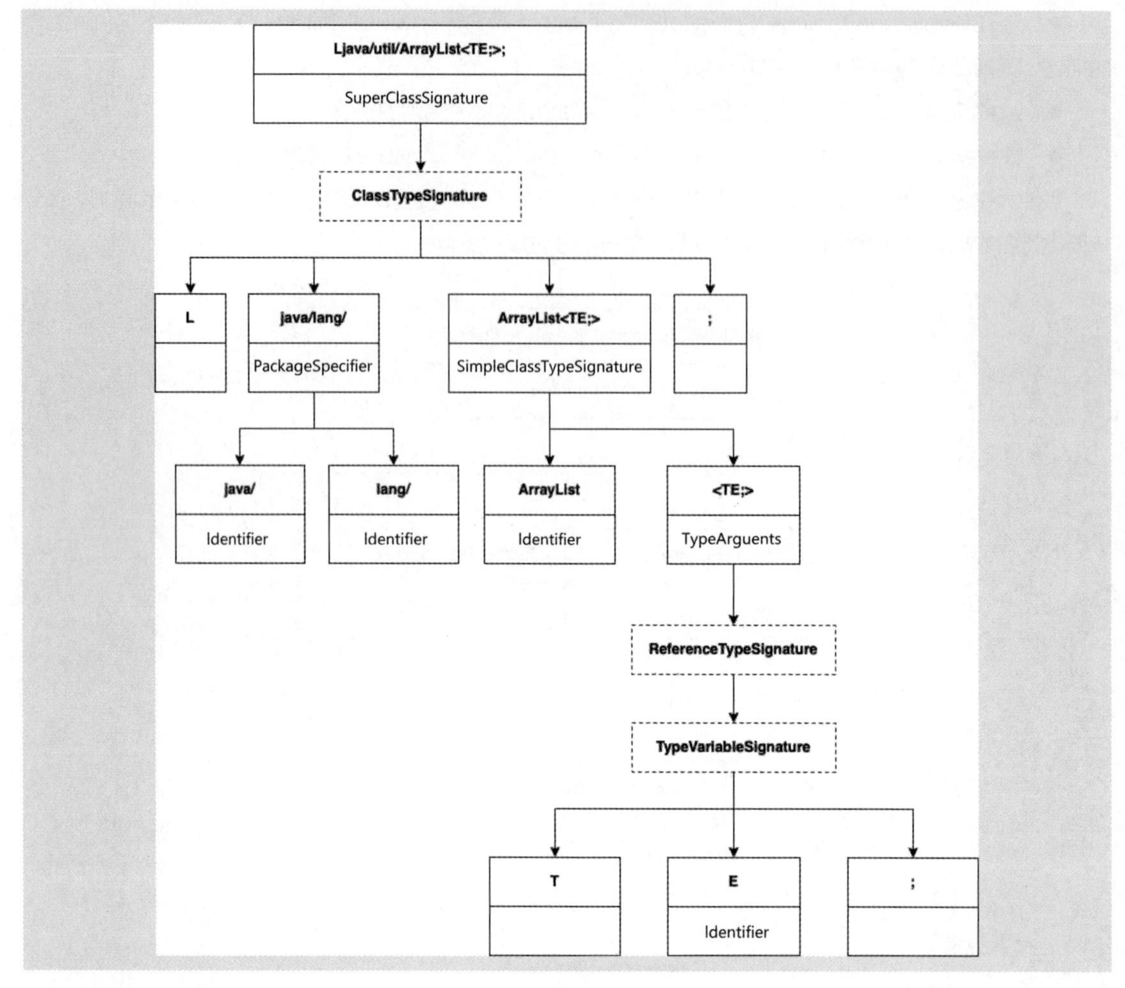

图 3-11　图解泛型 2

以上就是方法签名涉及的规则介绍，规则还是挺复杂的，一些特定的组合会使得签名变得几乎不可读，例如下面这个类的泛型签名。

```
public class SignatureDemo<U extends List<? super Number>, V extends HashMap<? extends String,?>>

//泛型签名:
<U::Ljava/util/List<-Ljava/lang/Number;>;V:Ljava/util/HashMap<+Ljava/lang/String;*>;>Ljava/lang/Object;
```

泛型层层嵌套会让具体的值变得异常复杂，不过这个还是交给程序来处理好了。介绍完了类签名规则，对于方法签名规则和字段签名规则就不再详细介绍了，接下来就是它们的规则，读者自行理解一下即可。

（2）方法签名规则。

```
MethodSignature: [TypeParameters] ( {JavaTypeSignature} ) Result {ThrowsSignature}
Result: JavaTypeSignature VoidDescriptor
ThrowsSignature: ^ ClassTypeSignature ^ TypeVariableSignature
VoidDescriptor: V
```

（3）字段签名规则。

```
FieldSignature: ReferenceTypeSignature
```

3.5 小结

本章介绍了 ClassFile 的结构并对其中的一些关键属性做了比较详细的介绍，主要目的是便于读者理解 ASM 中相关 API。ASM 中很多的 API 知识都需要读者有一定的了解，才能知道如何去使用。本章也涉及了一些字节码相关的知识，第 4 章将详细地介绍这部分内容。

4. 字节码指令

字节码指令是相对复杂但并非很难理解的内容，本章将通过图文并茂的方式介绍这部分知识，使读者能够相对轻松地入门。

Java 虚拟机指令由一个字节长度的操作码（opcode）和紧跟着的 0 至多个操作数（operand）组成，其形式如下。

```
<opcode> [operand1 operand2...]
```

对于操作码，直接去记它的值是比较难的，所以 Java 虚拟机的设计者提供了助记符的方式来代表操作码。例如操作码 0x1A 表示将整数 0 加载到操作数栈中，其对应的助记符是 iload_0。我们通常说的操作码，其实指的就是其对应的助记符。由于限制了操作码的长度只有一个字节，所以又称 Java 虚拟机指令为字节码指令。因为字节码指令的长度为一个字节，所以字节码指令的数量不会超过 256 个。截至目前，我们使用的字节码指令已经有 200 多个，大概占总数的 80%，另外约 20% 未使用，留作后续扩展。

对于操作数，因为 ClassFile 的结构放弃了编译后代码的操作数对齐，这就意味着当操作数的长度超过一个字节的时候，它将会以 big-endian 顺序存储，即高位在前，低位在后。例如将一个 16 位长度的无符号整数使用两个无符号整数存储起来（假设这两个整数分别为 byte1 和 byte2），那这个 16 位无符号整数的值就是 byte1 << 8 | byte2。

字节码指令不同于机器码指令，字节码指令是运行在 Java 虚拟机上的。为了能够理解字节码指令，先来了解一些 Java 虚拟机方面的知识。

4.1 Java 虚拟机栈

虚拟机的实现方法有基于栈和基于寄存器的两种，市面上较流行的是 Oracle 的 HotSpot 虚拟机，它是基于栈的，而 Android 的 Dalvik 虚拟机则是基于寄存器的。关于两者的区别，读者可以上网查找对应的资料了解，本书讨论的是基于栈的虚拟机。《Java 虚拟机规范》是 Oracle 为了规范虚拟机的设计推出的，描述了一种抽象化的虚拟机的行为，说白了就是实现 Java 虚拟机的一个指导方针。该规范中规定，每一个 Java 虚拟机线程都有自己的 Java 虚拟机栈（Java virtual machine stack），虚拟机栈会和线程一同被创建和销毁，此栈用于存储方法调用中的局部变量和一些尚未计算好的结果。在 Java 虚拟机栈中会有很多方法的调用，方法的数据和部分结果则使用栈帧这个数据结构来存储。

图 4-1 展示了 Java 虚拟机栈和栈帧的结构，表示每个线程都对应一个 Java 虚拟机栈，每个 Java 虚拟机栈可能存在多个栈帧，每进入一个方法就会创建一个栈帧，退出方法后销毁栈帧。

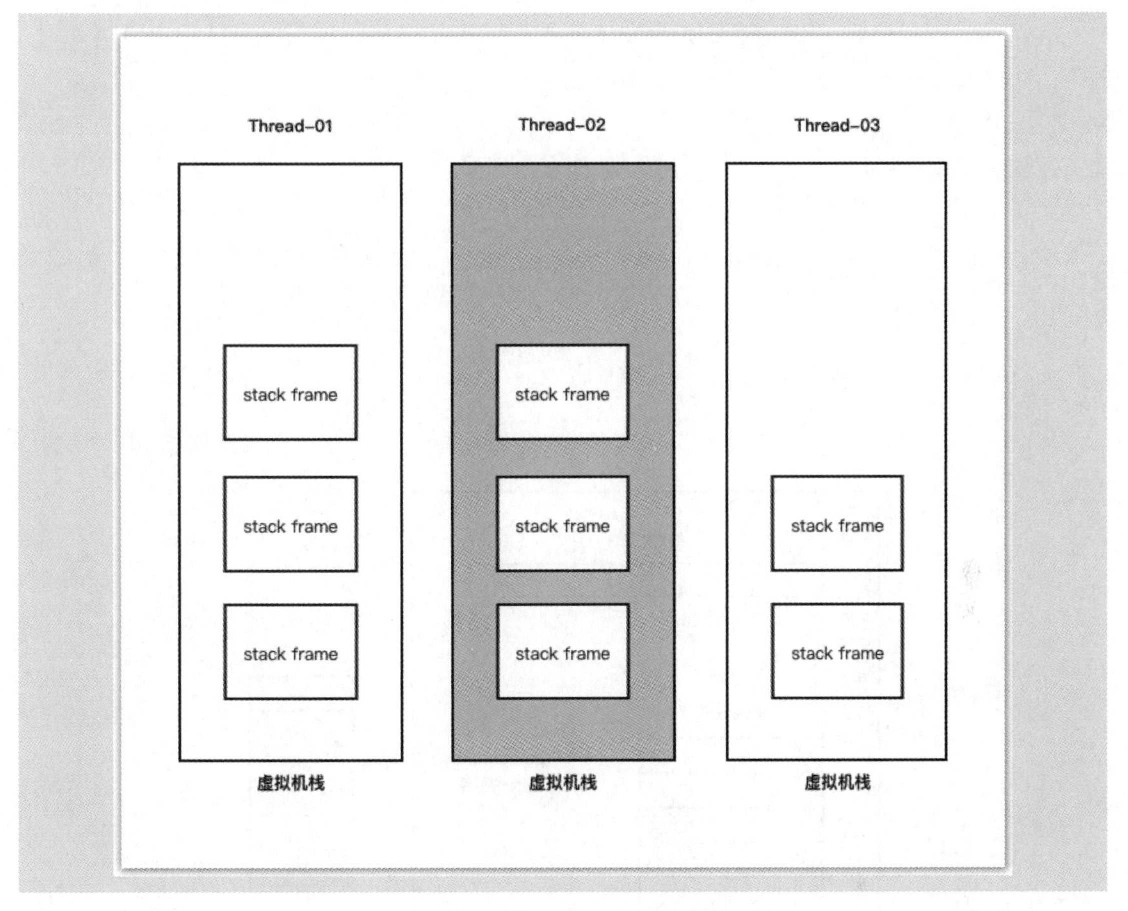

图 4-1　Java 虚拟机栈和栈帧的结构

4.2　栈帧

栈帧（stack frame）是用来支持虚拟机进行方法调用和方法执行的数据结构。栈帧随着方法调用而创建，随着方法结束（无论方法是正常完成还是异常完成都算作方法结束）而销毁。栈帧的存储空间由创建它的线程分配在 Java 虚拟机栈中，栈帧由局部变量（local variable）表、操作数栈（operand stack）以及指向当前方法所在类的运行时常量池的引用组成，如图 4-2 所示。其中局部变量表和操作数栈在编译期间确定，保存在 Code 属性中并提供给栈帧使用。

栈帧的大小取决于虚拟机的实现，虚拟机实现者可以提供对应的方式，用于配置栈帧的大小。例如递归调用一个方法的时候，如果没有设置退出条件，无限递归下去就会抛出如下异常。

```
public class RecursiveTest {
    // 模拟递归调用
    void foo(){
        foo();
    }
```

```
    public static void main(String[] args) {
        new RecursiveTest().foo();
    }
}

//将会抛出如下异常
Exception in thread "main" java.lang.StackOverflowError
    at cn.sensorsdata.RecursiveTest.foo(RecursiveTest.java:5)
    at cn.sensorsdata.RecursiveTest.foo(RecursiveTest.java:5)
    at cn.sensorsdata.RecursiveTest.foo(RecursiveTest.java:5)
    ...
```

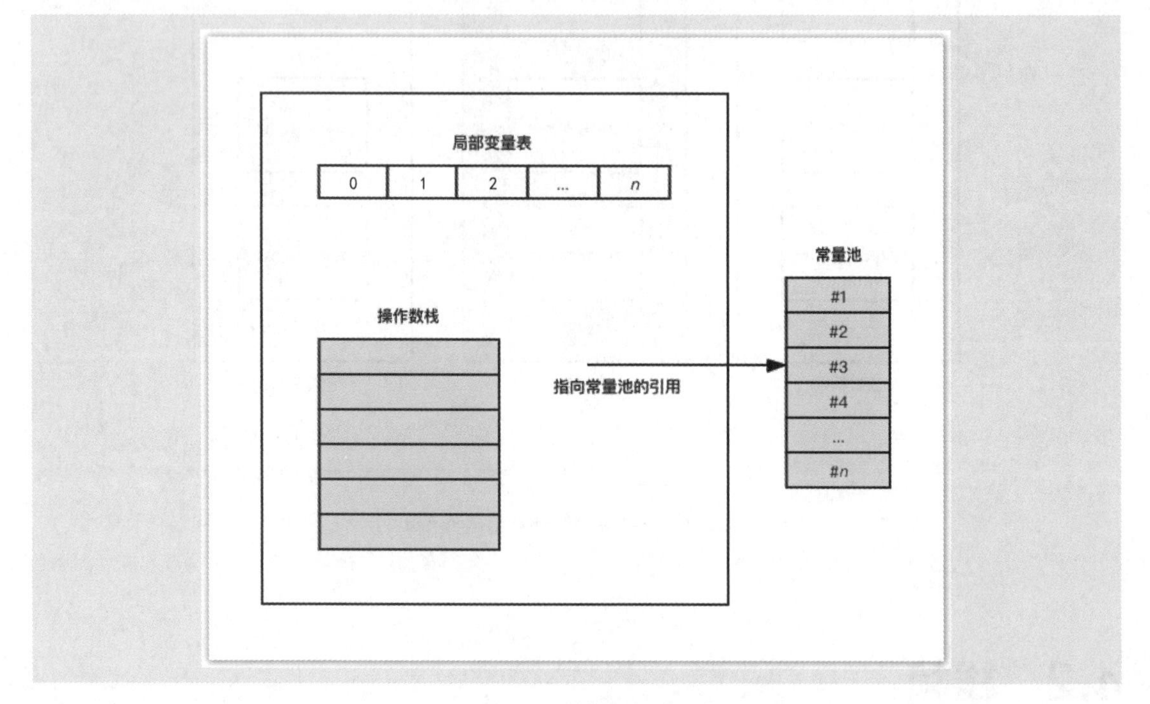

图 4-2　栈帧的组成

出现 StackOverflowError 异常的原因是线程在某一时刻只能执行一个方法，只有这个方法对应的栈帧是活动的，这个方法我们称为当前方法（current method），对应的栈帧称为当前栈帧（current stack frame）。如果当前方法未运行结束时（此时栈帧未销毁）又调用了其他方法，那么新的栈帧就会被创建，如此不断调用新方法就会不断创建新的栈帧，当累积的栈帧大小超过了设置的阈值就会抛出 StackOverflowError 异常。对于 HotSpot 虚拟机，可以通过 -Xss 来设定虚拟机的栈大小，例如 -Xss2m 表示将栈的大小设置为 2 兆字节（MB）。

4.3　局部变量表

每个栈帧的内部都包含一个局部变量表，局部变量表的长度在编译期确定，并存储在 Code 属性的

max_locals 中，虚拟机会根据 max_locals 的值分配方法运行时的变量列表的大小。可以将局部变量表理解为一个一个的卡槽，只不过有的类型的值使用一个槽就可以保存，有的则需要两个。在字节码中，虚拟机的数据类型（详细内容请参考 4.10 节）如 boolean、byte、char、short、int、float、reference 或 returnAddress 类型的值占用一个槽，而 long 和 double 类型存储的值比较大，要用两个槽。局部变量表中的索引是从 0 开始的，假如一个 long 类型的值存储在索引值为 *n* 的位置，那么实际上 *n* 和 *n*+1 两个槽都用来存储这个 long 类型的值。

Java 虚拟机使用局部变量表来完成方法调用时的参数传递。当调用方法时，方法中的参数将会依次存储在局部变量表中。也就是说进入方法的时候局部变量表中的值对应着方法中的参数，不过局部变量表中第 0 个位置的参数由方法修饰符决定。具体规则如下。

（1）如果是实例方法，那么第 0 个位置的局部变量是存储该实例方法所在的对象的引用，即 this。

（2）如果是类方法（static 修饰的方法），那么第 0 个位置的局部变量就是方法中的第一个参数。

我们来看看下面这个例子中的两个方法对应的局部变量表。

```
public class LocalsTest {

    public void foo(int num1, long num2){

    }

    public static void bar(int num1, long num2){

    }
}
```

使用如下命令编译 LocalsTest.java，获得 LocalsTest.class 文件，其中 -g 选项表示生成所有的调试信息，包括行号、局部变量信息等。

```
javac -g LocalsTest.java
```

再使用如下命令查看 foo() 和 bar() 方法的字节码。注意，javap 的 -l（L 的小写，用于输出行号和局部变量表）选项只有和 -g 选项一起使用才会起作用。

```
    javap -v -p -l LocalsTest
```

字节码信息如下。

```
  public void foo(int, long);
    descriptor: (IJ)V
    flags: ACC_PUBLIC
    Code:
      stack=0, locals=4, args_size=3
         0: return
      LineNumberTable:
        line 7: 0
      LocalVariableTable:
        Start  Length  Slot  Name   Signature
            0       1     0  this   Lcn/sensorsdata/LocalsTest; //(1)
            0       1     1  num1   I
            0       1     2  num2   J

  public static void bar(int, long);
    descriptor: (IJ)V
    flags: ACC_PUBLIC, ACC_STATIC
```

```
Code:
  stack=0, locals=3, args_size=2
     0: return
  LineNumberTable:
    line 11: 0
  LocalVariableTable:
    Start  Length  Slot  Name   Signature
        0       1     0  num1   I //(2)
        0       1     1  num2   J
```

观察上面 foo() 和 bar() 方法的 LocaLVariableTable 局部变量表属性信息，对比 Slot 这一列可以看到，虽然 foo() 和 bar() 在 Java 源码中方法的参数是一致的，但是局部变量表不一样。foo() 方法的 Slot 0 是 this，bar() 方法的 Slot 0 是方法的第一个参数。

表 4-1 所示为 foo() 方法的局部变量表的情况，其 locals（max_locals）的值为 4，表示局部变量表的最大长度是 4。

表4-1　foo()方法的局部变量表展示

Slot	0	1	2	3
局部变量	this	num1(int)	num2(long)	num2(long)

针对方法调用初始时的局部变量表情况，这里要注意两点：

（1）foo() 方法的参数 num2 是 long 类型，它占用了两个变量槽，即表 4-1 中的索引 2 和 3；

（2）对应槽中的数据并不是不变的，实际上，为了尽可能节省栈帧的内存，局部变量表中的变量槽是可以重用的。

4.4　操作数栈

每个栈帧的内部都包含一个称为操作数栈的后进先出（Last In First Out，LIFO）栈。同局部变量一样，操作数栈的最大深度也是在编译期确定的，并存储在 Code 属性的 max_stack 中。操作数栈的作用是保存计算过程的中间结果，同时作为计算过程中变量临时的存储空间。当一个方法刚执行时，操作数栈是空的，随着方法的执行和字节码指令的执行，对象实例的字段会从局部变量表中复制常量、变量压入操作数栈，再随着计算的进行将栈中元素出栈到局部变量表，或者消费栈中的元素返回给方法调用者，也就是出栈 / 入栈操作，一个完整的方法执行过程往往包含多个这样的出栈 / 入栈操作。

出栈和入栈操作是通过字节码指令来进行的，例如将局部变量表中的数据加载到操作数栈中、从操作数栈中取走数据等操作。接下来以下面这段代码对应的字节码为例，分析其中的 add() 方法。

```
// 方法
public int add(int a, int b){
      int sum = a + b;
      return sum;
}

// 上述方法对应的字节码
```

```
public int add(int, int);
    descriptor: (II)I
    flags: ACC_PUBLIC
    Code:
      stack=2, locals=4, args_size=3
         0: iload_1
         1: iload_2
         2: iadd
         3: istore_3
         4: iload_3
         5: ireturn
```

通过上述结果，可以看到操作数栈的最大深度是 2（stack=2），局部变量表的最大长度是 4（locals=4）。假设 add() 方法的参数 a =3、b=4，图 4-3 显示了 add() 方法的字节码执行的过程中局部变量表和操作数栈的变化情况。

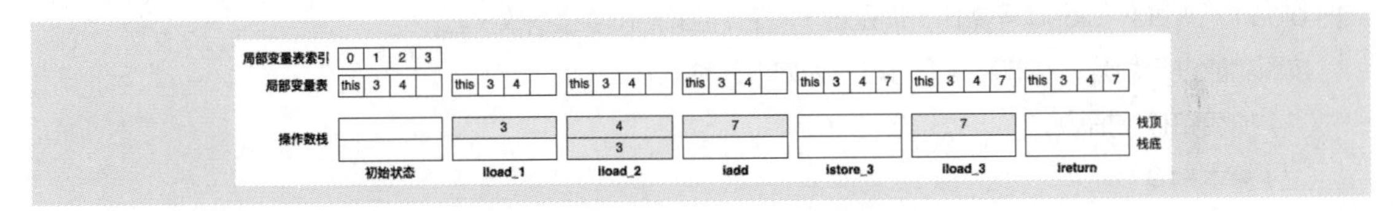

图 4-3　局部变量表和操作数栈的变化

图 4-3 展示的过程说明如下。

（1）初始状态：表示刚进入方法时，局部变量表中的内容就是方法的参数信息，因为 add() 方法是实例方法，所以局部变量表中的第一个内容是 this。

（2）iload_1 指令：表示将局部变量表索引 1 处的值 3 加载到操作数栈中，此时操作数栈顶的元素是 3。

（3）iload_2 指令：表示将局部变量表索引 2 处的值 4 加载到操作数栈中，此时栈顶的元素是 4。

（4）iadd 指令：表示去除栈顶位置的两个整型数据（此时操作数栈中的内容为空），执行加法计算，并将计算结果 7 压入栈顶中，所以指令执行后栈顶的值为 7。

（5）istore_3 指令：表示将栈顶的值存放在局部变量表索引为 3 的位置，此时操作数栈是空的，局部变量表索引 3 处的值是 7。

（6）iload_3 指令：表示将局部变量表索引 3 中的内容加载到操作数栈中，此时栈顶的值为 7。

（7）ireturn 指令：表示将栈顶的数据作为方法的返回值返回给方法调用者。

其实整个字节码指令的执行过程就是操作数栈和局部变量表之间不断加载与存储的过程。操作数栈支持的数据类型与局部变量表支持的数据类型一致，需要注意的是 32 位的数据类型所占的栈容量是 1，64 位的数据类型所占的栈容量是 2；当使用字节码指令对操作数栈进行操作的时候，要注意 long 和 double 类型的值所占的栈容量为 2，与局部变量表的情况类似。

至此介绍了栈帧、局部变量表以及操作数栈的知识，理解了这些知识以后再来学习字节码指令就容易一些。

4.5　字节码指令介绍

在 Java 虚拟机指令集中，大部分指令的助记符中包含特殊的字符来表明该指令可以操作哪种数据类型，如 i 开头（例如 iload）表示操作 int 数据类型（整型），l 开头（例如 lload）表示操作 long 数据类型（长整型），f 开头（例如 fload）表示操作 float（浮点型）数据类型，b 开头（例如 bipush）表示操作 byte 数据类型（字节型），c 开头（例如 caload）表示操作 char（字符型）数据类型，s 开头（例如 sipush）表示操作 short 数据类型（短整型），d 开头（例如 dload）表示操作 double 数据类型（双精度浮点型），a 开头（例如 aload）表示操作 reference 数据类型（引用型）。还有一些指令的助记符中没有明确的字符来表明具体操作的数据类型，例如 arraylength 指令。另外，有的指令与数据类型无关，例如 goto。

因为操作码的长度只有一个字节，假如每一种与类型相关的指令都支持 Java 虚拟机所有运行时的数据类型，那么肯定会超出一个字节所能表示的范围。因此，Java 虚拟机的指令集对于特定的操作只提供了有限的类型相关指令，也就是并非每种数据类型和每一种操作都有对应的指令。有一些指令可以在必要的时候用来将一些不支持的数据转换为可支持的类型，例如对于 byte、char、short 和 boolean 类型，没有对应的 bload、cload、sload 等操作，而是统一将它们当作 int 类型来处理。

按照功能进行划分，字节码指令可以分为如下几种。

（1）加载和存储指令。

（2）算术指令。

（3）类型转换指令。

（4）对象的创建和操作指令。

（5）操作数栈管理指令。

（6）控制转移指令。

（7）方法调用和返回指令。

（8）异常抛出指令。

（9）同步指令。

接下来详细介绍这些指令。

4.5.1 加载和存储指令

加载指令用于将数据从局部变量表中加载到操作数栈中，存储指令用于将操作数栈中的数据存储在局部变量表中。

（1）将一个局部变量加载到操作数栈中的指令（即加载指令）如表 4-2 所示。

表4-2 加载指令

操作码	操作数	描述
iload	index	用于将局部变量表索引为 index 的 int 值压入操作数栈中，例如 iload 1，是将局部变量表索引为 1 的值压入操作数栈的栈顶
lload	index	用于将局部变量表索引为 index 的 long 值压入操作数栈中
fload	index	用于将局部变量表索引为 index 的 float 值压入操作数栈中
dload	index	用于将局部变量表索引为 index 的 double 值压入操作数栈中
aload	index	用于将局部变量表索引为 index 的 reference 值压入操作数栈中

续表

<table>
<tr><th>操作码</th><th>操作数</th><th>描述</th></tr>
<tr><td>aaload</td><td>arrayref, index</td><td>用于从数组中加载一个reference类型的数据到操作数栈，这个指令有两个操作数,arrayref表示对数组的引用,index表示数组的索引。<pre>
public void foo(java.lang.String[]);
      descriptor: ([Ljava/lang/String;)V
      flags: ACC_PUBLIC
      Code:
        stack=2, locals=3, args_size=2
           0: aload_1
           1: iconst_0
           2: aaload
           3: astore_2
           4: return
</pre>上述字节码表示从foo()方法的String数组参数中，取数组索引为0的值。其中arrayref的值由aload_1引用，索引0由iconst_0产生。严格说此指令只是操作数组的内容，并非从局部变量表中加载数据，注意与其他加载指令的区别</td></tr>
<tr><td>lload_<n></td><td>无</td><td>用于将局部变量表索引为n的int值压入操作数栈中,n的值为 [0,3]。例如iload_1是将局部变量表索引为1的值压入操作栈的栈顶，效果与iload 1一致，不过iload_1没有操作数，效率更高</td></tr>
<tr><td>fload_<n></td><td>无</td><td>用于将局部变量表索引为n的float值压入操作数栈中,n的值为 [0,3]</td></tr>
<tr><td>dload_<n></td><td>无</td><td>用于将局部变量表索引为n的double值压入操作数栈中,n的值为 [0,3]</td></tr>
<tr><td>aload_<n></td><td>无</td><td>用于将局部变量表索引为n的reference值压入操作数栈中,n的值为 [0,3]</td></tr>
</table>

可以看到加载指令助记符中都带有 load 关键字，所以此类指令称为加载指令。另外，iload 相关的指令不仅能加载 int 类型，还可以加载 char、byte、boolean、short 类型。

（2）将值从操作数栈存储到局部变量表的指令（即存储指令）如表 4-3 所示。

表4-3　存储指令

<table>
<tr><th>操作码</th><th>操作数</th><th>描述</th></tr>
<tr><td>istore</td><td>index</td><td>用于将操作数栈栈顶的int值存放在局部变量表索引为index的位置，例如istore 1是将操作数栈栈顶的值存放在局部变量表的索引为1的位置</td></tr>
<tr><td>lstore</td><td>index</td><td>用于将操作数栈栈顶的long值存放在局部变量表索引为index的位置</td></tr>
<tr><td>fstore</td><td>index</td><td>用于将操作数栈栈顶的float值存放在局部变量表索引为index的位置</td></tr>
<tr><td>dstore</td><td>index</td><td>用于将操作数栈栈顶的double值存放在局部变量表索引为index的位置</td></tr>
<tr><td>astore</td><td>index</td><td>用于将操作数栈栈顶的reference值存放在局部变量表索引为index的位置</td></tr>
<tr><td>aastore</td><td>arrayref, index, value</td><td>用于向一个对象数组中添加一个值，这个指令有3个操作数,arrayref表示对数组的引用,index表示数组的索引,value表示具体的值：<pre>
public void foo(java.lang.String[]);
      descriptor: ([Ljava/lang/String;)V
      flags: ACC_PUBLIC
      Code:
        stack=3, locals=2, args_size=2
           0: aload_1
           1: iconst_0
           2: ldc   #2   // 字符串“hello”
           4: aastore
           5: return
</pre>上面的字节码表示将字符串“hello”存放在数组索引为0的位置。其中arrayref的值通过aload_1加载，索引是通过iconst_0表示，“hello”通过ldc指令加载，最后使用aastore指令消耗前面3个指令加载到操作数栈中的值</td></tr>
</table>

续表

操作码	操作数	描述
istore_<n>	无	用于将操作数栈栈顶的int值存放在局部变量表索引为n的位置，其中n的值为[0, 3]，例如istore_1是将操作数栈栈顶的值存放在局部变量表的索引为1的位置
lstore_<n>	无	用于将操作数栈栈顶的long值存放在局部变量表索引为n的位置，其中n的值为[0, 3]
fstore_<n>	无	用于将操作数栈栈顶的float值存放在局部变量表索引为n的位置，其中n的值为[0, 3]
dstore_<n>	无	用于将操作数栈栈顶的double值存放在局部变量表索引为n的位置，其中n的值为[0, 3]
astore_<n>	无	用于将操作数栈栈顶的reference值存放在局部变量表索引为n的位置，其中n的值为[0, 3]

可以看到存储指令助记符中都带有store关键字，所以此类指令称为存储指令。另外，istore相关的指令不仅能存储int类型，还可以存储char、byte、boolean、short类型。

（3）将常量加载到操作数栈中的指令（即加载常量指令）如表4-4所示，常见的加载常量相关的指令有const类、push类、ldc类。

表4-4　加载常量指令

操作码	操作数	描述
aconst_null	无	用于将null值压入操作数栈中
iconst_m1	无	将整型值 -1压入操作数栈中
iconst_<n>	无	用于将int类型的值n压入操作数栈中，n的值为[0, 5]，例如iconst_0表示将整数0压入栈顶，iconst_5表示将整数5压入栈顶
lconst_<n>	无	用于将long类型的值n压入操作数栈中，n的值为[0, 1]，例如lconst_1表示将1L压入栈顶
fconst_<n>	无	用于将float类型的值n压入操作数栈中，n的值为[0, 2]，例如fconst_1表示将1.0f压入栈顶
dconst_<n>	无	用于将double类型的值n压入操作数栈中，n的值为[0, 1]，例如dconst_1表示将1.0压入栈顶
bipush	value	用于将一个byte类型的有符号值，即−128～127的整数值压入栈顶中
sipush	value	用于将范围为−32768～32767的整型值压入栈顶
ldc	index	用于将int、float、string类型的常量压入栈顶，注意操作数index对应的是常量池中的有效索引
ldc_w	indexbyte1,indexbyte2	作用与ldc一致，不同点是ldc的操作数是一个字节，而ldc_w的操作数是两个字节。ldc指令指向的常量池索引的最大值为255，但是常量池的大小很轻松就能超过255，此时将无法使用ldc。ldc_w操作的是两个字节，可以覆盖到常量池中的所有值，例如ldc_w #300
ldc2_w	indexbyte1,indexbyte2	ldc2_w操作的也是两个字节，不过是将long和double类型的常量值压入栈顶

从表4-4可以看到，操作整型常量的指令有多种，不同范围的常量用不同的指令来操作，这么做的好处是对于一些字节码中常用的数字使用一个字节的操作码来表示可以使字节码更加紧凑，例如iconst_0指令比iconst 0好，因为iconst_0没有操作数。当需要将整型常量压入栈的时候可以参考表4-4中的数值范围选择对应的指令。如果在Java源码中取值不合理，编译器也会进行优化，选择合适的指令，例如下面这段代码。

```
public long add(int a, long b, int c) {
      short s = 4;
      return a + b;
}
```

上述代码 add() 方法中定义了一个 short 类型的变量 s，其值为 4。虽然 s 的类型是 short，但显然使用 iconst_4 指令更合理，而不是 sipush 指令。add() 方法对应的字节码如下。

```
public long add(int, long, int);
   descriptor: (IJI)J
   flags: ACC_PUBLIC
   Code:
     stack=4, locals=6, args_size=4
        0: iconst_4 //观察此处的字节码
        1: istore        5
        3: iload_1
        4: i2l
        5: lload_2
        6: ladd
        7: lreturn
```

4.5.2 算术指令

算术指令是对操作数栈上的值进行算术运算的指令，通常是取出（栈中的值出栈）操作数栈的值，运算后会将结果放入栈顶中。在算术运算中，一般需要关注不同数据类型之间的计算、计算结果的溢出、除数是否为 0 等情况。例如，当一个 int 型数据与一个 float 型数据进行相加时就会涉及精度问题。

与算术相关的指令如表 4-5 所示。

表4-5 算术指令

指令类别	具体指令				助记单词
	int类型	float类型	long类型	double类型	
加法指令	iadd	fadd	ladd	dadd	Add：加
减法指令	isub	fsub	lsub	dsub	Subtract：减
乘法指令	imul	fmul	lmul	dmul	Multiply：乘
除法指令	idiv	fdiv	ldiv	ddiv	Divide：除
求余指令	irem	frem	lrem	drev	Remainder：余数
取反指令	ineg	lneg	fneg	dneg	Negate：反面
位移指令	ishl、ishr、iushr	—	lshl、lshr、lushr	—	Shift Left、Shift Right
按位或指令	ior	—	lor	—	Boolean OR
按位与指令	iand	—	land	—	Boolean AND
异或指令	ixor	—	lxor	—	Boolean XOR
局部变量自增指令	iinc	—	—	—	Increment：增加
比较指令	—	fcmpg、fcmpl	lcmp	dcompg、dcmpl	Compare Float、Compare Double

接下来介绍部分有代表性的算术指令。

（1）iadd、fadd 指令。算术相关的指令都要求被操作的数据类型一致，例如 fadd，要求对应的两个操作数必须是 float 类型。如果不是 float 类型，例如 2 + 1.0f，就必须先将整数 2 转换成 float 类型。例如下面这个例子。

```
//Java 代码
public void bar() {
      int a = 2 + 3;
      float b = 1.0f + 2;
```

```
}

// 对应的字节码
public void bar();
      descriptor: ()V
      flags: ACC_PUBLIC
      Code:
        stack=1, locals=3, args_size=1
           0: iconst_5
           1: istore_1
           2: ldc          #3                  // 浮点数 3.0f
           4: fstore_2
           5: return
```

上面的 Java 代码中做了简单的加法运算，但是看字节码并没有相关的加法运算指令，这是因为编译器在编译的时候进行了优化。下面对这个例子进行修改，再看看结果。

```
//Java 代码
public void bar2() {
      int a = 2;
      int b = a + 2;
      float c = 1.0f ;
      c = a + c;
}

// 对应的字节码
public void bar2();
      descriptor: ()V
      flags: ACC_PUBLIC
      Code:
        stack=2, locals=4, args_size=1
           0: iconst_2
           1: istore_1
           2: iload_1
           3: iconst_2
           4: iadd
           5: istore_2
           6: fconst_1
           7: fstore_3
           8: iload_1
           9: i2f
          10: fload_3
          11: fadd
          12: fstore_3
          13: return
```

观察字节码指令偏移（关于什么是字节码指令偏移，单独在 4.9 节中讨论，读者务必先去了解这部分内容）索引位置为 8 ~ 12，即 c = a + c 对应的字节码，其中 iload_1 将局部变量中的 int 类型的值加载到局部变量后，使用 i2f 指令将 int 类型的值转换成 float 类型，即系统将 int 类型转换成了更大的数据类型，也就是 Java 中的隐式自动类型转换。与数据类型转换相关的字节码指令可以参考 4.6.3 小节。

（2）iinci 指令。这个指令比较特殊，它操作的对象不是操作数栈，而是局部变量表。其格式如下。

格式	iinc index value
操作数栈前后变化	无变化

该指令表示对局部变量表索引为 index 的局部变量的值加上 value。例如下面这段代码对应的字节码：

```
//Java 代码
public void bar() {
     int a = 100;
     a++;
     a--;
     float b =1.0f;
     b++;
}

// 对应字节码
public void bar();
      descriptor: ()V
      flags: ACC_PUBLIC
      Code:
        stack=2, locals=3, args_size=1
           0: bipush     100     // 将整数 100 压入栈中，此时栈中元素是 [100]
           2: istore_1           // 将栈顶的 100 存到局部变量表下标为 1 的位置，此时栈为 []
           3: iinc       1, 1    // 将局部变量表下标 1 位置的内容 +1
           6: iinc       1, -1   // 将局部变量表下标 1 位置的内容 +(-1)
           9: fconst_1
          10: fstore_2
          11: fload_2
          12: fconst_1
          13: fadd
          14: fstore_2
          15: return
```

上面的字节码，将整数值 100 存储在局部变量索引为 1 的位置，然后使用 iinc 分别对其进行加 1 和减 1 操作，而此时操作数栈没有任何变化。同时也可以看到对 float 类型的自增操作实际上对应的是使用 fadd 指令来对两个 float 类型操作。这里主要记住 iinc 操作的是局部变量表中的值，这一点与其他指令有很大的不同，也是字节码指令中唯一能对局部变量表中的值进行运算的指令。

（3）fcmpl、fcmpg 指令。比较两个 float 类型的值，并将比较的结果压入操作数栈栈顶中。其格式如下。

格式	fcmpl 或 fcmpg
操作数栈前后变化	..., value1,value2→..., result

假如有 float 类型的值 value1 和 value2，其比较规则如下。

- 如果value1大于value2，就将1压入栈顶中；
- 如果value1等于value2，就将0压入栈顶中；
- 如果value1小于value2，就将 -1压入栈顶中；
- 如果value1和value2中至少有一个是NaN（Not a Number），fcmpg返回1，fcmpl返回-1。

以下面这段代码对应的字节码为例。

```
//Java 代码
public void bar() {
     float f1 = 2.0f;
     float f2 =1.0f;
     boolean result = f1 > f2;
}

// 对应的字节码
```

```
public void bar();
      descriptor: ()V
      flags: ACC_PUBLIC
      Code:
        stack=2, locals=4, args_size=1
           0: fconst_2
           1: fstore_1
           2: fconst_1
           3: fstore_2
           4: fload_1        // 加载 f1 的值
           5: fload_2        // 加载 f2 的值
           6: fcmpl          // 比较 f1 和 f2，因为 f1 > f2, 1 会被压入栈顶
           7: ifle       14 // 比较栈顶的元素是否不大于 0，是则跳转到 14，否就顺序执行
          10: iconst_1
          11: goto        15
          14: iconst_0
          15: istore_3
          16: return
```

上述 Java 代码的 bar() 方法中定义了两个 float 类型变量 f1 和 f2（f1 大于 f2），并将最终结果保存到一个 boolean 类型变量中。对应字节码指令偏移 6: fcmpl 对 f1 和 f2 进行比较，因为 f1 值大于 f2，所以 1 被压入栈顶。字节码指令偏移 7: ifle 14 的作用是判断栈顶的元素是否不大于 0，是则跳转到偏移 14 的位置继续执行；否就顺序执行，即执行 10: iconst_1。

4.5.3 类型转换指令

前面介绍算术指令的时候提到，如果进行算术操作的两个值的类型不一致，需要先将数据转换成适当的类型，这就用到了类型转换指令。类型转换指令可以将不同的数据类型相互转换，这些转换操作一般用于实现用户代码中的显式类型转换，例如 int a = (int)9L，强制将一个 long 类型变量转换成 int 类型变量。

Java 虚拟机直接支持以下数据类型的宽化类型转换（widening numeric conversion，即小范围类型向大范围类型的安全转换）。

- 从int类型到long、float、double类型；
- 从long类型到float、double类型；
- 从float类型到double类型。

与之相对，处理窄化类型转换（narrowing numeric conversion）时，就必须显式地使用转换指令来完成。窄化类型转换可能会导致转换结果产生不同的正负号、不同的数量级的情况，转换过程很可能导致数值的精度丢失，比如将 long 类型转换成 int 类型。具体的类型转换指令如表 4-6 所示。

表4-6 类型转换指令

数据类型	int	long	float	double	byte	char	short
int	—	i2l	i2f	i2d	i2b	i2c	i2s
long	l2i	—	l2f	l2d	—	—	—
float	f2i	f2l	—	f2d	—	—	—
double	d2i	d2l	d2f	—	—	—	—

与转换相关的还有一个 checkcast 指令，此指令用于将一个 reference 类型强制转换成另一个类型，其格式如下。

格式	checkcast indexbyte1 indexbyte2
操作数栈前后变化	..., objectref→..., objectref

其中 indexbyte1 和 indexbyte2 构建成两个字节（构建方式是 indexbyte1 << 8 \| indexbyte2）的值，用于表示常量池中的索引。

以下面这段代码对应的字节码为例。

```
//Java 代码
public void bar() {
     Object a = "ss";
     String b = (String) a;
}

// 对应的字节码
Constant pool:
     #1 = Methodref          #6.#16       // java/lang/Object."<init>":()V
     #2 = String             #17          // hello
     #3 = String             #18          // ss
     #4 = Class              #19          // java/lang/String
     ...
{
    public void bar();
      descriptor: ()V
      flags: ACC_PUBLIC
      Code:
        stack=1, locals=3, args_size=1
           0: ldc            #3             // String ss
           2: astore_1
           3: aload_1
           4: checkcast      #4             // class java/lang/String
           7: astore_2
           8: return
}
```

上例中 checkcast 是将栈顶的引用类型强制转换成常量池索引 #4 所代表的类型。对于 checkcast，如果栈顶的值为 null，那么操作数栈中的值将不会有变化，如果强制类型转换失败将抛出异常。

与 checkcast 指令相似的另一个指令是 **instanceof**。**instanceof** 指令用于判断栈顶的元素是否是某个类型，并将判断的结果压入栈顶中。其判断规则是，如果是某个类型，就将 1 压入栈顶；如果不是，就将 0 压入栈顶；如果栈顶的类型是 null，也会将 0 压入栈顶。

4.5.4 对象的创建和操作指令

对象创建以后，就可以通过对象访问指令来获取对象实例中的字段和数组实例中的数组元素，这些指令汇总如下。

- 创建类实例的指令：new。
- 创建数组的指令：newarray、anewarray、multianewarray。
- 访问实例字段的指令：getfield、putf ield。

- 访问类字段（static字段）的指令：getstatic、putstatic。
- 将数组值从局部变量表加载到操作数栈的指令：baload、caload、saload、iaload、laload、faload、daload、aaload。
- 将操作数栈中的数组值存储到局部变量表中的指令：bastore、castore、sastore、iastore、lastore、fastore、dastore、aastore。
- 取数组长度的指令：arraylength。
- 检查类实例或数组类型的指令：instanceof、checkcast。

aload 和 astore 指令是用来操作和加载 reference 类型的，数组也是 reference 类型的，不过操作数组有单独的指令。数组类型最外面的一维元素的类型，叫作该数组类型的组件类型（component type），一个数组的组件类型也可以是数组。从任意一个数组开始，如果发现其组件类型也是数组类型，那就继续取这个数组的组件类型，不断执行这样的操作，最终可以遇到组件类型不是数组的情况，此时把这种类型称为本数组类型的元素类型（element type）。例如 int[ ][ ][ ] 的组件类型是 int[ ][ ]，元素类型是 int。因为数组实例也是引用类型的，所以数组的指令是由数组类型的"元素类型 + aload/astore"构成的。例如加载 int[ ] 数组的值的指令是 iaload，读者可以按照这种方式来记忆。

接下来介绍部分有代表性的对象相关指令。

（1）new 指令。该指令用于创建对象，其格式如下。

格式	new indexbyte1 indexbyte2
操作数栈前后变化	... →..., objectref

其中 indexbyte1 和 indexbyte2 用于构建（构建方式前面已介绍过）指向常量池的引用值，指令调用后会将创建类的 reference 结果压入操作数栈中。下面以创建一个 String 对象的字节码为例。

```
//Java 代码
public void bar() {
    String str = new String();
}

// 对应的字节码
Constant pool:
    ...
    #2 = Class        #14       // java/lang/String
    #3 = Methodref    #2.#13    // java/lang/String."<init>":()V
    #4 = Class        #15       // cn/sensorsdata/ArrayTest
    ...
{
   public void bar();
     descriptor: ()V
     flags: ACC_PUBLIC
     Code:
       stack=2, locals=2, args_size=1
          0: new           #2    // class java/lang/String  (1)
          3: dup                 // (2)
          4: invokespecial #3    // Method java/lang/String."<init>":()V
          7: astore_1
          8: return
}
```

观察上述位置（1）处的字节码，new 指令后面跟了 dup 和 invokespecial 两个指令，其中 dup 指令的作用是复制操作数栈栈顶的值并压入栈顶。invokespecial 在此处的作用是调用实例的构造函数，关于这个指令的介绍可以参考 4.6 节。这里用一个例子简单介绍使用方式，如 a.add(b)，表示调用 a 对象（add() 方法的拥有者）的 add() 方法，此方法需要的参数是 b。为正常执行这段代码，就需要将方法拥有者 a 和方法参数 b 压入栈中，当执行 invokespecial 指令的时候会消耗掉 a 和 b。

那 new 指令后面为什么要跟 dup 和 invokespecial 两个指令呢？那是因为 new 指令是创建类的一个引用，即在内存中创建了一块区域，但内存中的内容是未初始化的，所以需要 invokespecial 调用构造函数为其实例化。又因为 invokespecial 调用的构造函数没有返回值，而 invokespecial 会消耗栈顶的元素，如果没有 dup 指令对 objreference 进行复制，此时栈将变成空，即无法再对创建的对象进行操作，所以要先使用 dup 对 new 的结果进行复制。

具体操作流程如图 4-4 所示。

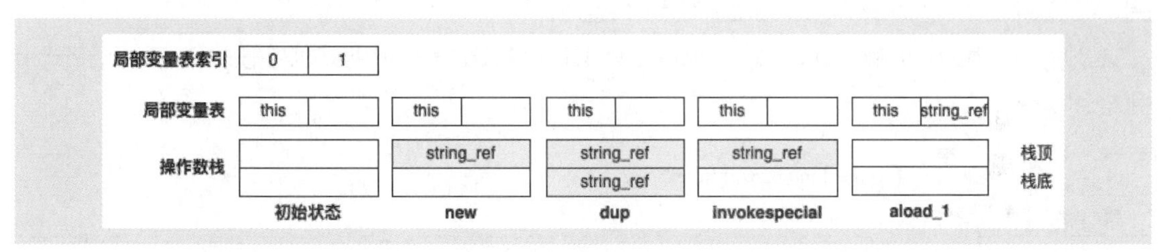

图 4-4 new 指令示例

（2）newarray 指令。该指令用于创建数组实例，其格式如下。

格式	newarray atype
操作数栈前后变化	..., count→..., arraryref

其中操作数 atype 指明了数组类型的元素类型，在操作数栈中需要先将数组的长度 count 压入栈顶。数组类型和 atype 值对照如表 4-7 所示。

表 4-7 数组类型和 atype 值对照

数组类型	atype 值
T_BOOLEAN	4
T_CHAR	5
T_FLOAT	6
T_DOUBLE	7
T_BYTE	8
T_SHORT	9
T_INT	10
T_LONG	11

newarray 指令的操作过程是：一个以 atype 为元素类型、以 count 值为长度的数组将会被分配在 GC（Garbage Collection，垃圾回收）堆中，并且一个代表该数组的 reference 类型的数据 arrayref 将被压入操作数栈中。这个新数组的所有元素将被分配为相应类型的初始值。

举例如下。

```
//Java 代码
public void bar() {
```

```
    int[] a = new int[2];
}

// 对应的字节码
public void bar();
      descriptor: ()V
      flags: ACC_PUBLIC
      Code:
        stack=1, locals=2, args_size=1
           0: iconst_2
           1: newarray   T_INT
           3: astore_1
           4: return
```

上述代码可创建一个元素类型是 int、长度为 2 的数组。类似地，anewarray 指令用于创建元素类型为接口类型的数组，multianewarray 则用于创建多维数组。

（3）getf ield、putf ield、getstatic、putstatic 指令。这几个指令可用于获取字段的值，它们的格式和用法分别如下。

格式	getfield indexbyte1 indexbyte2
操作数栈前后变化	..., objectref→..., value

getfield 指令用于获取实例中字段的值，该指令的操作数是一个指向常量池的索引值，会消费操作数栈中的 objectref 的值并将字段的结果压入栈顶。

格式	putfield indexbyte1 indexbyte2
操作数栈前后变化	..., objectref, value→...

putfield 指令用于给实例中的字段设置值，该指令的操作数同样是一个指向常量池的索引值，同时会消费操作数栈中的实例引用（objectref）和为字段设置的值（value）。

格式	getstatic indexbyte1 indexbyte2
操作数栈前后变化	... →..., value

getstatic 指令用于获取类变量的值，该指令的操作数同样是一个指向常量池的索引值，不会消费操作数栈中的值，会将 static 字段的值压入栈顶。

格式	putstatic indexbyte1 indexbyte2
操作数栈前后变化	..., value→...

putstatic 指令用于设置类变量的值，该指令的操作数同样是一个指向常量池的索引值，同时会消费操作数栈中为字段设置的值。

举例如下。

```
//Java 代码
public class FieldTest {

      int a = 1;
      static int B = 1;
      public void bar() {
```

```
        int c = this.a;
        this.a = 2;
    }

    public void foo() {
      int d = B;
      B = 2;
    }
}

//FieldTest 对应的字节码
Constant pool:
    #2 = Fieldref           #4.#19        // cn/sensorsdata/FieldTest.a:I
    #3 = Fieldref           #4.#20        // cn/sensorsdata/FieldTest.B:I

   int a;
     descriptor: I
     flags:

   static int B;
     descriptor: I
     flags: ACC_STATIC

   public void bar();
     descriptor: ()V
     flags: ACC_PUBLIC
     Code:
       stack=2, locals=2, args_size=1
          0: aload_0
          1: getfield      #2                  // Field a:I
          4: istore_1
          5: aload_0
          6: iconst_2
          7: putfield      #2                  // Field a:I
         10: return

   public void foo();
     descriptor: ()V
     flags: ACC_PUBLIC
     Code:
       stack=1, locals=2, args_size=1
          0: getstatic     #3                  // Field B:I
          3: istore_1
          4: iconst_2
          5: putstatic     #3                  // Field B:I
          8: return
```

上述代码 FieldTest 类中的 bar() 方法是对实例字段的获取和赋值操作，foo() 方法是对类变量的获取和赋值操作，读者对照上述理解即可。

4.5.5 操作数栈管理指令

Java 虚拟机也提供了一些能够直接对操作数栈中的元素进行管理的指令，例如前面提到的 dup 指令，这类指令如下。

- 使操作数栈栈顶的一个或两个元素的值出栈的指令：pop、pop2。
- 复制栈顶的一个或两个数值并将复制值或双份的复制值重新压入栈顶指令：dup、dup2、dup_x1、dup2_x1、dup_x2、dup2_x2。
- 交换栈顶的两个元素指令：swap。

（1）pop、pop2、dup、dup2 指令。pop 和 pop2 指令都是使栈顶的值出栈，那它们有什么不同呢？前文提到，不同 Java 虚拟机数据类型在操作数栈中占用的槽不一样，具体如表 4-8 所示。

表4-8　不同数据类型所占槽数和类别

实际数据类型	计算数据类型	类别	占用槽数 / 类别
boolean	int	一	1
byte	int	一	1
char	int	一	1
short	int	一	1
int	int	一	1
float	float	一	1
reference	reference	一	1
returnAddress	returnAddress	一	1
long	long	二	2
double	double	二	2

不同数据类型所占槽数和类别对于 pop 和 pop2 的区别，可以简单地看成让几个槽中的数据出栈，pop 是让一个槽中的数据出栈，而 pop2 是让两个槽中的数据出栈。假如栈顶的元素是 long 类型的，long 类型的值占用两个槽，使用 pop 指令将其出栈就会破坏 long 值的完整性，所以使用 pop2 指令来操作更合适。pop2 是使栈顶的一个或者两个元素的值出栈，具体是一个还是两个，由元素所占的槽数来决定。假如栈顶是两个 int 类型，那么使用 pop2 会使这两个 int 类型都出栈；如果是 long 类型，则使这一个 long 类型出栈。

同理，dup 是复制栈顶类元素并插入栈顶；dup2 是复制栈顶的一个或两个元素并插入栈顶，具体由元素所占的槽数决定。假如栈顶是两个 int 类型，那么使用 dup2 会复制这两个 int 类型，栈顶就有 4 个 int 类型；如果是 long 类型，使用 dup2 则复制这个 long 类型，栈顶就有两个 long 类型。

简单来说，使用操作数管理指令时既要考虑栈中的元素类型，也要考虑指令能够操作的槽数量。为了后续表述方便，表 4-8 根据元素类型所占的槽数不同，分成两个类别：类别一的元素在操作数栈中占一个槽，类别二的元素占两个槽。

（2）dup_x1 指令。该指令的作用是复制栈顶的元素并将复制的值向下插入栈顶元素后一个元素的后面（也可以理解为栈顶元素后面一个槽之后），使用前后的操作数栈变化如下。

```
..., value2, value1 → ..., value1, value2, value1
```

dup_x1 指令要求 value1 和 value2 的元素类型必须是类别一。例如，栈中的值是 int 类型的 1、2、3（其中 3 在栈顶位置），使用 dup_x1 后其结果如图 4-5 所示。

（3）dup_x2 指令。该指令的作用是复制栈顶的元素并将复制的值向下插入栈顶元素后一个或两个元素的后面（也可以理解为栈顶元素后面两个槽之后），dup_x2 要求栈顶的元素必须是类别一。根据其栈中元素的类别不同，操作数栈的变化有如下两种场景。

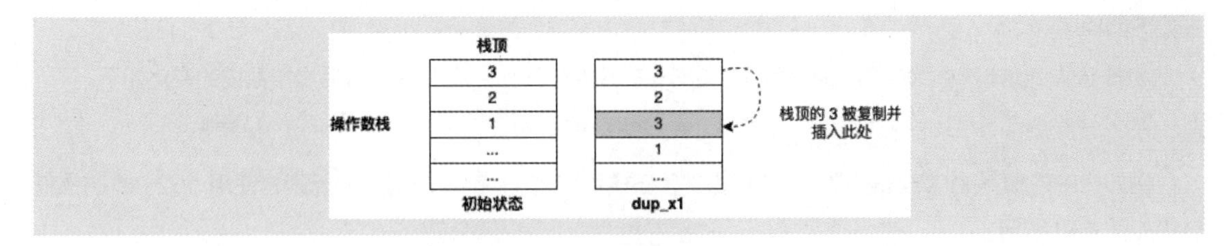

图 4-5 dup_x1 操作数栈的变化

场景一：value1、value2 和 value3 都为类别一的元素，dup_x2 执行后操作数栈的变化如下。

```
..., value3, value2, value1 → ..., value1, value3, value2, value1
```

例如栈中的值是 int 类型的 1、2、3（其中 3 在栈顶位置），都为类别一的元素，使用 dup_x2 的操作结果如图 4-6 所示。

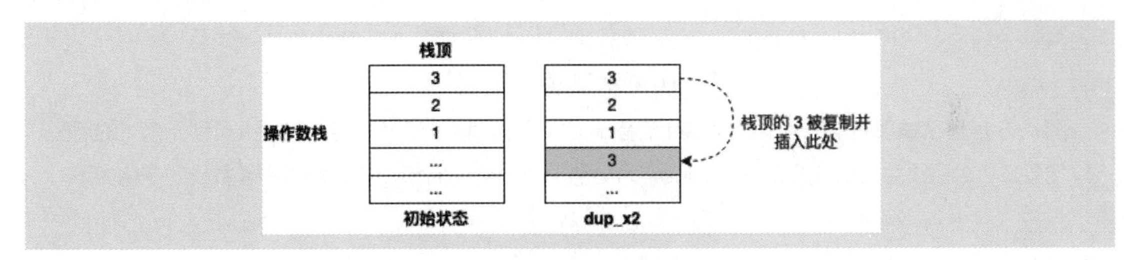

图 4-6 dup_x2 场景一操作数栈变化

图 4-6 描述了栈顶元素类别为一的 3 被复制后，插入 3 后面的两个槽之后，即 1 的后面。

场景二：value1 为类别一的元素，value2 为类别二的元素，dup_x2 指令执行后操作数栈的变化如下。

```
..., value2, value1 → ..., value1, value2, value1
```

例如栈中的元素分别为int类型的1、long类型的2L、int类型的1，使用dup_x2的操作结果如图4-7所示。

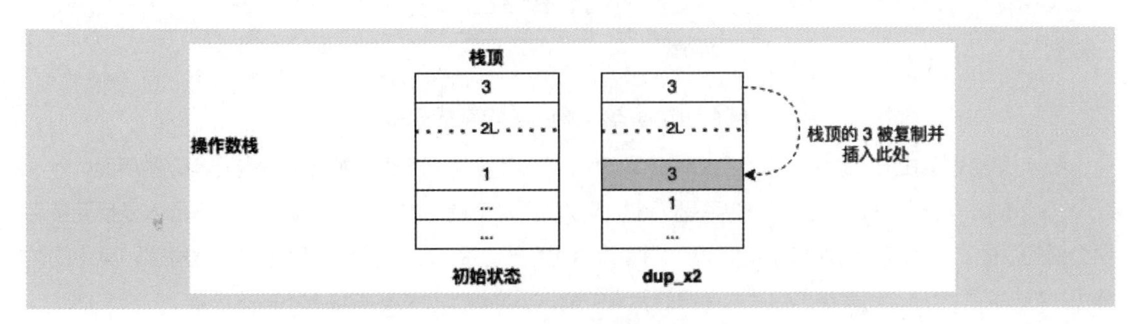

图 4-7 dup_x2 场景二操作数栈变化

图 4-7 描述了栈顶元素类别为一的 3 被复制后，插入 3 后面的两个槽之后 的 2L（占两个槽）之后。

对 dup、dup_x1、dup_x2 做一个总结：它们都要求栈顶的元素必须是类别一，即占一个槽；栈顶元素被复制后具体插入哪个位置由 dup 指令后面的参数决定，即 x1 表示栈顶元素后面一个槽之后、x2 表示栈顶元素后面两个槽之后。

（4）dup2_x1 指令。搞明白 dup、dup_x1、dup_x2 后，dup2(_x1) 就比较好理解了。dup2 是复制栈顶两个槽的内容，dup2_x1 指令的作用是复制操作数栈顶两个槽的内容并将复制的值向下插入栈顶元素后一个元素的后面。因为 dup2 指令复制栈顶两个槽的内容，根据其栈中元素的类别不同，操作数栈的变化有如下两种场景。

场景一：value1、value2 和 value3 都为类别一的元素，dup2_x1 指令执行后操作数栈的变化如下。

```
..., value3, value2, value1 → ..., value2,value1, value3, value2, value1
```

例如栈中的值是 int 类型的 1、2、3（其中 3 在栈顶位置），都为类别一的元素，使用 du2_x1 的操作结果如图 4-8 所示。

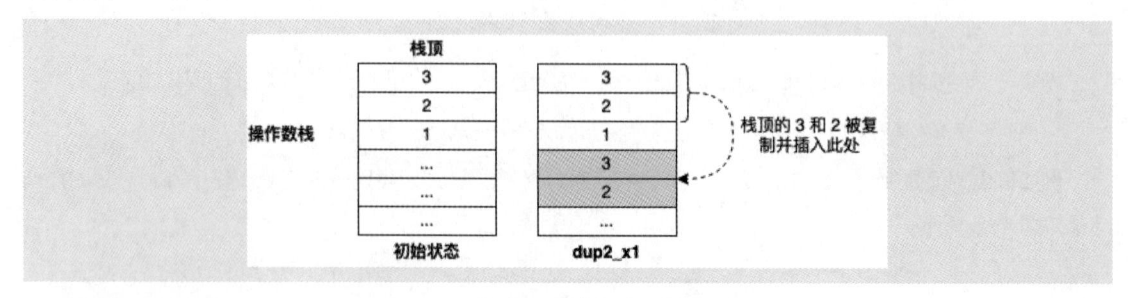

图 4-8　dup2_x1 场景一操作数栈变化

图 4-8 描述了栈顶元素是两个类别一的元素 3、2（即栈顶两个槽的内容）被插入类别元素 1 的后面。

场景二：value1 为类别二的元素，value2 为类别一的元素，dup2_x1 执行后操作数栈的变化如下。

```
..., value2, value1 → ..., value1, value2, value1
```

例如，栈中的元素是 1、2、3L，其中栈顶的元素 3L 是 long 类型的，1 和 2 是 int 类型的，使用 dup2_x1 后的结果如图 4-9 所示。

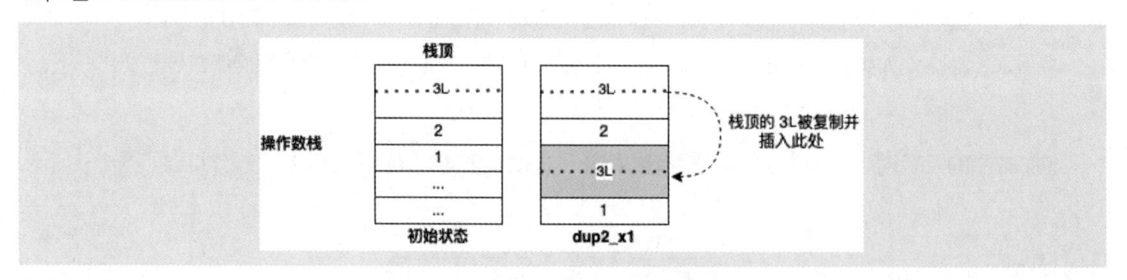

图 4-9　dup2_x1 场景二操作数栈变化

图 4-9 描述了栈顶元素是一个类别二的元素 3L（即占用栈顶两个槽）被插入类别元素 2 的后面。

（5）dup2_x2 指令。该指令的作用是复制操作数栈顶两个槽的内容并将复制的值向下插入栈顶元素后两个槽的后面。相信读者在明白了前面 dup* 指令的使用方式后，dup2_x2 也就很容易理解了，其中的关键点是栈顶前两个槽的元素类别和后两个槽的元素类别，下面给出针对这种情况不同的场景。

场景一：

```
..., value4, value3, value2, value1 → ..., value2, value1, value4, value3, value2, value1
```

其中 value1 ~ value4 都是类别一的元素。

场景二：

```
..., value3, value2, value1 → ..., value1, value3, value2, value1
```

其中 value1 为类别二的元素，value2 和 value3 为类别一的元素。

场景三：

```
..., value3, value2, value1 → ..., value2, value1, value3, value2, value1
```

其中 value1 和 value2 是类别一的元素，value3 是类别二的元素。

场景四：

```
..., value2, value1 → ..., value1, value2, value1
```

其中 value1 和 value2 都是类别二的元素。

可以看到本小节涉及的指令看起来很复杂，但是如果从操作数栈中槽的角度来看就比较容易理解了。

4.5.6 控制转移指令

控制转移指令可以让 Java 虚拟机重定向到指定字节码指令偏移位置执行，而不是按照顺序继续执行控制转移指令的下一条指令。从概念模型上理解，可以认为控制转移指令就是在有条件或无条件地修改 PC（Program Counter，程序计数器）寄存器的值。

控制转移指令包括以下几类。

- 条件分支指令：ifeq、iflt、ifle、ifne、ifgt、ifge、ifnull、ifnonnull、if_icmpeq、if_icmpne、if_icmplt、if_icmpgt、if_icmple、if_icmpge、if_acmpeq、if_acmpne。
- 复合条件分支指令：tableswitch、lookupswitch。
- 无条件分支：goto、goto_w、jsr、jsr_w、ret。

观察条件分支指令，这部分只提供了处理 int 和 reference 类型的指令，而并没有操作 boolean、byte、char、short、long、float 和 double 类型的指令。对于 boolean、byte、char 和 short 类型的条件分支比较操作，都使用 int 类型的比较指令操作即可；而对于 long、float、double 类型的条件分支比较操作，需要先使用比较运算指令（dcmpg、dcmpl、fcmpg、fcmpl、lcmp）计算结果，再根据结果使用条件分支指令即可。

接下来介绍部分有代表性的控制转移指令。

（1）if<cond>：ifeq、iflt、ifle、ifne、ifgt、ifge 指令。此类指令用于比较栈顶的 int 类型的值与 0 的大小关系，其格式如下。

格式	if<cond> branchbyte1 branchbyte2
操作数栈前后变化	..., value→...

其中操作数栈中的 value 会拿来与 0 比较，如果满足条件（succeed）就会跳转到以（branchbyte1 << 8 | branchbyte2）为结果的字节码指令偏移位置。if<cond> 各指令的比较规则如下。

- ifeq：如果value的值为0，则满足条件，跳转。
- ifne：如果value的值不等于0，则满足条件，跳转。
- iflt：如果value的值小于0，则满足条件，跳转。
- ifle：如果value的值不大于0，则满足条件，跳转。
- ifgt：如果value的值大于0，则满足条件，跳转。
- ifge：如果value的值不小于0，则满足条件，跳转。

举例如下。

```
//Java 代码
public void foo(int a) {
    if (a > 0) {
        a++;
    }
}
```

```
// 对应的字节码
public void foo(int);
      descriptor: (I)V
      flags: ACC_PUBLIC
      Code:
        stack=1, locals=2, args_size=2
           0: iload_1
           1: ifle          7
           4: iinc          1, 1
           7: return
```

上述 Java 代码的 foo() 方法比较入参 a 与 0 的大小关系，如果 a >0，就执行 a++。在 foo() 方法对应的字节码中，0: iload_1 表示从局部变量表中加载参数 a 的值到操作数栈中；1: ifle 指令判断 a 的值是否不大于 0，如果是就跳转到 7: return 继续执行，如果否就按顺序继续向下执行 4: iinc。

（2）if_icmp<cond>：if_icmpeq、if_icmpne、if_icmplt、if_icmple、if_icmpgt、if_icmpge 指令。此类指令用于比较操作数栈栈顶的两个 int 类型的值，如果满足条件就会跳转到指定的偏移位置继续执行，其格式如下。

格式	if_icmp<cond> branchbyte1 branchbyte2
操作数栈前后变化	..., value1, value2→...

其中操作数栈中的 value1 和 value2 相互比较，如果满足条件就会跳转到以（branchbyte1 << 8 | branchbyte2）为结果的偏移位置。if_icmp<cond> 各指令的比较规则如下。

- if_icmpeq：当value1等于value2时，则满足条件，跳转。
- if_icmpne：当value1不等于value2时，则满足条件，跳转。
- if_icmplt：当value1小于value2时，则满足条件，跳转。
- if_icmple：当value1不大于value2时，则满足条件，跳转。
- if_icmpgt：当value1大于value2时，则满足条件，跳转。
- if_icmpge：当value1不小于value2时，则满足条件，跳转。

举例如下。

```
//Java 代码
public void foo(int a, int b) {
      if (a > b) {
          a++;
      }
}

// 对应的字节码
public void foo(int, int);
      descriptor: (II)V
      flags: ACC_PUBLIC
      Code:
        stack=2, locals=3, args_size=3
           0: iload_1
           1: iload_2
           2: if_icmple     8
           5: iinc          1, 1
           8: return
```

上述 Java 代码的 foo() 方法比较入参 a 和 b，如果 a > b，就执行 a++。在 foo() 方法对应的字节码中，0: iload_1 和 0: iload_2 表示分别从局部变量表中加载参数 a 和 b 的值到操作数栈中; 2: if_icmple 比较操作数栈栈顶的两个元素，如果 a 不大于 b 就跳转到 8: return 继续执行，否则就按顺序继续向下执行 5: iinc。

注意，if<cond> 是与 0 做对比。

（3）if_acmp<cond>：if_acmpeq、if_acmpne 指令。此类指令用于比较操作数栈栈顶的两个 reference 类型的值，用法与 if_cmp<cond> 相似，其格式如下。

格式	if_acmp<cond> branchbyte1 branchbyte2
操作数栈前后变化	..., value1, value2→...

其中 value1 和 value2 都是 reference 类型的，如果满足条件就会跳转到以 (branchbyte1<<8) | branchbyte2 为结果的偏移位置。if_acmp<cond> 各指令的比较规则如下。

- if_acmpeq：当value1等于value2时，则满足条件，跳转。
- if_acmpne：当value1不等于value2时，则满足条件，跳转。

（4）ifnull、ifnonnull 指令。这两个指令的作用是弹出栈顶的 reference 类型元素，并判断是为 null（ifnull，null 表示空）还是不为 null（ifnonnull），然后跳转到相应位置，其指令格式如下。

格式	if_null 或 if nonnull branchbyte1 branchbyte2
操作数栈前后变化	..., value→...

比较规则如下。

- ifnull：当value1为null时，则满足条件，跳转。
- ifnonnull：当value不为null时，则满足条件，跳转。

举例如下。

```
//Java 代码
public void foo(String a) {
      if (a == null) {
          a = "Hi";
      }
}

// 对应的字节码
public void foo(java.lang.String);
      descriptor: (Ljava/lang/String;)V
      flags: ACC_PUBLIC
      Code:
        stack=1, locals=2, args_size=2
           0: aload_1
           1: ifnonnull     7
           4: ldc           #2          // String Hi
           6: astore_1
           7: return
```

上述 Java 代码的 foo() 方法判断字符串参数 a 是否为 null，满足条件则赋值“Hi”。对应的字节码 0: iload_1 表示从局部变量表中加载参数 a 的值到操作数栈; 1: ifnonnull 判断栈顶的值是否不为 null，如果是跳转到 7: return，否就继续向下执行 4: ldc。

（5）tableswitch、lookupswitch 指令。tableswitch 是通过索引访问跳转表并跳转到指定的偏移位置继续执行，lookupswitch 则是通过匹配键值访问跳转表并跳转。这两个指令主要用在 Java 的 switch-case 语句中，编译器会根据 switch-case 分支的实现情况，采取不同的指令，其中 tableswitch 适用于分支比较集中的情况，lookupswitch 适用于分支比较松散的情况。这么做主要是基于执行效率考虑的。

看下面的例子。

```
public void foo(int a){
      switch (a){
          case 10:
              break;
          case 11:
              break;
          case 14:
              break;
          default:
              break;
      }
}
```

上面这段 Java 代码对应的字节码如下。

```
public void foo(int);
      descriptor: (I)V
      flags: ACC_PUBLIC
      Code:
        stack=1, locals=2, args_size=2
           0: iload_1
           1: tableswitch    { // 10 to 14
                        10: 36
                        11: 39
                        12: 45
                        13: 45
                        14: 42
                   default: 45
              }
          36: goto          45
          39: goto          45
          42: goto          45
          45: return
```

观察上面的 Java 代码和对应的字节码可以发现，Java 代码中的 case 分支没有 12、13 这两项，但是观察字节码的 1: tableswitch 部分，它的索引值是 10 ~ 14 的连续值。这是因为编译器对 case 中的值做分析，如果 case 值存在连续或者少量断层，就会补齐断层并使用 tableswitch 指令来实现 switch-case 功能。当 JVM 执行到 tableswitch 指令时，它会检测 switch(a) 中的 a 值是否在 10 ~ 14：如果在范围内，a 的值就是 tableswitch 的索引值，直接跳转到指定的位置即可；如果不在范围内就跳到 default 处。可见 tableswitch 指令的时间复杂度是 O(1)。

假如 case 断层严重会怎样呢？看下面这个例子。

```
public void foo(int a){
      switch (a){
          case 20:
              break;
          case 10:
```

```
                break;
            case 30:
                break;
            default:
                break;
        }
}
```

上述 Java 代码对应的字节码如下。

```
public void foo(int);
      descriptor: (I)V
      flags: ACC_PUBLIC
      Code:
        stack=1, locals=2, args_size=2
           0: iload_1
           1: lookupswitch  { // 3
                        10: 36
                        20: 39
                        30: 42
                   default: 45
              }
          36: goto          45
          39: goto          45
          42: goto          45
          45: return
```

观察上述 Java 代码和字节码，如果还是使用 tableswitch 补齐 case 中分支的方式，那就需要补充非常多的分支。lookupswitch 指令对应的键值是按照升序来排列的，当 JVM 执行到 tableswitch 指令时，它会将 switch(a) 中的 a 值依次与 key 值进行比较，如果匹配就直接跳转，如果遇到比 a 大的 key 就跳转到 default。另外，键值通过升序排列也允许 JVM 实现这条指令时进行优化，比如采用二分搜索的方式取代线性扫描。

（6）goto、goto_w 指令。goto 指令会跳转到指定的地址位置，goto_w(wide) 表示能够跳转到更大的范围。goto 指令的操作数是两个字节，goto_w 的操作数是 4 个字节，可以跳转的范围更大，但是现在还是受到方法最大字节码长度为 65535 的限制，未来 Java 虚拟机版本可能会增大这个范围。

（7）jsr、jsr_w、ret 指令和 returnAddress 类型。jsr、jsw_w、ret 这几个指令和 returnAddress 类型的设计是用来实现 Java 语言中的 finally 语句块的，这种用法已经废弃。关于 finally 的实现原理请参考 4.7.4 小节，关于 jsr、jsw_w、ret 这几个指令以及 returnAddress 类型的介绍请参考 4.10 节。

4.5.7 方法调用和返回指令

在字节码指令中，与方法调用相关的指令有如下 5 种。

- invokestatic：该指令用于调用静态方法。
- invokeinterface：该指令用于调用接口方法。
- invokespecial：该指令用于调用一些需要特殊处理的实例方法，包括实例初始化方法、私有方法和父类方法。
- invokevirtual：该指令用于调用对象的实例方法。

- invokedynamic：该指令用于在运行时动态解析出调用点限定符所引用的方法，并执行该方法。

读者可能会有疑问，方法调用为什么需要 5 种不同的指令呢？这主要是由效率和方法调用的实现逻辑来决定的，关于这部分的内容，我们将在 4.6 小节展开来介绍。

方法返回指令根据返回值的类型进行区分，包括 ireturn（当返回值是 boolean、byte、char、short、int 类型时使用）、lreturn（当返回值是 long 类型时使用）、freturn（当返回值是 float 类型时使用）、dreturn（当返回值是 double 类型时使用）和 areturn（当返回值是 reference 类型时使用），还有一个 return 指令供声明为 void 的方法、实例初始化方法（<init>）、类和接口的类初始化方法（<clinit>）使用，由此可见所有的方法都需要有返回指令。

4.5.8 异常抛出指令

Java 语言中的 throw 语法由 athrow 指令来实现，另外还有一些异常会在字节码指令被检测到异常状况时自动抛出。例如对于 idiv 指令，当除数为 0 时，虚拟机会抛出 ArithmeticException 异常。关于 Java 异常（catch 语句）语法的实现，早期版本是通过前面介绍的 jsr、jsr_w、ret 等指令来实现的，现在是通过异常表来实现的，关于这部分内容的详细介绍请参考 4.7.4 小节。

4.5.9 同步指令

在 Java 中可以给方法或者方法内的代码片段添加 synchronized 关键字来实现同步，这两种同步的使用方式都是由同步锁来支持的。

对于方法级的同步是隐式的，虚拟机根据方法的结构中是否有 ACC_SYNCHRONIZED 访问标识符来区分一个方法是不是同步的。当调用方法时，调用指令会检查方法的 ACC_SYNCHRONIZED 访问标识符是否设置了，如果设置了，执行线程先获取同步锁，再执行方法，最后在方法完成时（不管是正常完成还是非正常完成）释放同步锁。如果一个同步方法在执行期间抛出了异常，并且在方法内部没有捕获此异常，那么这个同步方法所持有的锁将在异常抛出时自动释放。

对于方法中的同步代码，由 monitorenter 和 monitorexit 两条指令来支持 synchronized 关键字的语义。正确实现 synchronized 关键字需要编译器和 Java 虚拟机两者协作。

举例如下。

```
//Java 代码
public void foo() {
      synchronized (this){
          int a = 10;
      }
}

// 对应的字节码
public void foo();
      descriptor: ()V
      flags: ACC_PUBLIC
      Code:
        stack=2, locals=4, args_size=1
```

```
         0: aload_0
         1: dup
         2: astore_1
         3: monitorenter
         4: bipush        10
         6: istore_2
         7: aload_1
         8: monitorexit
         9: goto          17
        12: astore_3
        13: aload_1
        14: monitorexit
        15: aload_3
        16: athrow
        17: return
      Exception table:
         from    to  target type
             4     9    12   any
            12    15    12   any
```

上述 Java 代码中的 foo() 方法使用 synchronized(this) 获取 this 对象上的监视锁，monitorenter 指令用于获取对象上的监视锁，它的操作数是 object reference 类型的，所以字节码偏移 0 ~ 2 表示将 this 对象压入栈中，并复制一份 this 对象保存在局部变量表索引为 1 的位置，monitorenter 则获取 this 对象的监视锁。字节码偏移 4 ~ 6 表示 synchronized 代码块中的逻辑。7: aload_1 是加载局部变量表中索引 1 处的值 this 到操作数栈中，8: monitorexit 则用于释放 this 对象的监视锁。

另外编译器为了保证在方法异常发生完时能够正确地释放锁，会在每个方法退出的地方都添加 monitorexit 指令，也就是编译器在处理 synchronized 代码块的时候会为其添加 try-catch 功能。

上述字节码偏移 12 ~ 16 的意思如下。

- 12：当发生异常时将异常存放在局部变量表索引为3的位置。
- 13~14：使用monitorexit指令释放this对象的监视锁。
- 15~16：加载异常对象，再使用athrow指令将异常抛出。

4.6 方法调用

4.5.7 小节简单地介绍了 5 种与方法调用相关的指令的用法，但对它们的详细用法并没有多做介绍，本节将详细介绍这些内容。

4.6.1 invokevirtual指令

invokevirtual 指令用于调用普通的实例方法，其格式如下。

格式	invokevirtual indexbyte1 indexbyte2
操作数栈前后变化	..., objectref, [arg1, [arg2, ...]] → ..., [value]

其中 indexbyte1 << 8| indexbyte2 共同构成了指向常量池中结构类型为 CONSTANT_Methodref_info 的有效索引值。根据方法的签名信息，需要将方法的接收者，即将对象引用 objectref（方法的所有者），以及方法需要的参数 [arg1, [arg2, ...]] 依次压入栈中。如果方法有返回值，会将方法的返回值放入栈顶中。

举例如下。

```
//Java 代码
public void foo() {
      String str = "hello";
      int index = str.indexOf("ll");
}

// 对应的字节码
Constant pool:
     ...
     #4 = Methodref  #17.#18 // java/lang/String.indexOf:(Ljava/lang/String;)I
     ...
{
public void foo();
      descriptor: ()V
      flags: ACC_PUBLIC
      Code:
        stack=2, locals=3, args_size=1
       0: ldc           #2 // String hello
       2: astore_1
       3: aload_1
       4: ldc           #3 // String ll
       6: invokevirtual #4 // Method java/lang/String.indexOf: \
                                 (Ljava/lang/String;)I
       9: istore_2
      10: return
}
```

上述 Java 代码的 foo() 方法中定义了一个字符串，并调用了 String 的 indexOf() 方法。其中，indexOf() 的方法签名是“(Ljava/lang/String;)I”，它表示方法的参数是 java.lang.String 类型并且返回值是 int 类型的，对应的字节码按照偏移位置说明如下。

- 0: ldc，表示将字符串常量“hello”压入栈顶中。
- 2: astore_1，表示将字符串常量“hello”存放在局部变量表的索引1处。
- 3: aload_1，表示将字符串对象实例从局部变量表中加载到操作数栈中，它将作为indexOf()方法的接收者（方法的owner）。
- 4: ldc，表示将字符串常量“ll”压入操作数栈中，它将作为indexOf()方法的参数。
- 6: invokevirtual，用于确定方法的名称和所在类的信息（如常量池 #4 所示），方法执行后会将结果放入操作数栈中。
- 9: istore_2，表示将结果保存在局部变量表中。

4.6.2 invokestatic指令

invokestatic 指令用于调用静态方法，即使用 static 关键字修饰的方法，其格式如下。

格式	invokestatic indexbyte1 indexbyte2
操作数栈前后变化	..., [arg1, [arg2, ...]] →.., .[value]

其中 indexbyte1 << 8| indexbyte2 共同构成了指向常量池中结构类型为 CONSTANT_Methodref_info 的有效索引值。invokestatic 与 invokevirtual 指令的不同点在于它不需要将实例对象引用加载到操作数栈中，只需要将方法需要的参数入栈即可。如果方法有返回值，会将执行结果放入操作数栈中。

举例如下。

```
//Java 代码
public void foo() {
String str = String.valueOf(100);
}


// 对应的字节码
Constant pool:
     #1 = Methodref          #4.#12      // java/lang/Object."<init>":()V
     #2 = Methodref          #13.#14    // java/lang/String.valueOf:(I)Ljava/lang/String;
     ...
    public void foo();
      descriptor: ()V
      flags: ACC_PUBLIC
      Code:
        stack=1, locals=2, args_size=1
           0: bipush          100
           2: invokestatic   #2   // Method java/lang/String.valueOf:\
                                     (I)Ljava/lang/String;
           5: astore_1
           6: return
```

上述 Java 代码的 foo() 方法中使用 java.lang.String 类的静态方法 valueOf() 将整数 100 转换成字符串。对应的字节码指令偏移说明如下。

- 0: bipush，将整数100压入操作数栈中，将它作为静态方法String.valueOf()的参数。
- 2: invokestatic，用于调用静态方法的指令，其操作数用于确定方法的名称和所在类的信息（常量池 #2所示），方法执行后会将结果放入操作数栈中。
- 5: astore_1，将结果保存在局部变量表索引为1的位置。

4.6.3 invokespecial指令

invokespecial 指令用于调用一些“特殊”的实例方法，具体包括如下 3 种。

- 实例的构造方法，即 <init>。
- 实例的私有方法，即private。
- 父类方法，即使用super关键字调用的方法。

invokespecial 指令的格式和用法与 invokevirtual 指令的一致，只是使用场景不同。

举例如下。

```
//Java 代码
public class InvokeSpecialTest {
      @Override
      public String toString() {
          return super.toString();
      }

      public void foo() {
         String result = bar();
      }

      private String bar(){
          return new String("hello world");
      }
}

// 对应的字节码
Constant pool:
    ...
{
    public java.lang.String toString();
      descriptor: ()Ljava/lang/String;
      flags: ACC_PUBLIC
      Code:
        stack=1, locals=1, args_size=1
           0: aload_0
           1: invokespecial #2  // Method java/lang/Object.toString:\(1)
                                    ()Ljava/lang/String;
           4: areturn

    public void foo();
      descriptor: ()V
      flags: ACC_PUBLIC
      Code:
        stack=1, locals=2, args_size=1
           0: aload_0
           1: invokespecial #3  // Method bar:()Ljava/lang/String;(3)
           4: astore_1
           5: return

    private java.lang.String bar();
      descriptor: ()Ljava/lang/String;
      flags: ACC_PRIVATE
      Code:
        stack=3, locals=1, args_size=1
           0: new             #4   // class java/lang/String
           3: dup
           4: ldc             #5   // String hello world
           6: invokespecial #6   // Method java/lang/String."<init>":\(2)
                                    (Ljava/lang/String;)V
           9: areturn

}
```

上述 Java 代码部分的 InvokeSpecialTest 类中定义了 3 个方法。其中 toString() 重写了父类 Object 的 toString()；bar() 方法的访问修饰符是 private，并且使用 new String() 创建了一个字符串对象并返回，也就是调用了 String 的构造方法；foo() 方法的访问修饰符是 public，其内部调用私有方法 bar()。观察对应的字节码位置（1）（2）（3），可以看到使用的指令都是 invokespecial。

4.6.4 invokeinterface 指令

invokeinterface 指令，顾名思义，就是用于调用接口的指令。其格式和用法与 invokevirtual 指令一致，只是使用场景不同，示例如下。

```
//java 代码
List<String> list = new ArrayList<>();
int size = list.size();

// 对应的字节码
0: new            #2                    // class java/util/ArrayList
3: dup
4: invokespecial #3                     // Method java/util/ArrayList."<init>":()V
7: astore_1
8: aload_1
9: invokeinterface #4,  1               // InterfaceMethod java/util/List.size:()I
14: istore_2
```

List 是一个接口，调用接口的 size() 方法对应的指令是 invokeinterface。

4.6.5 方法调用指令的区别和方法分派

至此就介绍完了 invokevirtual、invokestatic、invokespecial、invokeinterface 这 4 种方法调用指令，接下来解释如下几个问题。

（1）为什么方法调用需要不同的指令？

（2）invokespecial 和 invokevirtual 有什么区别？

（3）invokevirtual 和 invokeinterface 有什么区别？

首先需要明白的是，所有的方法调用都只是对常量池的符号引用，并不是对应真正的物理内存中的地址。在 Java 中，一些方法如何执行在编译期就可以确定下来，另一些方法如何执行则需要在运行时才能确定。

举例如下。

```
public class InvokeTest {
      public static abstract class Color{
           public abstract void showColor();
}

      public static class Red extends Color{
          @Override
          public void showColor() {
              System.out.println("show red color");
          }
      }

      public static class Blue extends Color{
          @Override
```

```
        public void showColor() {
            System.out.println("show blue color");
        }
    }

    public static void main(String[] args) {
        Color color = new Random().nextBoolean()? new Red(): new Blue();
        color.showColor();
    }
}
```

上述 Java 代码 InvokeTest 类中定义了一个抽象类 Color 和 Color 的两个子类 Red 和 Blue，在 main() 方法中随机将 Red 和 Blue 的实例赋值给类型为 Color 的变量，然后调用这个 Color 对象的 showColor() 方法，color.showColor() 具体的输出是由 Color 的“实际类型”来决定的。什么是“实际类型”呢？用下面这段代码来引入两个概念。

```
Color red = new Red();
```

上述代码中 Color 称为变量的“静态类型”，Red 称为变量的“实际类型”，两者的区别是变量的静态类型不会被改变，并且最终的静态类型在编译期是确定的，而实际类型的结果在运行时才能确定。观察 main() 方法中的 color.showColor() 对应的字节码。

```
invokevirtual #9   // Method cn/sensorsdata/InvokeTest$Color.showColor:()V
```

从中可以看到方法的 owner 是 Color，而具体的输出则需要在运行时由 Color 的实际类型来决定，注意字节码中使用了 invokevirtual 指令。

不过，有些方法的调用在编译期就可以确定，例如下面这段代码。

```
//Java 代码
Red red = new Red();

// 对应的字节码
0: new            #2  // class cn/sensorsdata/InvokeTest$Red
3: dup
4: invokespecial #3  // Method cn/sensorsdata/InvokeTest$Red."<init>":()V
```

其中 new Red() 对应的 <init> 方法在编译期即可确定下来，也不存在运行时变成其他类型的可能，注意这里使用的是 invokespecial 指令。可以看到，不同的情况下，方法调用的指令是不一样的，这么做是为了针对不同的使用场景，使用不同的指令，用于增加效率。这就是为什么方法调用需要不同的指令。

那么，有哪些方法可以在解析阶段中确定唯一的调用版本呢？Java 语言里符合这个条件的方法有静态方法、私有方法、实例构造器方法、父类方法这 4 种，它们使用 invokestatic 和 invokespecial 指令来调用。其实还有一个 final() 方法，尽管 final() 方法使用 invokevirtual 指令（历史原因）调用，这 5 种方法调用会在类加载的时候就把符号引用解析为该方法的直接引用，我们称这 5 种方法为非虚方法（non-virtual method），与之相对的则是虚方法（virtual method）。

再来看第三个问题：invokevirtual 和 invokeinterface 有什么区别？为什么不使用 invokevirtual 指令来实现接口方法的调用呢？这就涉及“方法分派”这个概念。方法分派分为静态分派和动态分派，在正式介绍之前，读者可以思考 Java 中的重载和重写的实现原理是什么，它们在虚拟机中又是如何实现的。

1. 静态分派

先来看看什么是静态分派，想想下面这段代码的输出结果。

```
public class StaticDispatch {

    public static class Color{
    }

    public static class Red extends Color{
    }

    public static class Blue extends Color{
    }

    public void printColor(Color color){
        System.out.println("Colorful earth!");
    }

    public void printColor(Red redColor){
        System.out.println("Red flower!");
    }

    public void printColor(Blue redColor){
        System.out.println("Blue sky!");
    }

    public static void main(String[] args) {
        Color red = new Red();
        Color blue = new Blue();
        StaticDispatch dispatch = new StaticDispatch();
        dispatch.printColor(red);
        dispatch.printColor(blue);
    }
}
```

输出结果如下。

```
Colorful earth!
Colorful earth!
```

上述代码的 StaticDispatch 类有 3 个 printColor() 重载方法，具体使用哪一个重载方法是由方法参数的数量和类型决定的。main() 方法中故意定义了静态类型相同而实际类型不同的两个变量 red、blue，但编译时是根据方法参数的静态类型来选择重载方法的，而不是实际类型。由于静态类型在编译期是可知的，因此选择了 printColor(Color) 作为执行方法。通过观察 main() 方法中的 printColor() 方法对应的字节码，也可以看到字节码中的符号引用是指向 printColor(Color) 方法的。

```
26: invokevirtual #13 // Method printColor:\
                        (Lcn/sensorsdata/StaticDispatch$Color;)V
...
31: invokevirtual #13 // Method printColor:\
                        (Lcn/sensorsdata/StaticDispatch$Color;)V
```

所有依赖静态类型来决定方法执行版本的分派动作，都称为静态分派，其最典型的应用就是方法重载。当有多个重载方法的时候，编译器会根据静态类型去推断选择一个“合适”的版本，在《深入理解 Java 虚拟机：JVM 高级特性与最佳实践（第 3 版）》一书中给出了下面这样一个例子。

```
public class Overload {
    //1
```

```
        public static void sayHello(char arg){
            System.out.println("hello car");
        }

        //2
        public static void sayHello(Object arg){
            System.out.println("hello object");
        }

        //3
        public static void sayHello(int arg){
            System.out.println("hello int");
        }

        //4
        public static void sayHello(long arg){
            System.out.println("hello long");
        }

        //5
        public static void sayHello(Character arg){
            System.out.println("hello Character");
        }

        //6
        public static void sayHello(Serializable arg){
            System.out.println("hello Serializable");
        }

        //7
        public static void sayHello(char... args){
            System.out.println("hello char...");
        }

        public static void main(String[] args){
            sayHello('a');
        }
}
```

读者可以运行上述代码后，按照方法注释中的顺序依次注释对应的方法后再运行，看看每一次的输出结果（上述代码中的输出顺序就是注释中的顺序），感受一下何为“合适”的版本。

2. 动态分派

再来看看什么是动态分派，想想下面这段代码的输出结果。

```
public class DynamicDispatch {
      public static class Color{
          public void showColor(){
              System.out.println("base color");
          }
      }

      public static class Red extends Color{
          @Override
          public void showColor() {
              System.out.println("red color");
          }
```

```
        }

        public static class Blue extends Color{
            @Override
            public void showColor(){
                System.out.println("blue color");
            }
        }

        public static void main(String[] args) {
            Color red = new Red();
            Color blue = new Blue();
            red.showColor();
            blue.showColor();
        }
}
```

输出结果如下。

```
red color
blue color
```

上述代码中 main() 方法中定义的 red、blue 两个变量的静态类型都是 Color，但输出结果是实际类型中方法的输出，导致这个现象的原因是什么呢？先来看看 main() 方法对应的字节码。

```
public static void main(java.lang.String[]);
...
0: new     #2     //class cn/sensorsdata/DynamicDispatch$Red
3: dup
4: invokespecial #3  //Method cn/sensorsdata/DynamicDispatch$Red."<init>":()V
7: astore_1
8: new    #4      //class cn/sensorsdata/DynamicDispatch$Blue
11: dup
12: invokespecial #5  //cn/sensorsdata/DynamicDispatch$Blue."<init>":()V
15: astore_2
16: aload_1
17: invokevirtual #6 //cn/sensorsdata/DynamicDispatch$Color.showColor:()V
20: aload_2
21: invokevirtual #6 //cn/sensorsdata/DynamicDispatch$Color.showColor:()V
24: return
```

以 blue.showColor() 对应的字节码为例，字节码指令的偏移范围介绍如下。

- 8~15：调用了new Blue()方法，并将结果存储在局部变量表索引为2的位置。
- 20: aload_2：将局部变量表索引2处的Blue对象加载到操作数栈中。
- 21: invokevirtual：调用Color的showColor()方法。

可见 showColor() 方法的输出由运行时 blue 变量的实际类型来决定，而其对应的方法调用指令是 invokevirtual。《Java 虚拟机规范》中对 invokevirtual 指令的查找规则有如下描述。

- 假定方法接收者的实际类型是C，例如blue.showColor() 对应变量blue的实际类型是Blue。
- 如果C找到了与invokevirtual操作数所指向的常量池方法引用相同的方法，则选择这个方法指向，并且停止查找，例如会先在blue.showColor() 对应变量blue的实际类型Blue类中查找showColor()方法，如果找到了就使用它并停止查找。
- 否则，如果C有父类，就在其父类中查找，依次进行不断的递归查找。假如第二步中没有找到，就会在Blue的父类Color中查找，如果Color中没有会继续在Color的父类Object中查找。

- 如果在Object类中也没有找到，就会抛出AbstractMethodError。

正是因为 invokevirtual 指令的这种查找规则，才会有上面的输出结果，这个过程就是 Java 语言中方法重写的本质，我们把这种需要在运行时根据实际类型确定方法执行版本的分派过程称为动态分派。不过到这里为止仍然没有介绍 invokevirtual 和 invokeinterface 两个指令之间的区别，为了弄清楚这个问题，我们还需要了解一点方法分派的原理。

3. 方法分派的原理

不同的虚拟机在实现分派的原理上是有一些差别的，一种比较基础且高效的方式是为类型在方法区中建立一个虚方法表（virtual method table，也叫作 vtable）。使用虚方法表可以有效提高查找方法版本的速度。

举例如下。

```
public class Color {
      public void showColor(){}
      public void setName(){}
}
public class Red extends Color {
      public void setName(){}
      public void setAlpha(){}
}
```

上述代码中定义了一个 Color 类，Red 类继承了 Color 类，并重写了 setName() 方法，这两个类的虚方法表如图 4-10 所示。

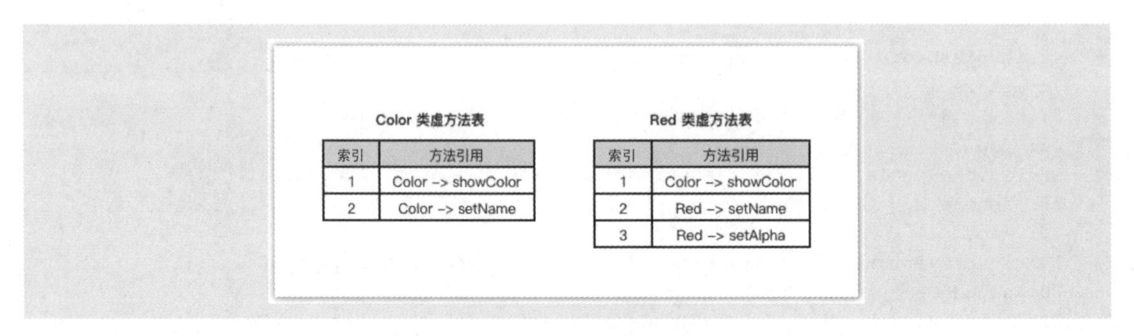

图 4-10　虚方法表示例

从图 4-10 中可以看到 Red 类的虚方法表保留了父类 Color 中虚方法的顺序，并且 Red 类覆盖了 setName 指向的方法链接。假如这个时候调用 Red 类的 setName() 方法，只需要找到索引 2 的方法引用就可以了。[注意上面例子中的 Color() 和 Red() 方法，都没有重写 Object 中的方法，Object 中的方法理应也写在虚方法表中，此处省略了。]

虚方法表中存放着各个方法的实际入口地址，如果某个方法在子类中没有被重写，那么子类的虚方法表中的地址和父类相同方法的地址入口是一致的，都指向父类的实现入口。如果子类重写了父类的方法，子类的虚方法表中的地址会被替换成指向子类实现版本的入口地址。虚方法表的特点是子类的虚方法表保持了父类的虚方法表中方法的编号顺序，所以这种查找是非常高效的。

不过 Java 语言的一个特性是一个类只能继承单个类，但可以实现多个接口，而每个接口的函数编号又都是和自己相关的，因此虚方法表无法解决多个对应接口的函数编号问题。为了支持 Java 接口的实现，Java 虚拟机还建立了一个接口方法表（interface method table，也叫作 itable）结构用于查找接口的实现，itable 是由偏移量表（offset table）和方法表（method table）组成的。当需要调用某个接口的方法时，虚拟机会先在偏移量表中找到接口对应的方法表位置，接着在方法表中找到具体的方法版本。

从上述虚拟机实现方法分派的原理可以发现，虚方法表的效率要比接口方法表的高一些，因此分别使用 invokevitual 和 invokeinterface 两个指令来应对不同的方法调用，这就是它们的区别。

4.6.6 invokedynamic指令

invokedynamic 指令是 JDK 7 新增的一个指令，用于实现动态类型语言（dynamically typed language），那么什么是动态类型语言呢？动态类型语言是指运行时才去做类型检测的语言，例如下面这段代码。

```
obj.println("Hello World!")
```

上面这段代码输出的结果是什么？你可能会说“Hello World!”，但对计算机来说却不是，因为没有给出这段代码的上下文。在 Java 语言中，这段代码对应的字节码如下。

```
0: getstatic      #2    // Field java/lang/System.out:Ljava/io/PrintStream;
3: astore_1
4: aload_1
5: ldc            #3    // String Hello World!
7: invokevirtual #4    // Method java/io/PrintStream.println:\
                           (Ljava/lang/String;)V
```

可以看到在字节码中要求调用 println() 方法的接收者是 java.io.PrintStream 对象、方法参数是 java.lang.String 类型、返回值是 void 类型。如果在 obj 对应的类中找不到符合条件的函数，会在其父类中查找。如果 obj 所属的类和 java.io.PrintStream 没有继承关系，就算 obj 所属的类有符合前面条件的方法也无法调用成功，因为类型不符合。但是上面这段代码在 JavaScript 中就不存在这个问题，JavaScript 语言不用关心 obj 的具体类型，只要 obj 中有 println() 方法并且其参数是字符串即可，这就是动态类型语言和静态类型语言的区别。

Java 为支持动态类型语言的功能在 JDK 7 中新增了 invokedynamic 指令，另外还新增了 java.lang.invoke 包，这个包的作用是解决之前只能依靠符号引用来确定调用的目标方法的问题，提供了一种新的确定调用方法的机制，称为“方法句柄”。

例如下面的 Kotlin 代码中定义了一个 compare() 方法用于比较两个值，具体的比较逻辑由方法类型的参数 method 来决定。

```
fun max(a: Int, b: Int): Int {
      return if (a > b) a else b
}

fun compare(num1: Int, num2: Int, method: (Int, Int) -> Int): Int {
      return method(num1, num2)
}

fun main() {
      compare(1, 2, ::max)
}
```

虽然 Java 中不能完全实现这种语法，但可以借助方法句柄，委托实现这个功能。下面通过方法句柄来实现上述 Kotlin 代码的功能。

```
public class MethodHandleTest {
    public static int max(int a, int b) {
        return a > b ? a : b;
    }

    public static int min(int a, int b) {
        return a < b ? a : b;
    }

    public static int compare(int num1, int num2, MethodHandle methodHandle)
        throws Throwable
    {
        return (int) methodHandle.invokeExact(num1, num2);
    }

    public static void main(String[] args) throws Throwable {
        MethodType methodType =
            MethodType.methodType(int.class, int.class, int.class);
        MethodHandles.Lookup lookup = MethodHandles.lookup();

        //max
        MethodHandle maxMethodHandle =
            lookup.findStatic(MethodHandleTest.class, "max", methodType);
        int maxResult = compare(1, 2, maxMethodHandle);

        //min
        MethodHandle minMethodHandle =
            lookup.findStatic(MethodHandleTest.class, "min", methodType);

        // 执行
        int minResult = compare(1, 2, maxMethodHandle);
    }
}
```

上述 Java 代码中定义了一个 compare() 方法，另外还定义了两个方法签名都是 (II)I 的 max() 方法和 min() 方法。我们期望 compare() 方法不需要关注具体的比较逻辑，类似于传递了一个方法指针 [compare() 方法中执行方法的调用，可对比前面的 Kotlin 代码]。再来看看 main() 方法中的逻辑。

- MethodType用于指定方法的签名信息，静态方法MethodType.methodType的第一个参数是返回值类型，其余参数对应方法签名中入参的类型。
- MethodHandles.Lookup类的作用是在指定的类中寻找匹配的方法，除了上面例子中使用到的findStatic()方法，还有findVirtual()、findConstructor()、findspecial()等方法，这些方法返回方法句柄（MethodHandle）。
- 构建好方法句柄以后就可以调用具体的方法了，将方法参数传入MethodHandle的invoke()或者invokeExact()方法就可以进行方法调用。

可以看出 compare() 方法中的 MethodHandle 具体调用什么方法可以灵活定义，相当于我们把 invokestatic 指令从字节码层面转移到了 Java 层面。那 MethodHandle 跟 invokedynamic 又有什么关系呢？从某种意义上说，invokedynamic 指令与 MethodHandle 的目的是一样的，之前介绍的 4 个 invoke* 指令将方法的分派固定在虚拟机中，invokedynamic 指令则是将方法的分派转移到用户的代码中。

invokedynamic 指令的调用流程如下。

- 首先，执行invokedynamic的时候会调用引导方法。
- 引导方法会返回一个调用点CallSite对象，通过调用点CallSite对象的getTarget()方法获取到MethodHandle对象。
- 在CallSite没有变化的情况下，MethodHandle可以一直被调用；如果CallSite有变化，重新查找即可。

下面来看看 invokedynamic 的格式和在字节码中的使用。

格式	invokedynamic indexbyte1 indexbyte2 0 0
操作数栈前后变化	..., [arg1, [arg2, ...]] →...

invokedynamic 的操作数有 4 个字节，其中前两个字节组合指向常量池中类型为 CONSTANT_InvokeDynamic_info 的常量，后两个字节固定为 0。另外观察操作数栈的变化，invokedynamic 指令需要额外的参数会从操作数栈中获取。CONSTANT_InvokeDynamic_info 类型的结构已经在 3.4.5 小节中介绍过，它包括引导方法的位置，以及方法名和方法描述符等信息。

下面从实际案例来理解 invokedynamic。

```
public class LambdaMain {
      Object obj = "hello";

      public void foo() {
          Consumer<String> consumer = o -> {
          };

          Consumer<String> consumer2 = o -> {
              Object tmpObj = this.obj;
          };
      }
}
```

上面这段 Java 代码中使用 Java 8 中 Lambda 的写法定义了两个 Consumer 对象，对应的字节码如下。

```
Constant pool:
    ...
     #5 = InvokeDynamic #1:#30  // #1:accept:(Lcn/sensorsdata/LambdaMain;)\
                                    Ljava/util/function/Consumer;
    ...
{

    ...
    public void foo();
      descriptor: ()V
      flags: ACC_PUBLIC
      Code:
        stack=1, locals=3, args_size=1
           0: invokedynamic #4,  0  // InvokeDynamic #0:accept: \
                                        ()Ljava/util/function/Consumer;
           5: astore_1
           6: aload_0
           7: invokedynamic #5,  0  // InvokeDynamic #1:accept: \
                (Lcn/sensorsdata/LambdaMain;)Ljava/util/function/Consumer;
          12: astore_2
          13: return
```

```
    ...
}

BootstrapMethods:
    ...
    1: #24 invokestatic java/lang/invoke/LambdaMetafactory.metafactory:\(1)
                (Ljava/lang/invoke/MethodHandles$Lookup;\
                Ljava/lang/String;Ljava/lang/invoke/MethodType;\
                Ljava/lang/invoke/MethodType;\
                Ljava/lang/invoke/MethodHandle;\
                Ljava/lang/invoke/MethodType;)Ljava/lang/invoke/CallSite;
      Method arguments:
        #25 (Ljava/lang/Object;)V
        #29 invokespecial cn/sensorsdata/LambdaMain.lambda$foo$1:\
                        (Ljava/lang/String;)V
        #27 (Ljava/lang/String;)V
```

注意：javap 的输出结果以单行表示过长，为方便阅读和印刷，本书会对字节码做换行拼接。

观察 foo() 方法对应的字节码，其中定义的 Lambda 被翻译成了 invokedynamic，以其中的 7: invokedynamic #5, 0 为例。

- #5表示CONSTANT_InvokeDynamic_info结构在常量池中的位置，后面为固定两个字节的0。
- 常量池中 #5的内容是InvokeDynamic #1:#30，其中 #1是引导方法在引导方法属性表BootstrapMethods_attribute（BootstrapMethods_attribute表是ClassFile中的一个属性）中的位置，#30是对应引导方法需要的动态调用名称、参数和返回类型信息。

这里补充一下引导方法属性表的结构。

```
BootstrapMethods_attribute {
      u2 attribute_name_index;
      u4 attribute_length;
      u2 num_bootstrap_methods;
      {   u2 bootstrap_method_ref;
          u2 num_bootstrap_arguments;
          u2 bootstrap_arguments[num_bootstrap_arguments];
      } bootstrap_methods[num_bootstrap_methods];
}
```

说明如下。

- attribute_name_index为固定的值“BoostrapMethods”。
- attribute_length为属性表的长度。
- num_bootstrap_methods为引导方法的数目。
- bootstrap_methods为引导方法数组。

数组item结构中的bootstrap_method_ref为指向常量池中结构为CONSTANT_MethodHandle_info的索引；num_bootstrap_arguments和bootstrap_arguments为引导方法需要的参数。引导方法的参数类型必须是CONSTANT_String_info、CONSTANT_Class_info、CONSTANT_Integer_info、CONSTANT_Long_info、CONSTANT_Float_info、CONSTANT_Double_info、CONSTANT_MethodHandle_info或CONSTANT_MethodType_info结构中的一个。

LambdaMain 字节码中位置（1）处就是引导方法的结果展示。本书将在 8.4 节对 Lambda 相关的知识做进一步的探讨，包括引导方法的详细介绍，读者暂且了解这么多即可。

4.7 案例分析

经过前面大量理论知识的学习，读者可能有些“累了”，接下来通过一些轻松、有趣的内容，了解字节码是如何与 Java 源码相对应的。

4.7.1 System.out.println

读者在学习了 ASM 之后，首先想到的是在方法中插入一段代码来验证结果，而这段代码通常是 System.out.println（“Hello World”），本小节就来看看这段代码被翻译成字节码以后是怎样的。首先，要调用 println() 方法，需要知道 println() 方法的签名、访问标识等信息。查看源码可以得到 println() 是在 PrintStream 类中定义的方法，代码如下。

```
public class PrintStream {
      public void println(String x) {
          ...
      }
}
```

可以看到它是一个实例方法，调用它就需要获取 PrintStream 对象，获取的方式是调用 System 类中的静态字段 out，即 System.out。所以，要想输出“Hello World”，就需要先获取到 System 类中的静态字段 out，然后调用 out 的 println() 方法，理顺了这个逻辑之后再来看字节码就很好理解了。

```
0: getstatic      #2 // Field java/lang/System.out:Ljava/io/PrintStream;
3: ldc            #3 // String Hello World
5: invokevirtual #4 // Method java/io/PrintStream.println:(Ljava/lang/String;)V
```

使用 invokevirtual 指令调用 println() 方法，需要将方法接收者和方法参数值依次压入栈中，对应字节码操作如下。

- 0: getstatic：通过getstatic指令，将System类中类型是PrintStream的静态字段out压入栈中。
- 3: ldc：将println()需要的参数值“Hello World”通过ldc指令从常量池中压入操作数栈中。
- 5: invokevirtual：调用PrintStream对象的println()方法。

4.7.2 switch-case与String

在 Java 7 之前要想实现比较字符串条件流的唯一方式是使用 if-else 语句，不过 Java 7 扩展了 switch-case，使其也能支持 string 类型的判断。接下来让我们通过字节码一探其原理。

```
public void chooseFruit(String type) {
      int a = 1;
      switch (type) {
          case "Apple":
              a = 2;
              break;
          case "Durian":
              a = 3;
```

```
                break;
            default:
                a = 4;
                break;
        }
}
```

上述代码对应的字节码如下。

```
public void chooseFruit(java.lang.String);

   0: iconst_1
   1: istore_2
   2: aload_1
   3: astore_3
   4: iconst_m1
   5: istore        4
   7: aload_3
   8: invokevirtual #2  // Method java/lang/String.hashCode:()I
11: lookupswitch  { // 2
        63476538: 36
      2058334421: 51
         default: 63
    }
36: aload_3
37: ldc           #3  // String Apple
39: invokevirtual #4  // Method java/lang/String.equals:(Ljava/lang/Object;)Z
42: ifeq          63
45: iconst_0
46: istore        4
48: goto          63
51: aload_3
52: ldc           #5  // String Durian
54: invokevirtual #4  // Method java/lang/String.equals:(Ljava/lang/Object;)Z
57: ifeq          63
60: iconst_1
61: istore        4
63: iload         4
65: lookupswitch  { // 2
               0: 92
               1: 97
         default: 102
    }
92: iconst_2
   93: istore_2
   94: goto          104
   97: iconst_3
   98: istore_2
   99: goto          104
102: iconst_4
103: istore_2
104: return
```

在解读上述字节码之前可以回顾一下 lookupswitch 指令和 tableswitch 指令的区别。上述字节码的说明如下。

- 0~1：将int型的值1保存到局部变量表索引2处，也就是变量a。

- 2~3：将chooseFruit()方法的参数保存到局部变量表索引3处，假设这里的type值为“Apple”。
- 4~5：将 -1保存在局部变量表索引4处。
- 7~8：将局部变量表索引3处的“Apple”压入栈顶，并调用其hashCode()方法，并将结果压入栈顶中。
- 11：根据hashCode()的结果，调用lookupswitch指令寻找匹配的值，这里“Apple”的hashCode值为63476538，所以对应的字节码偏移位置是36。
- 36~39：将局部变量表索引3处的“Apple”压入栈顶，再将常量池中的“Apple”压入栈顶，调用equals()方法比较它们的值。
- 42~48：ifeq指令会判断equals()返回的结果，如果为0，就跳转到字节码偏移63的位置；因为传入的是“Apple”，所以equals()返回为true，继续执行ifeq后面的指令，45~48中的指令是将0存储在局部变量表索引4处并跳转到63继续执行。
- 63：将局部变量表索引4处的值加载到操作数栈中，此时值为0。
- 65：lookupswitch指令寻找匹配的值，此时操作数栈中的值为0，所以跳转到92。
- 92~94：将int型的值2压入操作数栈中，并存储在局部变量表索引4处，然后执行goto语句跳转到104。
- 104：执行return指令，程序结束。

通过以上指令解读可以发现，switch-case 为支持 string 类型，实际上是先比较字符串的 hashCode 值。因为 hashCode 可能存在碰撞冲突（例如“Ab”和“BC”的 hashCode 值都是 2113），所以还需要调用 String.equals() 方法再进行一次比较得到一个 int 值，接着比较这个 int 值，跳转到字符串 case 分支中代码块的具体逻辑。字节码中使用了两个 lookupswitch 指令，相当于有两个 switch-case 语句，我借助工具反编译上面生成的 .class 文件（做了部分调整，添加了返回值），可以看到两个 switch-case 的结果。

```
public int chooseFruit(String var1) {
      boolean var2 = true;
      byte var4 = -1;
      switch(var1.hashCode()) {
      case 63476538:
          if (var1.equals("Apple")) {
              var4 = 0;
          }
          break;
      case 2058334421:
          if (var1.equals("Durian")) {
              var4 = 1;
          }
      }
      byte var5;
      switch(var4) {
      case 0:
          var5 = 2;
          break;
      case 1:
          var5 = 3;
          break;
      default:
          var5 = 4;
      }
```

```
        return var5;
    }
```

4.7.3 for循环原理

下面是一个简单的 for 循环，大家可以想一想，根据我们已经掌握的字节码知识，是否可以从字节码层面来实现呢？

```
public int foo() {
        int a = 0;
        for (int index = 0; index < 10; index++) {
            if (index / 2 == 0) {
                a++;
            }
        }
        return a;
}
```

对上面的代码进行分解，不考虑 for 循环体中的内容。首先需要判断 index 的值是否小于 10，因为 index 是 int 型的，可以考虑使用 if_icmpge，当不小于 10 的时候使用 goto 跳转到指定的字节码偏移位置，即跳出 for 循环；否则就执行 for 循环体里面的代码，并在结尾执行 index++ 操作，然后使用 goto 指令跳转到 for 循环开始的地方。对应的字节码如下。

```
public int foo();
     descriptor: ()I
     flags: ACC_PUBLIC
     Code:
       stack=2, locals=3, args_size=1
          0: iconst_0
          1: istore_1
          2: iconst_0
          3: istore_2
          4: iload_2
          5: bipush        10
          7: if_icmpge     25
         10: iload_2
         11: iconst_2
         12: idiv
         13: ifne          19
         16: iinc          1, 1
         19: iinc          2, 1
         22: goto          4
         25: iload_1
         26: ireturn
```

这里有一个知识点，iinc 指令操作的是局部变量表，不是操作数栈，可不要忘记哟！

4.7.4 try-catch-finally 原理

作为 Java 程序员，读者可能遇到过类似下面这样的面试题，请问它的输出结果是什么？

```
public static int foo() {
```

```
        int i = 1;
        try {
            i = i / 0;
            return i;
        } catch (Exception e) {
            i++;
            return i;
        } finally {
            i = 10;
            //return i; 去掉注释，输出结果又是什么
        }
}
```

输出结果如下。

```
2
```

要解释为什么输出结果是 2，先来了解在 Java 中 try-catch-finally 背后的原理。

在第 3 章介绍 Code 属性时提到，Code 属性中有一个异常表可记录 Java 中的异常信息，其结构如下。

```
Code_attribute {
        ...
        u2 exception_table_length;
        {   u2 start_pc;
            u2 end_pc;
            u2 handler_pc;
            u2 catch_type;
        } exception_table[exception_table_length];
        ...
}
```

当发生异常的时候，从异常表头部遍历，找到合适的异常处理器进行处理，每个异常处理器由 start_pc、end_pc、handler_pc、catch_type 组成，其中 [start_pc, end_pc) 构成异常处理器处理的字节码偏移范围; catch_type 表示异常的类型，如果其值非 0，那么 catch_type 指向常量池中的 CONSTANT_Class_info 结构，如果其值为 0，表示可以处理任意的异常类型；当满足 [start_pc, end_pc) 和 catch_type 时，就跳转到字节码偏移位置为 handler_pc 的地方继续执行。

举例如下。

```
public int bar() {
        int i = 1;
        try {
            i = i / 0;
        } catch (ArithmeticException exception) {
            i = 2;
        } catch (Exception e) {
            i = 3;
        }
        return i;
}
```

对应的字节码如下。

```
public int bar();
        descriptor: ()I
        flags: ACC_PUBLIC
```

```
    Code:
      stack=2, locals=3, args_size=1
         0: iconst_1
         1: istore_1
         2: iload_1
         3: iconst_0
         4: idiv
         5: istore_1
         6: goto          18
         9: astore_2
        10: iconst_2
        11: istore_1
        12: goto          18
        15: astore_2
        16: iconst_3
        17: istore_1
        18: iload_1
        19: ireturn
      Exception table:
         from    to  target type
             2     6     9   Class java/lang/ArithmeticException
             2     6    15   Class java/lang/Exception
```

我们来分析上面的这段字节码。

- 0~1：是将i=1的值存在局部变量表索引1处的位置。
- 2~6：将局部变量表索引1处的值加载到操作数栈，并将常量0也压入栈中，使用idiv指令计算两者的值，并将结果存储在局部变量表索引1处，如果没有出现异常会使用goto指令跳转到字节码指令偏移18的位置继续执行。
- 观察异常信息表，其中from = start_pc、to = end_pc、target = handle_pc、type = catch_type，可以看到在字节码偏移2~6匹配了两个异常类型，并且它们的target不相同。以bar()方法的实际运行结果为例，当执行到4位置的idiv指令时会发生ArithmeticException异常，此时会从异常表中查找发生异常的字节码偏移位置是否在 [from, to) 范围中，并且type类型与抛出的异常匹配的异常表中的项，跳转到target处继续执行。本例中抛出的异常就是ArithmeticException，所以会跳转到9的位置继续执行。
- 9：这条指令的意思是将触发的异常值存储在局部变量表索引2处。
- 10~12：是将常量2压入操作数栈中并保存在局部变量表索引为1的位置，然后跳转到18的位置继续执行，这部分就是catch (ArithmeticException exception) 部分的代码。
- 15：这条指令的意思也是将触发的异常值存储在局部变量表索引2处。
- 16~17：是将常量3压入操作数栈中并保存在局部变量表索引为1的位置，然后跳到18的位置继续执行，这部分是catch (Exception e) 部分的代码。
- 18~19：是返回局部变量表索引1处的值。

以上就是 try-catch 异常表的基本介绍，如果添加上 finally 语句块，那对应的字节码又是怎样的呢？Java 程序员都知道 finally 语句块一定会执行，其背后的原理是什么呢？继续看下面这段 Java 代码对应的字节码。

```
public void foo1() {
    try {
        doSomething();
```

```
        } catch (Exception e) {
            doException();
        } finally {
            doFinally();
        }
}
```

对应的字节码如下。

```
public void foo1();
      descriptor: ()V
      flags: ACC_PUBLIC
      Code:
        stack=1, locals=3, args_size=1
           0: aload_0
           1: invokespecial #13                  // Method doSomething:()V
           4: aload_0
           5: invokespecial #14                  // Method doFinally:()V
           8: goto          30
          11: astore_1
          12: aload_0
          13: invokespecial #15                  // Method doException:()V
          16: aload_0
          17: invokespecial #14                  // Method doFinally:()V
          20: goto          30
          23: astore_2
          24: aload_0
          25: invokespecial #14                  // Method doFinally:()V
          28: aload_2
          29: athrow
          30: return
        Exception table:
           from    to  target type
               0     4    11   Class java/lang/Exception
               0     4    23   any
              11    16    23   any
```

可以看到字节码中doFinally()方法被调用了3次，并且被调用的位置是在正常return或者throw之前，下面对上述字节码进行介绍。

- 0~5：代表try中的内容，结合异常表，在执行doSomething()方法时，如果没有发生异常就紧接着执行finally语句块中的内容，即执行doFinally()。如果执行doSomething()时发生了异常，就查找异常表，确定匹配的异常项，然后跳转到target位置。本例中[0~4)对应着两个异常，其中type=any表示可以匹配任意的异常。假如doSomething()抛出了java.lang.Exception异常，就会匹配到异常表的第一个项，跳转到偏移位置11继续执行；假如抛出的异常是java.lang.Throwable，就会匹配到异常表的第二个项，跳转到偏移位置23继续执行。
- 11~17：代表catch中的内容，这里会执行doException()方法。在执行此方法时也需要注意，如果此方法没有抛出异常就紧接着执行finally语句块中的内容，即执行doFinally()方法；如果执行doException()方法时抛出了异常，通过异常表可以跳转到23位置。
- 23~29：此部分指令的意思是将异常存储在局部变量表中，然后执行finally中的方法doFinally()，接着将异常加载到操作数中，通过throw指令抛出异常。

通过上述字节码介绍，可以得出如下结论。

（1）finally 中的代码一定会执行，字节码中会构建一些异常表来保证当未捕获异常发生时也能够执行 finally 中的代码。

（2）字节码实现 finally 的方式是复制 finally 中的代码块，并将内容插入到 try 和 catch 代码块中所有正常退出和异常退出之前。

有了这些基础知识后，对于 try-catch-finally 中存在 return 的情况就比较容易理解了，本节开头举例的代码对应的字节码如下。

```
...
14: iinc          0, 1
17: iload_0
18: istore_2
19: bipush        10
21: istore_0
22: iload_2
23: ireturn
...
```

以上指令是截取了 catch 部分的结果，int 型的值存储在局部变量表索引 0 处，14 ~ 18 是执行 i++ 操作，即将局部变量表索引 0 处的值 +1 后（此时这个位置的值为 2），重新加载局部变量表索引 0 处的值并保存在了局部变量表索引 2 处，然后执行 finally 中的代码，并将 int 型的值 10 保存在局部变量表索引 0 处，最后局部变量表索引 2 处的值被压入操作数栈中进行返回，这就是为什么其结果是 2。

从字节码中可以看出其执行流程是，在返回 i 之前，会先将 i 的值单独保存在局部变量表中新的位置，再执行 finally 中的代码块，最后将新位置的 i 值返回。假如 finally 中也存在 return 那又会怎样？我们知道在 catch 中语句执行 return 或者 throw 之前先执行 finally 中的代码，而且 finally 中的代码是被整体复制的，所以可以想象，finally 中的 return 会在 catch 中的 return 或者 throw 指令之前执行。看下面这段代码。

```
public static int foo() {
          int i = 1;
          try {
              i = i / 0;
              return i;
          } catch (Exception e) {
              throw e;
          } finally {
              return 10;
          }
}
```

输出结果如下。

```
10
```

上面这段代码是否会抛出异常呢？根据前面的分析，这段代码并不会抛出异常，因为在执行 throw 之前会先执行 finally 中的代码块，finally 中的 return 10 会执行方法返回，并不会执行 throw 部分的代码。

关于 try-catch-finally 部分的内容就讲到这里，大家可以思考下面这几个问题。

（1）如果将本节开始的例子中的 int i 的类型改成 reference 类型，那结果是什么样的呢？

（2）try-catch-resource 自动关闭资源这个语法糖又是如何实现的呢？

这两个问题，读者可以先写一个小 Demo，然后查看字节码来分析。

4.8 加载、链接、初始化

本节我们来探讨类加载和初始化方面的知识，为什么要了解这方面的知识呢？原因是：这部分知识本身就比较重要，特别是类初始化的流程，理解了这部分知识，对后面学习和理解 ASM 也很有帮助。

关于这部分知识，将以 Java 中的 static 为切入点来讲解。Java 的 static 关键字大家应该都不陌生，网上也有很多介绍 static 的文章，本节带大家从虚拟机类加载机制的角度详细认识一下 static。

在正式开始之前，先看看两个小例子。

案例一

思考下面的代码输出什么。

```
class Base{

      static{
          System.out.println("base static");
      }

      public Base(){
          System.out.println("base constructor");
      }
}

public class Main extends Base{

      static{
          System.out.println("main static");
      }

      public Main(){
          System.out.println("main constructor");
      }


      public static void main(String[] args) {
          new Main();
      }
}
```

结果如下。

```
// 输出结果
basic static
main satic
base constructor
main constructor
```

案例二

思考下面的代码又输出什么。

```
class Food{
      static{
          System.out.println("food static");
      }
      public Food(String str) {
```

```
            System.out.println(str +"'s food");
        }
}

public class Person {
        Food foo = new Food("person");
        static{
            System.out.println("person static");
        }

        public Person() {
            System.out.println("person constructor");
        }
}

class Student extends Person {
        Food bar = new Food("Student");

        static{
            System.out.println("student static");
        }

        public Student() {
            System.out.println("student constructor");
        }

        public static void main(String[] args) {
            new Student();
        }
}
```

结果如下。

```
//输出结果
person static
student static
food static
person's food
person constructor
student's food
student constructor
```

上面这两个案例是非常经典的 Java 笔试题，你是否能准确理清案例的输出结果呢？尤其是案例二，如果不了解类加载机制、字节码等相关知识，很可能会陷入思绪混乱的状态。本节将带大家一起从源头上剖析上面的案例，彻底搞明白这部分知识。

4.8.1 加载时机

一个类从被加载到虚拟机内存中开始，到从内存中卸载为止，它的生命周期如图 4-11 所示。

对于初始化阶段，《Java 虚拟机规范》严格规定有且只有遇到下面 6 种情况必须对类立即进行初始化（而加载、验证、准备自然需要在此之前开始）。

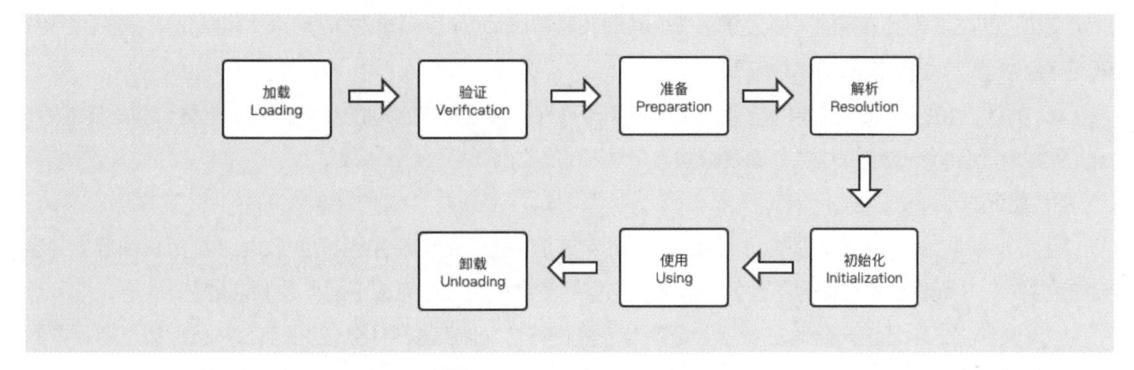

图 4-11　类加载的 7 个阶段

（1）遇到 new、getstatic、putstatic 或 invokestatic 这 4 条字节码指令时，如果类没有进行过初始化，则需要先触发其进行初始化。生成这 4 条指令最常见的 Java 代码场景如下。

- 使用 new 关键字实例化对象。
- 读取或者设置类的静态字段（被 final 修饰、已在编译器把结果放入常量池的静态字段除外）。
- 调用类的静态方法（类方法）。

（2）使用 java.lang.reflect 包的方法对类进行反射调用的时候，如果类没有进行过初始化，则需要先触发其进行初始化。

（3）当初始化类的时候，如果发现其父类还没有进行过初始化，则需要触发父类进行初始化。

（4）当虚拟机启动时，用户需要指定执行的主类（包含 main() 方法的类），虚拟机会先初始化这个类。

（5）当使用 JDK 7 新加入的动态语言（*invokedynamic*）时，如果 java.lang.invoke.MethodHandle 实例最后的解析结果为 REF_getStatic、REF_putStatic、REF_newInvokeSpecial、REF_invokeStatic 这 4 种类型的方法句柄，并且方法句柄对应类没有进行过初始化，则需要先触发其进行初始化。

（6）当一个接口中定义了 JDK 8 新加入的默认方法（被 default 关键字修饰的接口方法）时，如果这个接口的实现类发生了初始化，那么接口要在其之前被初始化。

4.8.2　加载过程

4.8.1 小节中提到了类从加载到内存以及从内存卸载的整个生命周期，并且介绍了类的 6 种加载时机，类的加载时机对我们理解 static 是很关键的。接下来简单介绍一下类加载的几个主要过程，其中类加载过程中的准备和初始化阶段都跟 static 有关，需要关注一下。

1. 加载

首先要说明的是加载（loading）只是类加载（class loading）过程的一个阶段，不要混淆了这两个概念。在加载阶段，虚拟机需要完成以下 3 件事。

（1）通过一个类的全限定名来获取定义此类的二进制字节流。

（2）将这个字节流所代表的静态存储结构转化为方法区的运行时数据结构。

（3）在 Java 堆中生成一个代表这个类的 java.lang.Class 对象，作为方法区这个类的各种数据的访问入口。

这里有如下几个注意点。

- 《Java 虚拟机规范》中对这 3 件事的要求其实并不是特别具体。例如：对第一件，开发人员可以通过定义自己的类加载器来完成（通过重写一个类加载器的loadClass()方法），可以实现从JAR、ZIP、WAR等压缩包中读取，也可以从网络中获取。

- 方法区是各个线程共享的内存区域，它用于存储已被虚拟机加载的类型信息、常量、静态变量等，因此加载阶段会将类的结构信息存储在这里。

2. 验证

验证（verification）阶段是验证ClassFile的字节流中包含的信息是否符合《Java虚拟机规范》的全部约束要求，从而保证这些信息不会危害虚拟机自身的安全。大致会完成下面4个验证动作。

（1）文件格式验证这一阶段主要是验证字节流是否符合.class文件格式的规范，并且能被当前版本的虚拟机处理。例如是否以魔数0xCAFEBABE开头、主次版本号是否在虚拟机介绍范围等。

（2）元数据验证这一阶段主要是对字节码描述的信息进行语义分析，以保证其描述的信息符合Java语言规范的要求。例如这个类是否有父类、这个类的父类是不是final类型的，以及如果这个类不是抽象类，那么是否实现了父类或接口中要求实现的方法等，总的来说就是验证语法是否正确。

（3）字节码验证这一阶段是整个验证阶段中最复杂的，主要工作是进行数据流和控制流分析。在第二阶段对元数据信息中的数据类型做完校验后，这一阶段将对类的方法体进行校验、分析。这一阶段的任务是保证被校验类的方法在运行时不会做出危害虚拟机安全的行为。

（4）符号引用验证主要是在虚拟机将符号引用转化为直接引用时进行校验，这个转化动作发生在解析阶段。符号引用可以看作对类自身以外（常量池的各种符号引用）的信息进行匹配性的校验。

验证阶段对虚拟机的类加载机制来说，是一个非常重要但不一定是必要的阶段。如果所运行的全部代码都已经被反复使用和验证过，在实施阶段就可以考虑使用-Xverify:none参数来关闭大部分的类验证措施，从而缩短虚拟机类加载的时间。

3. 准备

准备（preparation）阶段是正式为类变量（即static变量）分配内存并设置类变量初始值的阶段。例如下面定义的类中的变量。

```
public static int value = 1;
```

注意在初始化阶段结束后其值为0，而不是1，这是因为这个时候还未执行任何Java方法（静态变量会在<cinit>()方法中初始化）。那么有没有办法在准备阶段为value赋值呢？答案是有的，可以将其定义为常量类型。因为定义成常量后它的信息会被存储在字段属性表中的ConstantValue属性中。修改后代码如下：

```
public static final int value = 1;
```

那么在类中定义的成员变量是在什么时候进行赋值的呢？稍后会介绍。

4. 解析

解析（resolution）阶段是虚拟机将常量池内的符号引用替换为直接引用的过程。

5. 初始化

前面介绍了类初始化的6个时机，下面再进一步介绍初始化（initialization）阶段做了哪些事情。

在准备阶段，类变量已经赋值过一次系统要求的初始零值。而在初始化阶段，则会根据程序员通过程序编码制定的主观计划去初始化类变量和其他资源。简单来说，就是初始化阶段执行类构造器<cinit>()方法的过程。关于类构造方法和实例构造方法的说明如下。

- 类构造器<cinit>()方法：是由编译器自动收集类中的所有类变量的赋值动作和静态语句块

（static{}）中的语句合并产生的，编译器收集的顺序是由语句在源文件中出现的顺序决定的。在初始化阶段会调用一次，因此只加载一次。

- 实例构造 <init>()方法：是在实例创建出来的时候调用，包括调用new指令、调用Class或java.lang.reflect.Constructor对象的newInstance()方法、调用任何现有对象的clone()方法以及通过 java.io.ObjectInputStream类的getObject()方法反序列化。

注意：类构造器方法和实例构造方法不同，它不需要调用父类构造方法，Java 虚拟机会保证在子类的 <cinit>() 方法执行前，父类的 <cinit>() 方法已经执行完毕。

至此，我们了解了类加载的时机、类加载的过程，可以得出如下结论。

（1）在类的初始化时机中，当初始化一个类时，如果发现其父类还没有进行过初始化，则需要触发父类的初始化。

（2）在准备阶段，会给类变量赋零值，如果是常量，其值是保存在 ConstantValue 属性中的。

（3）在初始化阶段，JVM 会调用类构造器方法，并且 Java 虚拟机会保证在子类的类构造方法执行前，父类的类构造方法已经执行完毕，这一点也是我们理解 static 的关键。

（4）编译期会收集静态代码块，其顺序是按照在源码中定义的顺序决定的，并最终生成一个类构造器方法。

这些都是跟 static 相关的知识点，接下来我们再看一下 Java 成员变量在字节码中是如何初始化的。

4.8.3 字节码剖析

下面举一个例子来说明，代码如下。

```
public class Test {
      private static final int age = 99;// 类变量，常量
      private float pi;// 成员变量，未初始化
      private long la = 9999;// 成员变量，已初始化
      private final Date date = new Date();// 非 static 类型
      private static List<String> list = new ArrayList<>();//static 类型

      // 静态代码块
      static {
          System.out.println("static part one");
      }

      // 静态代码块
      static {
          System.out.println("static part two");
      }

      // 无参构造方法
      public Test(){
          System.out.println("test constructor");
      }

      // 构造方法
      public Test(String sr){

      }

      public static void main(String[] args) {
```

```
        new Test();
    }

    private static int name = 10;// 静态成员变量
    private int level = 100;// 非静态成员变量
}
```

上面的这些成员变量在字节码中是如何体现的呢？来看对应的字节码。

```
// 字节码信息
public class cn.curious.asm.Test
...
Constant pool:
     #1 = Methodref          #20.#47    // java/lang/Object."<init>":()V
     ...
    #68 = Utf8               println
{
    private static final int age;
      descriptor: I
      flags: ACC_PRIVATE, ACC_STATIC, ACC_FINAL
      ConstantValue: int 99 // 基本类型的常量会保存在 ConstantValue 属性中

    private float pi;
      descriptor: F
      flags: ACC_PRIVATE

    private long la;
      descriptor: J
      flags: ACC_PRIVATE

    private final java.util.Date date;
      descriptor: Ljava/util/Date;
      flags: ACC_PRIVATE, ACC_FINAL

    private static java.util.List<java.lang.String> list;
      descriptor: Ljava/util/List;
      flags: ACC_PRIVATE, ACC_STATIC
      Signature: #34      // Ljava/util/List<Ljava/lang/String;>;

    private static int name;
      descriptor: I
      flags: ACC_PRIVATE, ACC_STATIC

    private int level;
      descriptor: I
      flags: ACC_PRIVATE
public cn.curious.asm.Test();// 无参构造方法 (1)
      descriptor: ()V
      flags: ACC_PUBLIC
      Code:
        stack=3, locals=1, args_size=1

    0: aload_0
    1: invokespecial#1// Method java/lang/Object."<init>":()V// 隐式调用父类构造方法       (2)
    4: aload_0
    5: ldc2_w         #2      // long 99991
    8: putfield       #4      // Field la:J        (3)
```

```
    11: aload_0
    12: new             #5     // class java/util/Date
    15: dup
    16: invokespecial #6     // Method java/util/Date."<init>":()V
    19: putfield      #7     // Field date:Ljava/util/Date;        (4)
    22: aload_0
    23: bipush         100
    25: putfield      #8     // Field level:I// 初始化 level     (5)
    28: getstatic      #9     // Field java/lang/System.out:Ljava/io/PrintStream;
    31: ldc            #10    // String test constructor
    33: invokevirtual #11    // Method java/io/PrintStream.println:(Ljava/lang/String;)V
    36: return

public cn.curious.asm.Test(java.lang.String);
      descriptor: (Ljava/lang/String;)V
      flags: ACC_PUBLIC
      Code:
        stack=3, locals=2, args_size=2

     0: aload_0
     1: invokespecial #1  // Method java/lang/Object."<init>":()V
     4: aload_0
     5: ldc2_w          #2  // long 99991
     8: putfield       #4  // Field la:J
    11: aload_0
    12: new             #5  // class java/util/Date
    15: dup
    16: invokespecial #6   // Method java/util/Date."<init>":()V
    19: putfield       #7  // Field date:Ljava/util/Date;    /// 初始化 date
    22: aload_0
    23: bipush          100
    25: putfield       #8       // Field level:I
    28: return

public static void main(java.lang.String[]);
      descriptor: ([Ljava/lang/String;)V
      flags: ACC_PUBLIC, ACC_STATIC
      Code:
        stack=2, locals=1, args_size=1
           0: new             #12    // class cn/curious/asm/Test
           3: dup
           4: invokespecial #13    // Method "<init>":()V
           7: pop
           8: return

static {};//<clinit>类构造器方法 (6)
      descriptor: ()V
      flags: ACC_STATIC
      Code:
        stack=2, locals=0, args_size=0

    0: new             #14  // class java/util/ArrayList
    3: dup
    4: invokespecial #15  // Method java/util/ArrayList."<init>":()V
    7: putstatic      #16  // Field list:Ljava/util/List;// 初始化 list
  10: getstatic      #9   // Field java/lang/System.out:Ljava/io/PrintStream;
  13: ldc            #17  // String static part one // 静态代码块中代码
```

```
15: invokevirtual #11  // Method java/io/PrintStream.println:(Ljava/lang/String;)V
18: getstatic     #9   // Field java/lang/System.out:Ljava/io/PrintStream;
21: ldc           #18  // String static part two //静态代码块中代码
23: invokevirtual #11  // Method java/io/PrintStream.println:\
                     (Ljava/lang/String;)V
26: bipush        10
28: putstatic     #19  // Field name:I //写在类末尾的 name
31: return
}
```

观察字节码，可以得出如下信息。

（1）首先可以看到位置（1）的无参构造方法在开始的位置[即位置（2）]调用了父类中的无参构造方法，这个是编译器自动加上的，即在 Test 的构造方法中调用 super()。

（2）在调用了父类的无参构造方法后，可以看到非静态成员变量 la [位置（3）]、date [位置（4）]以及在源码尾部定义的 level [位置（5）] 依次进行了初始化。

（3）除了常量 age 外，其他 static 变量以及静态代码块按照在源码中定义的顺序会合并到类构造器方法中 [位置（6）]，类构造器方法会在类初始化阶段调用。

（4）类中定义的非 static 变量的初始化操作是在实例构造方法中进行的，跟定义顺序有关系，不过一定是在构造方法中的源码执行之前。

（5）实例构造方法会隐式调用父类的构造方法。

现在我们得到了如上的结论，再来分析一下前面的两个例子。直接看第二个例子，根据上面的结论，按照字节码中的代码执行顺序来修改代码。

```
class Food{
      static{
          System.out.println("food static");
      }
      public Food(String str) {
          super();
          System.out.println(str +"'s food");
      }
}

public class Person {
      Food foo;
      static{
          System.out.println("person static");
      }

      public Person() {
          super();
          foo = new Food("person");
          System.out.println("person constructor");
      }
}

class Student extends Person {
      Food bar;

      static{
          System.out.println("student static");
```

```
    }

    public Student() {
        super();
        bar = new Food("Student");
        System.out.println("student constructor");
    }

    public static void main(String[] args) {
        new Student();
    }
}
```

现在我们来分析上面代码执行的流程。

- 首先运行Student类中的main()方法。根据前面介绍的类加载时机，我们知道这个类有一个main()方法，是一个入口类，会立即初始化Student类。在类初始化阶段会先调用类构造器方法 <clinit>()，JVM会保证子类的 <clinit>()方法执行之前先执行完父类的 <clint>()方法，会先调用父类Person中的静态代码块。因此，Person直接执行静态代码块，输出person static。
- 上一步输出父类的static代码块后，会调用自己的静态代码块输出student static。至此，JVM初始化了Student和Person这两个类。
- 然后执行Student的main()方法。在此方法中出现了new，按照我们前面介绍的类加载时机，遇到new会立刻初始化该类。又因为Student类已经初始化过了，所以不需要再初始化，然后执行Student的构造方法。
- 在执行Student构造方法时，会隐式调用父类的构造方法。调用父类构造方法结束后会给非static成员变量进行初始化操作，按照这个顺序继续往下，我们先看调用父类构造方法。
- 在Student的构造方法中调用父类Person的构造方法，Person的构造方法中也会隐式调用父类的构造方法。因为它的父类是Object，所以没有内容输出。接着会给Person中的food变量进行初始化，当遇到new的时候，JVM马上初始化Food这个类。因此，会先调用Food的静态代码块，输出food static。
- 在Food类初始化完毕后，继续调用Food的构造方法，这个时候输出：person's food。接着调用Person构造方法中的其他代码，输出person constructor。
- Person的构造方法执行完毕后，又回到Student的构造方法继续执行。此时又遇到了new Food，因为Food类已经初始化过了，所以这里直接调用它的构造方法，输出Student's food。
- 最后，执行Student构造方法中其余代码，输出student constructor。

经过上述分析，就得到了例子中的最终结果。

4.9 字节码指令偏移

本节以及4.10节从本章的结构上来说相对独立，但是读者又必须了解这两节的内容，所以放在本章的尾部来介绍。

观察这段代码对应的字节码。

```
//Java 代码
public String foo(int a) {
     String str = String.valueOf(a);
```

```
        return str;
}

// 对应的字节码
public java.lang.String foo(int);
     Code:
       stack=1, locals=3, args_size=2
          0: iload_1
          1: invokestatic #2 //java/lang/String.valueOf:(I)Ljava/lang/String;
          4: astore_2
          5: aload_2
          6: areturn
```

可以看到 foo() 方法对应的字节码对应的序号是 0、1、4、5、6，这些序号并不是连续的，这些序号实际上表示的是 Code 属性中每条指令出现在第几个字节中，示例如下。

- 0: iload_1：表示iload_1出现的位置是第一个字节。
- 1: invokestatic：因为操作码iload_1没有操作数，所以第二个字节就是invokestatic指令。
- 4: astore_2：因为invokestatic指令的操作数会占两个字节，所以astore_2在第四个字节，按照索引从0开始计数，对应的就是4。

这就是字节码前面序号产生的原因，我们称为字节码指令偏移。

4.10 Java虚拟机中的数据类型

Java 虚拟机进行计算时需要依靠特定的数据类型，这些数据类型与 Java 语言中的数据类型有些类似，JVM 中的数据类型如图 4-12 所示。

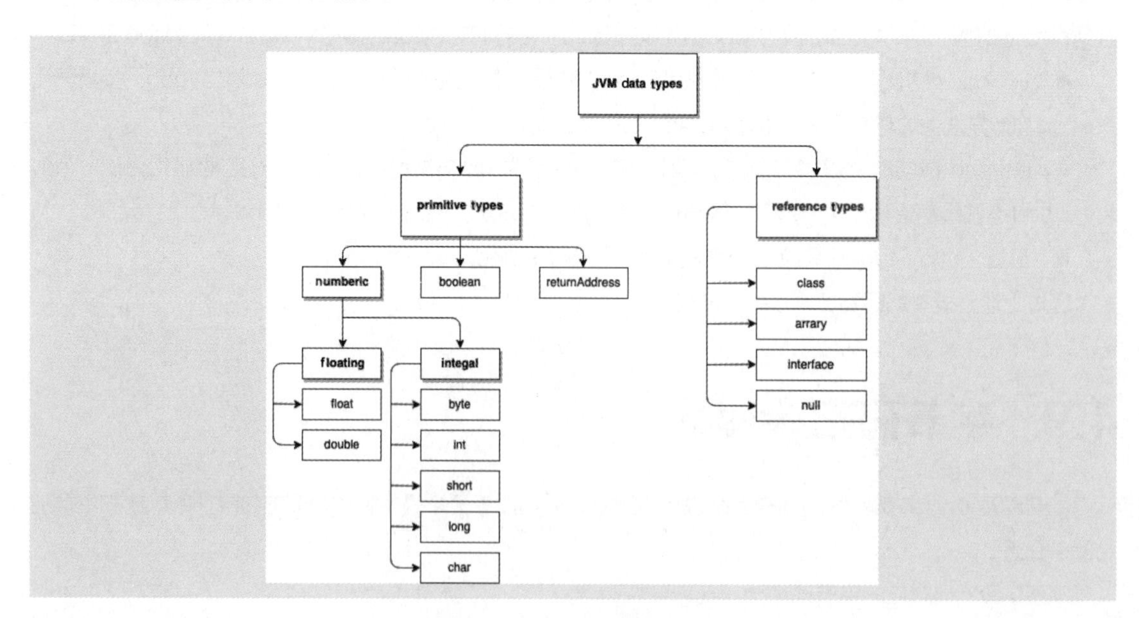

图 4-12　JVM 中的数据类型

从图 4-12 中可以看出 JVM 中的数据类型分为：primitive types（基本数据类型）和 reference types

（引用数据类型）。基本数据类型用于存储值，而引用数据类型用于指向对象的引用。

4.10.1 基本数据类型

JVM 中的基本数据类型包括 numberic（数字类型）、boolean（布尔类型）和 returnAddress 类型，其中 numberic 包括 floating（浮点型）和 integal（整数类型）。浮点型具体包括 float 和 double、整数类型包括 byte、int、short、long、char。可以看出 JVM 中的基本数据类型与 Java 语言中的基本数据类型很相似，只有细微的差别，具体如下。

- 虽然JVM中提供了boolean类型，但是对它的支持是很有限的。JVM中没有专门处理boolean类型的指令，相反，JVM会将Java语言中的boolean类型转换成虚拟机中int类型。JVM中的boolean类型中的false使用0来替代，true使用非0来替代。
- returnAddress类型是JVM中的特有类型，它与指令jsr、jsr_w、ret一起用于实现Java中的finally语句块。关于jsr、jsr_w、ret指令，从ClassFile版本50，即JDK 7开始就禁用了这些指令，而且javac编译器从JDK 1.4.2开始就不生成这几个指令了，所以在现实世界中读者基本上不会遇到这几个指令。

JVM 中有很多与 returnAddress 相关的内容，为方便读者理解，这里用一个具体例子来介绍。

```
//Java 代码
static int foo(boolean value) {
      try {
          if (value) {
              return 1;
          }
          return 0;
      } finally {
          System.out.println("Hello");
      }
}

// 对应的字节码
   0 iload_0
   1 ifeq 11
   4 iconst_1
   5 istore_3
   6 jsr 24
   9 iload_3
10 ireturn
11 iconst_0
12 istore_3
13 jsr 24
16 iload_3
17 ireturn
18 astore_1
19 jsr 24
22 aload_1
23 athrow
24 astore_2
25 getstatic #8
28 ldc #1
30 invokevirtual #7
33 ret 2
```

上述 Java 代码的 foo() 方法中定义了一个 try-finally 语句块，假如 foo() 方法的参数 value = true，分析其中部分字节码内容。

- 0 iload_0：表示将局部变量表索引0处的整型值加载到操作数栈中，foo() 方法是静态方法，所以局部变量表索引0处的值就是参数value，此时其值为1。
- 1 ifeq 11：将操作数栈中的值与0做比较，如果等于0就跳转到字节码指令偏移位置11处继续执行，因为栈中值为1，所以继续顺序执行。
- 4 iconst_1：表示将常量1压入栈顶中。
- 5 istore_3：表示将操作数栈顶的常量1存储到局部变量表索引3处。在4.7.4小节介绍try-catch-finally原理时提到，在return之前，会将return的值保存在新的位置。
- 6 jsr 24：jsr指令的格式是jsr branchbyte1 branchbyte2，其中（branchbyte1<<8）| branchbyte2为具体指令偏移量。注意此处是偏移量，不是指具体的位置。此处jsr指令的实际内容是A8 00 12，A8是jsr操作码对应的十六进制值，0x12对应的十进制值是18，所以此处的24由当前字节码的位置6 + 18得出。类似goto指令跳转到指定的字节码指令偏移位置，不过jsr指令是跳转到指定的字节码指令偏移位置处（本例中是24），并且将下一条指令的地址（本例对应的指令是9 iload_3）压入操作数栈顶中，栈顶中的地址就是returnAddress类型。
- 24 astore_2：将操作数栈顶的returnAddress类型的值保存在局部变量表索引2处。
- 25～30：执行System.out.println(“hello”) 代码。
- 33 ret 2：表示获取到局部变量表索引2处的returnAddress类型的值，并跳转到其值的位置，即9 iload_3处。
- 9 iload_3：加载局部变量表索引3处的值到操作数栈的栈顶，本例中是1。
- 10 ireturn：返回操作数栈顶的值。

相信读者现在对 returnAddress 类型应该有所了解了。

4.10.2 引用数据类型

JVM 中的引用数据共有 4 种：class（类类型）、array（数组类型）、interface（接口类型）、null。其中 class 的值为指向类实例的引用。array 的值为指向数组的引用。interface 的值为指向该接口实现类的引用，例如类 B 实现了接口 A，那接口 A 的引用数据类型为 A 的实现类 B 的实例的引用。null 是一个特殊的引用类型，它不指向任何引用。

4.11 小结

本章为读者详细地介绍了字节码指令以及相关用法，并通过一些例子介绍了 Java 中常见的语法在字节码世界中是如何展现的。读者也可以发挥奇思妙想，使用字节码来生成一些 Java 代码中无法实现的逻辑，甚至可以创造一些语法。

5. ASM基础

本书名为《ASM 全埋点开发实战》，直至本章才正式开始介绍 ASM 相关的内容，其原因是前面章节是为后续章节做铺垫的：本书关于 Gradle 的两章内容是为将 ASM 应用在 Android 中打下基础，关于字节码的两章内容则是为学习 ASM 打下基础。ASM 与字节码是强相关的，ASM API 中的很多方法和参数的命名都是基于字节码中的知识，所以有了字节码相关的知识再来学习 ASM 就会容易很多。

5.1 ASM简介

在学习了 ClassFile 结构和字节码相关的知识后，应该如何根据这些知识去修改 ClassFile 呢？比如添加一个方法，读者会发现即使对 ClassFile 的结构已经很熟悉，但这依然是非常难做到的事情，更不要说还需要手动计算 max_stack、max_locals 以及维护常量池等内容。这时候就需要一些第三方工具库帮我们简化这些操作，ASM 库就是一个通用的 Java 字节码操作和分析框架，使用它可以很容易地实现程序分析、代码生成和转换等功能。

ASM 项目最初由 Eric Bruneton 于 2000 年在 INRIA（法国国家信息与自动化研究所）攻读博士学位期间创建。在当时包含 java.lang.reflect.Proxy 包的 JDK 1.3 还没有发布，ASM 被用作代码生成器来生成动态代理类。当时的 ASM 还没有 ClassReader 和 Vistor 的概念，这两种概念是后来为了能够对已有的 ClassFile 进行修改而添加的。ASM 于 2002 年开源，自此之后 ASM 得到快速的发展并且被用在非常多的框架中，例如 Java 8 Lambda 表达式、Gradle、Fastjson、Kotlin、Groovy、MyBatis 等。注意，“ASM”并不是一些单词的简写，它取之于 C 语言中的 __asm 关键字。此关键字的作用是允许程序可以使用汇编语言来实现，其实通过对比可以发现字节码指令与汇编语言指令很相似。

ASM 并不是唯一的操作字节码的库，类似的比较出名的有 Javassist、Byte Buddy 等。与这些库比较，ASM 具有如下优点。

- 架构设计精巧,API设计良好。
- 体积小、速度快、性能好，而且非常稳定。
- 更新及时，支持最新的Java版本。

当然缺点是学习难度比较大，不过不用担心，这正是本书存在的意义。

5.2 ASM组成

ASM 库提供了两套用于生成和转换 ClassFile 的 API：一套是基于事件模型的 Core API，一套是基于对象的 Tree API。ASM 的 Core API 针对 ClassFile 的结构定义了一系列的事件回调方法，而 Tree API 则构建了包括整个 ClassFile 内容的对象，不过 Tree API 是基于 Core API 构建的。当然对于这两种模型，ASM 也相应地提供了 API 用于在这两种方式之间灵活地切换。

这两种方式的区别和优缺点，完全可以对比 XML 的两种解析方式：SAX（Simple API for XML，XML 简单应用程序接口）和 DOM（Document Object Model，文档对象类型）。SAX 方式按照 XML 文档的顺序读取，遇到 XML 节点会触发事件回调。SAX 自顶向下解析，边扫描边解析，因此这种方式的优点是内存占用小、速度快，缺点是无法对已经解析过的内容进行修改。DOM 方式是将 XML 内容整体加载到内存中，构建整个 XML 文档的对象树，所以其缺点是内存占用大、速度较慢，优点是可以随时对整个 XML 文档内容做修改。

整个 ASM 库由如下部分组成。

- asm.jar：Core API的核心包，该包中包含基于事件的API以及分析和编辑ClassFile的API，它们分别定义在org.objectweb.asm和org.objectweb.asm.signature这两个包中。
- asm-util.jar：提供了一些基于Core API的工具类，这些工具类可以用于调试ASM程序，这部分的API定义在org.objectweb.asm.util包中。
- asm-commons.jar：提供了几个有用的transformer，同样是基于Core API，这部分的API定义在org.objectweb.asm.commons包中。
- asm-tree.jar：Tree API所在的包，并且提供了与Core API相互转换的工具，这部分的API定义在org.objectweb.asm.tree包中。
- asm-analysis.jar：提供了class分析框架并且预置了一些class分析器工具，这部分的API定义在org.objectweb.asm.tree.analysis包中。

ASM 的包名都是以 org.objectweb 开头的，所以其官网网址中 ow 的意思就是 objectweb。读者在搜索 ASM 官网的时候最好加上 objectweb 关键字，否则在 Google 中搜索 asm 关键字时，显示结果的前几项并不是我们要找的。另外，ASM 的核心库是 asm.jar，其他库都是基于这个包的，本书关于 ASM 的介绍也会分成 Core API 和 Tree API 两部分，首先来看看 Core API。

ASM Core API 中最重要的 3 个类是 ClassReader、ClassVisitor、ClassWriter。

- ClassReader用于加载和解析ClassFile中的内容，其accept()方法接收ClassVisitor参数，ClassReader解析的结果会调用ClassVisitor中的visit*()方法，该类可以看作事件的生产者。
- ClassVisitor是一个抽象类，是ASM中生成和转换类的基础，该类中的每个visit*()方法都代表着ClassFile中的一个结构，它是事件的消费者。
- ClassWriter是ClassVisitor的一个具体的实现，它的toByteArray()方法用于输出ClassFile的内容。

接下来详细介绍这 3 个类。

5.3 ClassReader API介绍

顾名思义，ClassReader 用于加载、解析 ClassFile 的内容，内容包括包名、类名、访问标识符、继承、接口实现等。

5.3.1 构造方法

ClassReader 提供了 4 个构造方法用于初始化，具体如表 5-1 所示。

表5-1　ClassReader的构造方法

构造方法	说明
ClassReader(byte[] classFile)	将ClassFile对应的字节数组内容作为参数构建ClassReader对象
ClassReader (byte[] classFileBuffer, int classFileOffset, int classFileLength)	从包含ClassFile内容的字节数组内容的指定classFileOffset偏移处开始，取出classFileLength个字节作为参数构建ClassReader对象
ClassReader(inputStream inputStream)	将包含ClassFile内容的输入流作为参数传递给此构造方法
ClassReader(String className)	根据类名来构建ClassReader对象。其原理是通过ClassLoader.getSystemResourceAsStream(String)方法来获取className所对应类的数据，此处的className必须是全限定名，例如java/lang/Runnable。注意这种方式必须保证类在classpath中能够找到

例如，使用ClassReader输出一个类的基本信息，示例代码如下。

```
public static void main(String[] args) {
      try {
          // 初始化 ClassReader，通过全限定名来加载类
          ClassReader classReader = new ClassReader("java/lang/String");
          // 获取类名信息
          String className = classReader.getClassName();
          // 获取接口信息
          String[] interfaces = classReader.getInterfaces();
          // 获取父类信息
          String superName = classReader.getSuperName();
          // 获取访问标识符信息
          int access = classReader.getAccess();

          // 定义一个 StringBuilder，用于拼装类的基本信息
          StringBuilder builder = new StringBuilder();
          // 此处简单地判断是否是用 public 修饰的
          if((access & Opcodes.ACC_PUBLIC) != 0){
              builder.append("public").append(" ");
          }
          builder
              .append(className)
              .append(" ")
              .append("extends").append(" ").append(superName);
          if(interfaces != null && interfaces.length != 0){
              builder.append(" ")
                      .append("implements")
                      .append(" ")
                      .append(Arrays.toString(interfaces)
                              .replace("[","")
                              .replace("]", ""))
                      .append(" ");
          }
          // 输出结果
          System.out.println(builder.toString());
      } catch (IOException e) {
          e.printStackTrace();
      }
}
```

上述代码的输出结果如下。

```
public java/lang/String extends java/lang/Object implements \
     java/io/Serializable, java/lang/Comparable, java/lang/CharSequence
```

通过 ClassReader 只能获取一个类的基本信息，而无法获取到类中的字段、方法等内容。若想获得更多内容需要借助 ClassReader 的 accept() 方法。

5.3.2 accept()方法

accept() 方法的定义如下。

```
public void accept(ClassVisitor classVisitor, int parsingOptions)
```

accept() 方法有两个参数，其中 classVisitor 是事件消费者，ClassReader 将解析的结果回调给 classVisitor 消费；parsingOptions 的作用是配置 ClassReader 如何解析 ClassFile，可选值如下。

- ClassReader.SKIP_CODE：如果设置此值，将不解析也不访问Code属性中的内容。
- ClassReader.SKIP_DEBUG：如果设置此值，将不解析也不访问ClassFile中的调试信息（如SourceFile、LineNumberTable、LocalVariableTable等属性）。
- ClassReader.SKIP_FRAMES：如果设置此值，将跳过StackMapTable属性信息。如果跳过StackMapTable属性信息，输出结果没有该属性，将可能导致校验错误，此时需要配合ClassWriter.COMPUTE_FRAMES配置（该配置将在5.5节介绍），让ASM重新生成StackMapTable属性信息。
- ClassReader.EXPAND_FRAMES：设置计算栈映射帧的方式。在3.4.10小节介绍StackMapTable时提到，StackMapTable是JDK 7新添加的属性，用于加快字节码校验的速度，我们称这种方式为“压缩”方式；而JDK 7之前是根据类型来推断的，我们称这种方式为“扩展”方式。默认情况下，ASM将根据ClassFile的版本来选择其默认支持的计算方式，如果设置为EXPAND_FRAMES，ASM将按照JDK 1.6以及JDK 1.6之前的“扩展”方式来计算。注意，使用此配置时，ASM会进行解压和压缩操作，因此会降低性能。

接下来实现输出 ClassFile 更多信息的例子，示例如下。

```
public static class ParsingVisitor extends ClassVisitor {
        public ParsingVisitor() {
            super(Opcodes.ASM6);
        }

        @Override
        public void visit(
            int version, int access, String name, String signature,
            String superName, String[] interfaces)
        {
            super.visit(version, access, name, signature,
                     superName, interfaces);
            System.out.println(
                "class " + name + " extends " + superName + " {");
        }

        @Override
        public FieldVisitor visitField(
            int access, String name, String descriptor,
```

```
                String signature, Object value)
        {
            System.out.println("    " + name + ", " + descriptor);
            return super.visitField(
                access, name, descriptor, signature, value);
        }

        @Override
        public MethodVisitor visitMethod(
            int access, String name, String descriptor, String signature,
            String[] exceptions)
        {
            System.out.println("    " + name + ", " + descriptor);
            return super.visitMethod(
                access, name, descriptor, signature, exceptions);
        }

        @Override
        public void visitEnd() {
            super.visitEnd();
            System.out.println("}");
        }
}
```

上述定义的 ParsingVisitor 类实现了 ClassVisitor 类，ClassVisitor 类中定义了一系列的 visit() 方法，这些方法会在解析 ClassFile 时触发。关于 ClassVisitor 会在 5.4 节详细介绍。同时为方便后续内容的演示，这里再定义一个 SensorsData 类，其内容如下。

```
package cn.sensorsdata.asm;

public class SensorsData {
      private String address;
      private static final int age = 6;

      @Deprecated
      public String getInfo(){
          String info = "Address: "+ address +" \nAge: "+ age;
          return info;
      }
}
```

最后将 ClassReader 和 ParsingVisitor 结合起来，代码如下。

```
public static void parsingClass() throws IOException {
      String filePath = "src/main/java/cn/sensorsdata/asm/";
      File file = new File(filePath + "SensorsData.class");
      // 读取 ClassFile 中的内容
      byte[] bytes = FileUtils.readFileToByteArray(file);(1)
      // 创建 ClassReader 对象
      ClassReader classReader = new ClassReader(bytes);(2)
      // 此处 accept() 的 parsingOptions 的值为 0，表示不对 ClassFile 做任何处理
      classReader.accept(new ParsingVisitor(), 0);(3)
  }
```

上述 parsingClass() 方法中首先获取 SensorsData.class 的 byte[] 数据 [位置（1）]，并将其作为参数构建 ClassReader 对象 [位置（2）]，最后调用 accept() 方法，将 ParsingVisitor 对象整合在一起 [位

置（3）]。执行上述方法后，输出的结果如下。

```
class cn/sensorsdata/asm/SensorsData extends java/lang/Object {
    address, Ljava/lang/String;
    age, I
    <init>, ()V
    getInfo, ()Ljava/lang/String;
}
```

可以看到输出了 SensorsData 类中的字段名和方法等信息。假如又想要获取 SensorsData 中 getInfo() 方法的注解信息、访问标识符、局部变量表和操作数栈的最大值信息、字段 age 的值等，应该如何做呢？这就需要详细地介绍 ClassVisitor。

5.4 ClassVisitor API 介绍

ClassVisitor 是事件的消费者，它的 API 如下。

```
public abstract class ClassVisitor {

    public ClassVisitor(final int api) {
      this(api, null);
    }

    public ClassVisitor(final int api, final ClassVisitor classVisitor)

    public void visit(
        final int version,
        final int access,
        final String name,
        final String signature,
        final String superName,
        final String[] interfaces)

    public void visitSource(final String source, final String debug)

    public ModuleVisitor visitModule(
        final String name, final int access, final String version)

    public void visitNestHost(final String nestHost)

    public void visitOuterClass(
        final String owner, final String name, final String descriptor)

    public AnnotationVisitor visitAnnotation(
        final String descriptor, final boolean visible)

    public AnnotationVisitor visitTypeAnnotation(
        final int typeRef, final TypePath typePath,
        final String descriptor, final boolean visible)

    public void visitAttribute(final Attribute attribute)

    public void visitNestMember(final String nestMember)
```

```
    public void visitPermittedSubclass(final String permittedSubclass)

    public void visitInnerClass

    public RecordComponentVisitor visitRecordComponent(
        final String name, final String descriptor, final String signature)

    public FieldVisitor visitField(
        final int access,
        final String name,
        final String descriptor,
        final String signature,
        final Object value)

    public MethodVisitor visitMethod(
        final int access,
        final String name,
        final String descriptor,
        final String signature,
        final String[] exceptions)

    public void visitEnd()
}
```

以上除去构造方法共有 15 个方法，这些方法在使用时按照如下顺序调用。

```
visit
[ visitSource ] [ visitModule ][ visitNestHost ][ visitOuterClass ]
( visitAnnotation | visitTypeAnnotation | visitAttribute )*
( visitNestMember | [ * visitPermittedSubclass ] | visitInnerClass | visitRecord
Component | visitField | visitMethod )*
visitEnd.
```

其中“[ ]”部分表示可选，例如当 ClassFile 中包含源码文件信息时就会调用 visitSource()，没有就不调用；“*”部分表示这些方法会执行 0 次或多次，例如类中定义了多个方法就会多次调用 visitMethod()；“|”部分表示并列关系，没有先后顺序，例如遇到字段时就调用 visitField()，遇到方法时就调用 visitMethod()，解析时遇到哪个就回调对应的方法。从这里的执行顺序可以看出，visit() 方法将会最先执行，visitEnd() 方法会在最后执行。接下来详细介绍其中的每一个方法。

5.4.1 ClassVisitor()构造方法

这里主要看下面这个构造方法。

```
public ClassVisitor(final int api, final ClassVisitor classVisitor)
```

其中 api 的取值必须是 Opcodes.ASM4、Opcodes.ASM5、Opcodes.ASM6、Opcodes.ASM7、Opcodes.ASM8 或 Opcodes.ASM9 中的一个。这里 api 的版本与 JDK 版本没有对应关系，例如本地使用 JDK 8，api 值不需要设置成 Opcodes.ASM8。api 的作用是表示 ClassVisitor 的实现版本，简单地说 ASM 中有一些 API 是高版本中才有的，如果 api 设置的值比较小，那么这些高版本中才有的 api 将不会被触发。关于 api 为什么要这么设计以及 ASM 是如何处理兼容性的，将会在 6.7.2 小节展开探讨。

classVisitor 参数表示当前 ClassVisitor 处理后继续向下委托处理的对象，可以设置为 null。ASM 框架采用 visitor 模式设计，具体将在 6.7.4 小节介绍 ASM 框架的原理。

5.4.2 visit()

visit() 方法用于访问 ClassFile 的一些基本信息，其定义如下。

```
public void visit(
      int version, int access, java.lang.String name,
      java.lang.String signature, java.lang.String superName,
      java.lang.String[] interfaces)
```

说明如下。

- version 表示 ClassFile 的版本信息，例如 Java 8 对应的 version 为 52。
- access 表示 ClassFile 的访问标识信息，例如接口的访问标识通常是 Opcodes.ACC_PUBLIC + Opcodes.ACC_ABSTRACT，这里的 ACC 就是 access 的简写。
- name 表示类的全限定名，例如 Runnable 接口的值为 java/lang/Runnable。
- signature 表示类的泛型签名信息，关于泛型格式，可以参考 3.4.10 小节介绍的 Signature 属性，如果没有就设置为 null。
- superName 表示当前类的父类名，在 Java 中所有的类都有父类，如果没有明确指定，则父类默认都是 java/lang/Object。
- interfaces 表示当前类实现的接口，一个类可以实现多个接口，所以这里是字符串数组，如果没有实现接口就设置为 null。

5.4.3 visitSource()

该方法用于访问 ClassFile 中与源码相关的属性信息，其定义如下。

```
public void visitSource(java.lang.String source, java.lang.String debug)
```

其中，source 参数对应的是 ClassFile 中 SourceFile 属性的值，用于表示当前 ClassFile 对应的源文件。例如，使用 javap 工具查看其结果。

```
Classfile .../src/main/java/cn/sensorsdata/asm/SensorsData.class
    Last modified 2021-8-12; size 605 bytes
    MD5 checksum 5893c69678d1e09e3e23e4b85a8bc389
    Compiled from "SensorsData.java"
```

debug 参数对应的是 ClassFile 中 SourceDebugExtension 属性的值，该属性是 JDK 1.6 中新增的，用于存储额外的调试信息。譬如 JSP 文件调试，是无法通过 Java 堆栈来定位到 JSP 文件的行号的。为解决此类问题，JSR-45 规定了那些用非 Java 语言编写，却需要编译成字节码并运行在 Java 虚拟机中的程序能够进行调试的标准，SourceDebugExtension 属性就是为实现这个标准而引入的。

5.4.4 visitModule()

该方法用于访问 ClassFile 中 Java 9 新添的模块化信息，其定义如下。

```
public ModuleVisitor visitModule(
      java.lang.String name, int access, java.lang.String version)
```

注意此方法的返回结果是一个 ModuleVisitor 对象。模块化是 Java 9 的一个重要特性，本书在此不多做介绍，如果想用此方法，则 ClassVisitor 构造参数中 api 的值不能低于 Opcodes.ASM6。

5.4.5 visitNestHost()

该方法用于访问 ClassFile 的宿主类，其定义如下。

```
public void visitNestHost(java.lang.String nestHost)
```

其中 nestHost 表示宿主类的类名，为了方便对该内容以及后面相关 API 进行介绍，我们先来了解如下几个概念。

- OuterClass。
- InnerClass。
- NestedClass。
- LocalClass。
- EnclosingClass。
- AnonymousClass。
- NestHost。
- NestMate。
- NestMember。

Java 中允许将一个类定义在另一个类的内部，例如下面的用法。

```
package cn.sensorsdata.asm.nested;

public class OuterClass {

      class InnerClass{

      }

      static class StaticNestedClass{

      }
}
```

上述的 OuterClass 类中定义了两个内部类 InnerClass 和 StaticNestedClass，根据内部类是否使用 static 修饰，又分为内部类（inner class）和嵌套类（nested class）。如果内部类使用 static 关键字修饰，那该类就被称为嵌套类（关于这两种内部类使用上的区别，本书不做介绍，此处旨在搞清楚这些称呼实际表示什么）。

Java 中还允许将类定义在方法中，例如下面的用法。

```
public class LocalTestClass {

      public void foo(){
          class LocalClass{
```

```
            }
        }
}
```

将定义在方法中的类称为 Local Class。当然还有另外一种定义类的方式，不需要声明类的名称，称为匿名类（anonymous class），其用法如下。

```
Runnable runnable = new Runnable() {
        @Override
        public void run() {

        }
};
```

上面代码中定义了一个 Runnable 接口的匿名类。其实不管一个类中定义了多少个内部类，一个 ClassFile 总是只代表一个类。例如我们编译 OuterClass 后会产生 3 个 ClassFile，如图 5-1 所示。

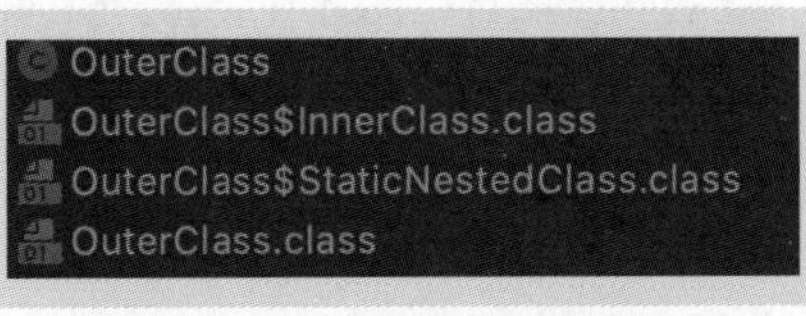

```
OuterClass
OuterClass$InnerClass.class
OuterClass$StaticNestedClass.class
OuterClass.class
```

图 5-1　OuterClass 编译后的结果

反编译 OuterClass$InnerClass 后显示的结果如图 5-2 所示。

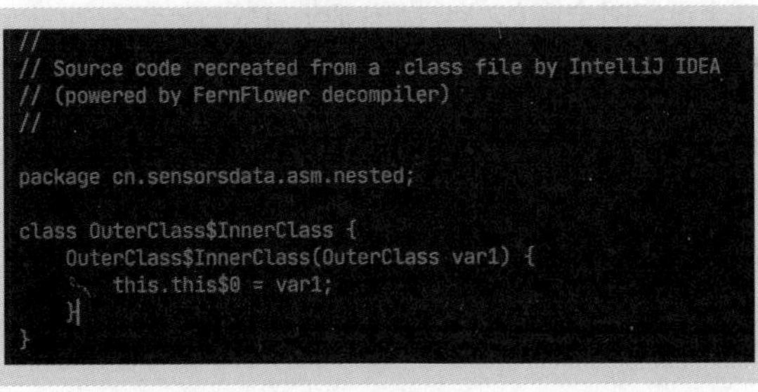

```
//
// Source code recreated from a .class file by IntelliJ IDEA
// (powered by FernFlower decompiler)
//

package cn.sensorsdata.asm.nested;

class OuterClass$InnerClass {
    OuterClass$InnerClass(OuterClass var1) {
        this.this$0 = var1;
    }
}
```

图 5-2　反编译 OuterClass$InnerClass 后显示的结果

可以看到 OuterClass$InnerClass 中有了外部类 OuterClass 的引用，这就是内部类 InnerClass 可以访问外部类私有变量的原因。

那么什么是 Enclosing Class 呢，这个不太好翻译，举例说明如下。

```
public class EnclosingTest {
        class InnerClass {
            class InnerA{
            }
        }
}
```

上述例子中 InnerA 的 Enclosing Class 是 InnerClass，InnerClass 的 Enclosing Class 是 EnclosingTest。

那 NestHost 和 NestMember 又是什么呢？再来看看下面这个例子。

```
public class NestHostTest {

    static class A {
        private int flag = 0;
    }

    static class B {
        public void foo() throws Exception {
            A a = new A();
            a.flag = 1;
            System.out.println(a.flag);// (1)

            Field f = A.class.getDeclaredField("flag");
            f.setInt(a, 2);
            System.out.println(a.flag);// (2)
        }
    }

    public static void main(String[] args) throws Exception {
        B b = new B();
        b.foo();
    }
}
```

上述 NestHostTest 中定义了两个类型为 Nested Class 的 A 和 B，在 B 的 foo() 方法中使用了 A 类，并且给 A 中的 private 属性 flag 设置了值，这种用法是没有问题的，所以位置（1）可以正常输出。接着通过反射的方式来给 A 对象的 flag 设置值，在位置（2）输出结果，理论上位置（1）可以，位置（2）也不会有问题，可实际在 Java 8 上运行时位置（2）会抛出 IllegalAccessException 异常，这显然是不合理的。因此 Java 11 对这个问题进行了修正，在 Java 11 上运行不会报错。

这就是 Nest、Nest Host 在 Java 11 中的定义，Java 11 还提供了 3 个方法用于输出这些值，它们在 Class 类中的定义如下。

```
public Class<?> getNestHost()
public Class<?>[] getNestMembers()
public boolean isNestmateOf(Class<?> c)
```

举例如下。

```
public class NestHostTest {

    static class A {
    }

    static class B {
    }

    public static void main(String[] args) throws Exception {
        // 输出 Class A、B 和 NestHostTest 的 Nest Host
        System.out.println("====Nest Host====");
        System.out.println("A class's nest host: " + A.class.getNestHost());
        System.out.println("B class's nest host: " + B.class.getNestHost());
        System.out.println("NestHostTest class's nest host: "
                          + NestHostTest.class.getNestHost());

        // 输出 Class A 的所有 Nest Members
```

```
            System.out.println("\n====A Class Nest Members====");
            Arrays.stream(A.class.getNestMembers()).forEach(System.out::println);

    //输出Class B的所有Nest Members
            System.out.println("\n====B Class Nest Members====");
            Arrays.stream(B.class.getNestMembers()).forEach(System.out::println);

    //输出Class NestHostTest的所有Nest Members
            System.out.println("\n====NestHostTest Class Nest Members====");
            Arrays.stream(NestHostTest.class.getNestMembers())
                .forEach(System.out::println);
        }
    }
```

输出结果如下。

```
====Nest Host====
A class's nest host: class cn.sensorsdata.asm.NestHostTest
B class's nest host: class cn.sensorsdata.asm.NestHostTest
NestHostTest class's nest host: class cn.sensorsdata.asm.NestHostTest

====A Class Nest Members====
class cn.sensorsdata.asm.NestHostTest
class cn.sensorsdata.asm.NestHostTest$B
class cn.sensorsdata.asm.NestHostTest$A

====B Class Nest Members====
class cn.sensorsdata.asm.NestHostTest
class cn.sensorsdata.asm.NestHostTest$B
class cn.sensorsdata.asm.NestHostTest$A

====NestHostTest Class Nest Members====
class cn.sensorsdata.asm.NestHostTest
class cn.sensorsdata.asm.NestHostTest$B
class cn.sensorsdata.asm.NestHostTest$A
```

上述 NestHostTest 类中定义了 Nested Class A 和 B，首先输出 Class A、B 和 NestHostTest 的 Nest Host 结果，可以看出它们的 Nest Host 都是指 NestHostTest Class。也就是说只要是 NestHostTest 中定义的类，它们的 Nest Host 都是 NestHostTest Class。

接着输出 Class A、B 和 NestHostTest 的 Nest Member 或者叫 Nestmate，从结果中可以看到它们的 Nest Members 都是一样的，这一点结合前面的翻译内容就很容易理解。

所以使用 Java 11 编译 NestHostTest 类，对于 Class NestHostTest$A，visitNestHost() 的值就是 cn/sensorsdata/asm/NestHostTest。

5.4.6 visitNestMember()

该方法用于访问 ClassFile 的 nestMember，方法定义如下。

```
public void visitNestMember(String nestMember)
```

根据 5.4.5 小节介绍的内容，对于 Class NestHostTest，visitNestMember 的值为 cn/sensorsdata/asm/NestHostTest$A 和 cn/sensorsdata/asm/NestHostTest$B。

5.4.7 visitInnerClass()

该方法用于访问 ClassFile 中内部类的相关信息，内部类的信息存储在 InnerClasses 属性中，如果有多个内部类会多次调用此方法，方法定义如下。

```
public void visitInnerClass(
      java.lang.String name, java.lang.String outerName,
      java.lang.String innerName, int access)
```

以前面的 OuterClass.class 类为例来解释 visitInnerClass() 方法中的各参数。

- name表示内部类的全限定名，例如cn/sensorsdata/asm/OuterClass$InnerClass。
- outerName为外部类名，例如cn/sensorsdata/asm/OuterClass。
- innerName为内部类名，例如InnerClass，如果是匿名类则为null。
- access为内部类的访问标识符。

5.4.8 visitOuterClass()

该方法的定义如下。

```
public void visitOuterClass(
      final String owner, final String name, final String descriptor)
```

该方法的名称非常具有误导性，它并非是获取内部类的外部类信息，它实际上是用来访问 LocalClass 外的方法的，以下面代码为例。

```
public class OuterClass {
      void someMethod(String s) {
          class InnerClass {
          }
      }

      static class StaticNestedClass{

      }
}
```

代码编译后，其中的 InnerClass 被编译成“OuterClass$1InnerClass”。对该类进行解析，visitOuterClass() 方法中各参数的值的结果如下。

- owner的结果为“OuterClass”。
- name的结果为“someMethod”。
- descripter的结果为“(Ljava/lang/String;)V”。

5.4.9 visitField()

该方法用于访问 ClassFile 中的字段信息，方法定义如下。

```
public FieldVisitor visitField(
      int access, java.lang.String name, java.lang.String descriptor,
```

```
        java.lang.String signature, java.lang.Object value)
```

我们以 private static final int A = 10 为例，来介绍 visitField() 方法中各参数的作用。

- access 比较好理解，就是字段的访问标识符，以上面的 A 为例，其值为 Opcodes.ACC_PRIVATE + Opcodes.ACC_STATIC + Opcodes.ACC_FINAL。
- name 为字段名，以上面的 A 为例，其值为 "A"。
- descriptor 为字段的描述符，以上面的 A 为例，其值为 I。
- signature 为泛型签名信息，如果没有就是 null。
- value 为字段的初始值，以上面的 A 为例，其值为 10，这里要注意只有常量才会有初始值。例如对于 private int A = 10，其初始值是 null，而并非 10，因为其初始化是在构造方法中进行的，这部分内容已在 4.8 节做过介绍。

注意，visitField() 方法的返回值是 FieldVisitor，此类提供了给字段添加属性、注解等相关的 API，FieldVisitor 中的 API 列表如下。

```
public FieldVisitor(final int api)
public FieldVisitor(final int api, final FieldVisitor fieldVisitor)

/**
  * 访问字段上定义的 Annotation
  */
public AnnotationVisitor visitAnnotation(
      final String descriptor, final boolean visible)

/**
  * 访问字段上定义的 Type Annotation
  */
public AnnotationVisitor visitTypeAnnotation(
      final int typeRef, final TypePath typePath, final String descriptor,
      final boolean visible)

/**
  * 访问非标准属性
  */
public void visitAttribute(final Attribute attribute)

/**
  * FieldVisitor 最后调用的方法，用于通知 visitor，字段的所有注解和属性已处理完毕
  */
public void visitEnd()
```

5.4.10 visitMethod()

该方法用于访问 ClassFile 中的方法，方法定义如下。

```
public MethodVisitor visitMethod(
      int access, java.lang.String name, java.lang.String descriptor,
      java.lang.String signature, java.lang.String[] exceptions)
```

以代码 private void foo(String) throws Exception 为例，来介绍 visitMethod() 方法中的各参数的作用。

- access 是方法的访问修饰符，以 foo() 方法为例，其值为 Opcodes.ACC_PRIVATE。
- name 为方法名，以 foo() 方法为例，其值为 “foo”。
- descriptor 为方法描述符信息，以 foo() 方法为例，其值为 “(Ljava/lang/String;)V”。
- signature 为泛型签名信息，如果没有就是 null。
- exceptions 为异常信息，以 foo() 方法为例，其值为 [“Ljava/lang/Exception;”]。

注意，此方法的返回为 MethodVisitor()，方法体中的内容由此方法生成。MethodVisitor 也是很重要的一个类，放在 5.7 节单独介绍。

5.4.11 visitAnnotation()

该方法用于访问声明在类上的注解，其方法定义如下。

```
public AnnotationVisitor visitAnnotation(
        java.lang.String descriptor, boolean visible)
```

下面通过例子来介绍 visitAnnotation() 方法中个参数的意义。

```
//1. 定义注解
package cn.sensorsdata.asm.annotation;

import java.lang.annotation.ElementType;
import java.lang.annotation.Retention;
import java.lang.annotation.RetentionPolicy;
import java.lang.annotation.Target;

@Retention(RetentionPolicy.RUNTIME)
@Target({ElementType.FIELD, ElementType.TYPE})
public @interface AnnotationOne {
}

//2. 定义测试类
package cn.sensorsdata.asm.annotation;

@AnnotationOne
public class AnnotationTest {

}
```

上面的代码定义了一个名为 AnnotationOne 的注解和一个使用此注解的 AnnotationTest 类。对 visitAnnotation() 来说，其中，descriptor 为注解的类描述符。在此例中，其值为 Lcn/sensorsdata/asm/annotation/AnnotationOne。visible 表示注解的 Rentention 设置的值是否可见，如果值为 RetentionPolicy.RUNTIME 表示可见，否则不可见。在此例中，其值为 true。

5.4.12 visitTypeAnnotation()

该方法用于访问声明在类的泛型上定义的注解。Java 8 扩展了注解功能，新增了如下两种类型。

```
public enum ElementType {
    ...
```

```
        /**
         * Type parameter declaration
         *
         * @since 1.8
         */
        TYPE_PARAMETER,

        /**
         * Use of a type
         *
         * @since 1.8
         */
        TYPE_USE,
}
```

这两种类型被称为类型注解（type annotation），具体内容如下。

- TYPE_PARAMETER允许注解被用在类型变量（即泛型）的声明中，比如MyClass<@Test T>，表示给泛型T添加Test注解。
- TYPE_USE则允许注解可以被用在任何类型使用的地方，比如new @NonEmpty @Readonly List<String>(myNonEmptyStringSet)，表示注解声明在构造方法前。

类型注解用来支持在Java的程序中做强类型检查，配合第三方插件工具（如Checker Framework），可以在编译期检测出runtime error（如UnsupportedOperationException、NullPointerException异常），避免异常延续到运行期才发现，从而提高代码质量，这就是类型注解的主要作用。

visitTypeAnnotation()方法的定义如下。

```
public AnnotationVisitor visitTypeAnnotation(
        int typeRef, TypePath typePath, java.lang.String descriptor, boolean visible)
```

说明如下。

- typeRef：注解的类型，其值必须是TypeReference.CLASS_TYPE_PARAMETER（0x00）、TypeReference.METHOD_TYPE_PARAMETER（0x01）、TypeReference.CLASS_EXTENDS（0x10）中的一种。
- typePath：类型注解使用的位置，例如数组元素、通配符、泛型参数等，此值可能为null。
- descriptor：为注解类描述符。
- visible：表示注解的Rentention设置的值是否可见。

下面我们定义一个具体的类型注解，用于观察visitTypeAnnotation()的输出结果。

```
//1.定义注解
package cn.sensorsdata.asm.annotation;

import java.lang.annotation.ElementType;
import java.lang.annotation.Retention;
import java.lang.annotation.RetentionPolicy;
import java.lang.annotation.Target;

@Retention(RetentionPolicy.RUNTIME)
//注意这里使用了Type_PARAMETER
@Target({ElementType.FIELD, ElementType.TYPE_PARAMETER})
public @interface AnnotationTwo {
```

```
}

//2. 定义测试类
package cn.sensorsdata.asm.annotation;
// 将注解定义在泛型上
public class AnnotationTest<@AnnotationTwo T> {
}
```

对 AnnotaionTest 进行解析，visitTypeAnnotaion() 方法输出的值如下：

```
typeRef:0, typePath:null, descriptor: Lcn/sensorsdata/asm/annotation/
AnnotationTwo;, visible:true
```

5.4.13 visitPermittedSubclass()

该方法用于访问允许的子类，方法定义如下。

```
public void visitPermittedSubclass(final String permittedSubclass)
```

举例，定义如下的密封类。

```
// 定义密封类 Shape
public sealed class Shape permits Circle, Square {
}

public final class Circle extends Shape {
      public float radius;
}
public non-sealed class Square extends Shape {
     public double side;
}
```

密封类 Shape 允许使用在 Circle、Square 类上，对 Class Shape 来说，其 permittedSubClass 就是 Circle 和 Shape。密封类是 Java 15 引入的特性，本书对此不多做介绍。

5.4.14 visitRecordComponent()

该方法用于访问类的 Record 部分。Java 14 引入了一个新的关键字 record，用于解决模板代码问题，类似于 Kotlin 中的 data class。例如，下面定义的 Java Bean:

```
final class Rectangle implements Shape {
      final double length;
      final double width;

      public Rectangle(double length, double width) {
          this.length = length;
          this.width = width;
      }

      double length() { return length; }
      double width() { return width; }
}
```

使用 record 关键字，则可以使用如下方式声明：

```
record Rectangle(double length, double width) { }
```

可以看到使用 record 可以大大简化代码、提高开发效率。关于 record 关键字的用法，本书亦不多做介绍。

5.4.15 visitEnd()

visitEnd() 在 ASM 中的多个 API 中出现，例如 FieldVisitor、MethodVisitor、AnnotationVisitor 等，它一般是各 visitor 最后调用的一个方法。它的作用一般是通知 visitor 已处理完相关元素的一个信号。

至此，我们对 ClassVisitor 中定义的所有方法都进行了介绍，关于其中的每个方法的作用，大家可以动手写一些测试例子验证一下。接下来介绍 ASM 中另外一个重要的类 ClassWriter。

5.5 ClassWriter API 介绍

ClassWriter 类，顾名思义，其作用是写入、生成 ClassFile。它可以单独使用，专门用来生成一个 Java class；也可以与 ClassReader、ClassVisitor 一起使用，对已有的 ClassFile 做修改。

5.5.1 构造方法

ClassWriter 继承自 ClassVisitor，它除了拥有 ClassVisitor 的所有功能，它的构造方法如下。

```
public ClassWriter(final int flags)
public ClassWriter(final ClassReader classReader, final int flags)
```

（1）参数classReader用于读取原始ClassFile数据，它会将整个常量池和引导方法以及其他合适的字节码片段直接复制到新的ClassFile对应的信息后面，新的ClassFile原本的数据不会被删除。该参数用于提高转换ClassFile的效率。

（2）flags用于设置修改ClassFile的默认行为，对直接复制的方法将不起作用，它的取值如下。

- 0表示ASM什么都不做，用户需要自己去计算栈映射帧、局部变量表和操作数栈的最大值。
- ClassWriter.COMPUTE_MAXS表示ASM将自动计算局部变量表和操作数栈的最大值。
- ClassWriter.COMPUTE_FRAMES表示ASM将自动计算栈映射帧、局部变量表和操作数栈的最大值。

5.5.2 toByteArray()

toByteArray() 是 ClassWriter 中的一个重要方法，用于返回获取 ClassFile 内容的数组。

关于 ClassWriter 的使用，举例如下。

```
package cn.sensorsdata.asm.cw;

public abstract class SensorsDataAPI {
      public static final String VERSION = "9.0.0";
      protected String serverUrl = "http://www.sensorsdata.cn";

      public SensorsDataAPI(String serverUrl){
          this.serverUrl = serverUrl;
      }

      public abstract void track(String jsonStr);
}
```

上述代码中定义了一个 SensorsDataAPI 抽象类，并且定义了两个字符串字段，一个是常量，另一个是普通变量；另外还定义了一个构造方法和一个抽象方法。下面来看看如何使用 ClassWriter 来生成这样一个类。

示例代码如下。

```
ClassWriter cw = new ClassWriter(0);(1)
cw.visit((2)
      Opcodes.V1_8,
      Opcodes.ACC_PUBLIC + Opcodes.ACC_ABSTRACT,
      "cn/sensorsdata/asm/cw/SensorsDataAPI", null, "Ljava/lang/Object;", null);
cw.visitField((3)
      Opcodes.ACC_PUBLIC + Opcodes.ACC_FINAL + Opcodes.ACC_STATIC,
      "VERSION", "Ljava/lang/String;", null, "9.0.0")
      .visitEnd();
cw.visitField((4)
      Opcodes.ACC_PROTECTED,
      "serverUrl",
      "Ljava/lang/String;", null,
      "http://www.sensorsdata.cn")
      .visitEnd();
cw.visitMethod((5)
      Opcodes.ACC_PUBLIC,
      "<init>",
      "(Ljava/lang/String;)V", null, null)
      .visitEnd();
cw.visitMethod((6)
      Opcodes.ACC_PUBLIC + Opcodes.ACC_ABSTRACT,
      "track", "(Ljava/lang/String;)V", null, null)
      .visitEnd();
cw.visitEnd();(7)

//保存数据
String filePath = "src/main/java/cn/sensorsdata/asm/cw/";
File file = new File(filePath + "SensorsDataAPI.class");
byte[] data = cw.toByteArray();(8)
FileUtils.writeByteArrayToFile(file, data);(9)
```

下面分析上述代码。

（1）位置（1）处创建 ClassWriter 对象，此处传值为 0，表示不对 ClassFile 的操作数栈和局部变量表等内容做自动计算操作，所有操作都必须通过手动调用相关方法计算。

（2）位置（2）处使用 visit() 方法添加 ClassFile 的基本信息。

- 方法的第一个参数 Opcodes.V1_8 表示 ClassFile 的版本号。
- 第二个参数表示类的访问标识符。
- 第三个参数表示类名。
- 第四个参数表示泛型信息，此类没有泛型，所以传值为 null。
- 第五个参数表示父类信息，Java 中所有的类都有父类，默认父类是 java/lang/Object。
- 第六个参数表示此类实现的接口，因为此类未实现任何接口，所以传 null。

（3）位置（3）和位置（4）使用 visitField() 方法创建字段信息，注意此方法的第五个参数，表示常量的默认值，即使用 static final 修饰的字段。因此对 serverUrl 来说，其设置第五个参数的默认值是不起作用的。还需要注意的是 visitField() 的返回值是 FieldVisitor，因此还需要调用它的 visitEnd() 方法。

（4）位置（5）和位置（6）使用 visitMethod() 方法创建方法信息，注意构造方法的名字是 <init>。另外 visitMethod 方法的返回值是 MethodVisitor，因此还需要调用其 visitEnd() 方法。

（5）位置（7）调用 visitEnd() 方法通知 ClassWriter 对象所有操作已结束。

（6）位置（8）调用 toByteArray() 方法获取生成 .class 文件的数据内容。

（7）位置（9）将通过 toByteArray() 获取的数据保存到文件中（此处的工具类 FileUtils 用的是 Apache commons-io 库）。

反编译上述代码输出的 .class 文件，结果如图 5-3 所示。

```
//
// Source code recreated from a .class file by IntelliJ IDEA
// (powered by FernFlower decompiler)
//

public abstract class SensorsDataAPI {
    public static final String VERSION = "9.0.0";
    protected String serverUrl;

    public SensorsDataAPI(String var1) {
    }

    public abstract void track(String var1);
}
```

图 5-3　SensorsDataAPI.class 的结果展示

可以看到最终结果与源码略有出入：其中 serverUrl 的值未设定，构造方法中的内容也没有。首先关于 serverUrl 的初始值的设定并非在定义 Field 的地方，而是在构造方法中对其赋值，ASM 也遵从这样的设定，具体可以参考 4.8 节的介绍。另外如何生成方法中的内容，需要先了解 5.7 节内容。

5.6　类的转换和修改

5.6.1　转换类的方式

截至目前，我们已经比较分散地介绍了 ClassReader、ClassVisitor、ClassWriter 这 3 个类的 API 的使用，还未介绍如何将这三者结合在一起的用法。根据这 3 个类的作用可以有如下两种组合方式。

第一种方式是将 ClassReader 产生的事件直接传递给 ClassWriter，形式如下。

```
byte[] classData = ...;
ClassWriter classWriter = new ClassWriter(0);
ClassReader classReader = new ClassReader(classData);
classReader.accept(classWriter, 0);
byte[] newClassData = classWriter.toByteArray();
```

此种方式的作用就是将 ClassReader 读取的数据再次构建一遍，没有产生任何改变，可以看出这是一个无意义的操作。

第二种方式是在 ClassReader 和 ClassWriter 中间使用 ClassVisitor，其形式如下。

```
byte[] classData = ...;
ClassWriter cw = new ClassWriter(0);
// 创建 ClassVisitor 的匿名实现类，将 cw 作为参数传递给它
ClassVisitor cv = new ClassVisitor(Opcodes.ASM8, cw) {};
ClassReader cr = new ClassReader(classData);
cr.accept(cv, 0);
byte[] newClassData = cw.toByteArray();
```

上述代码中定义了一个 ClassVisitor 的匿名类，因为 ClassWriter 继承自 ClassVisitor，所以可以将 ClassWriter 对象作为参数传递给 ClassVisitor 的构造方法。当需要对 CassFile 做转换时，通常会采用这种方式，并通过 toByteArray() 方法获取最终的结果。

第二种方式的处理过程可以用图 5-4 来表示。

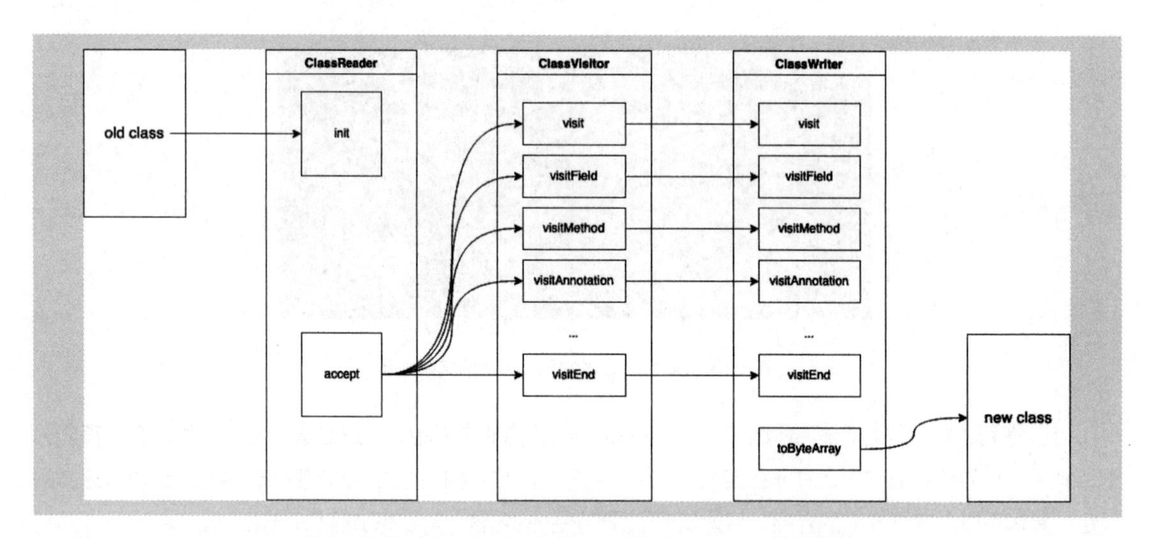

图 5-4　ClassReader、ClassVisitor 和 ClassWriter 处理流程

除了上述两种方式外，ASM 还有提供了一些优化方式，例如下面这个例子。

```
byte[] classData = ...;
ClassWriter cw = new ClassWriter(0);
ClassVisitor cv = new ClassVisitor(Opcodes.ASM8, cw)
{
      @Override
      public void visit(
           int version, int access, String name, String signature,
           String superName, String[] interfaces)
      {
```

```
            super.visit(Opcodes.V1_8, access, name, signature, superName, interfaces);
        }
};
ClassReader cr = new ClassReader(classData);
cr.accept(cv, 0);
byte[] newClassData = cw.toByteArray();
```

上述代码的作用是将 ClassFile 的版本号修改为 Java 8，假如为了修改这个版本号需要将 ClassFile 整体进行解析和重构，这样做显然是比较低效的。如果能将 ClassFile 中未发生改变的方法直接复制到新的 ClassFile 中，这样效率就会高很多，实际上 ASM 也确实是这么做的。当满足如下条件时，ASM 将会对其进行优化。

如果 ClassReader 检测到其 accept() 接收的 ClassVisitor 对应的 visitMethod() 返回的值是 MethodWriter 对象，这就意味着方法内容未改变，ClassReader 会将内容直接复制过去。

我们通过源码来理解 ASM 的这个优化逻辑，下面是 ClassReader 中读取 method 信息的部分源码。

```
private int readMethod(
       final ClassVisitor classVisitor, final Context context,
      final int methodInfoOffset)
{
      ...

      // If the returned MethodVisitor is in fact a MethodWriter
      // it means there is no method adapter between the reader and the writer
      // In this case, it might be possible to copy
      // the method attributes directly into the writer
      // If so, return early without visiting the content of these attributes
      if (methodVisitor instanceof MethodWriter) {
        MethodWriter methodWriter = (MethodWriter) methodVisitor;
        if (methodWriter.canCopyMethodAttributes(
            this,
            synthetic,
            (context.currentMethodAccessFlags & Opcodes.ACC_DEPRECATED) != 0,
            readUnsignedShort(methodInfoOffset + 4),
            signatureIndex,
            exceptionsOffset))
        {
          methodWriter.setMethodAttributesSource(
              methodInfoOffset, currentOffset - methodInfoOffset);
          return currentOffset;
        }
      }
      ...
}
```

上述代码注释中介绍，如果 MethodVisitor 对象的实际类型是 MethodWriter（MethodWriter 是 MethodVisitor 的子类），就意味着不需要读取方法中的具体内容，也不需要做修改。在这种情况下可以直接对原有方法信息做复制操作。上述代码中的 canCopyMethodAttributes() 负责进行数据复制。具体可以通过下面这个例子来验证。

```
byte[] bytes = ...
ClassReader cr = new ClassReader(bytes);
ClassWriter cw = new ClassWriter(0);
ClassVisitor cv = new ClassVisitor(Opcodes.ASM8, cw)
```

```
{
    @Override
    public void visit(
        int version, int access, String name, String signature,
        String superName, String[] interfaces)
    {
        super.visit(Opcodes.V1_8, access, name, signature, superName,
                interfaces);
    }
    @Override
    public MethodVisitor visitMethod(
        int access, String name, String descriptor,
        String signature, String[] exceptions)
    {
        MethodVisitor mv =
            super.visitMethod(access, name, descriptor, signature, exceptions);
        mv = new MethodVisitor(Opcodes.ASM8, mv) {}; //注释后则能进行优化转换
        return mv;
    }
};
cr.accept(cv, 0);
byte[] newClassData = cw.toByteArray();
```

上述这个例子将无法满足“方法内容未改变”这个条件，因此也就无法达到优化的目的，注释以后则能进行优化。

ASM 还提供了另外一种优化方式，其使用形式如下。

```
byte[] classData = null;
ClassReader cr = new ClassReader(classData);
ClassWriter cw = new ClassWriter(cr, 0);// 注意这里使用的构造方法 (1)
ClassVisitor cv = new ClassVisitor(Opcodes.ASM8, cw)
{
    @Override
    public void visit(
        int version, int access, String name, String signature,
        String superName, String[] interfaces)
    {
        super.visit(Opcodes.V1_8, access, name, signature, superName,
                interfaces);
    }
};
cr.accept(cv, 0);
byte[] newClassData = cw.toByteArray();
```

上述代码位置（1）中使用的是 ClassWriter(final ClassReader classReader, final int flags) 这个构造方法来接收 ClassReader 对象，即 ClassReader 和 ClassWriter 相互引用对方。通过这种方式来处理，效率上比前一种方式快一倍，具体特点如下。

- ClassWriter 的构造方法会将整个常量池和引导方法以及其他合适的字节码片段直接复制到新的 ClassFile 对应的信息后面，所以速度上要快很多。
- 因为会直接复制常量池数据，所以对增加字段、方法等操作来说影响不大，原因是这些操作涉及的常量池数据会在原有的技术上进行添加；但是如果删除或者修改字段、方法会使得 ClassFile 的体积增大，因为这种方式不会对常量池进行优化。

- 这种方式适合做一些“新增”性质的操作。

本小节介绍了转换类的方式，针对不同的使用场景可以选择不同的方式，不过通常情况下第二种方式使用得最多，它最通用，其他优化方式根据具体业务场景来做调整即可。

5.6.2 删除Class成员

对于需要做删除操作的 Class 成员，基本上都可以让 ClassVisitor 的 visit*() 方法返回 null 或者删除对应 visit() 方法的方法体即可。

举例如下。

```
byte[] bytes = ..
ClassReader cr = new ClassReader(bytes);
ClassWriter cw = new ClassWriter(0);
ClassVisitor cv = new ClassVisitor(Opcodes.ASM9,cw) {

      @Override
      public FieldVisitor visitField((1)
          int access, String name, String descriptor,
          String signature, Object value)
      {
          //删除 address 字段
          if("address".equals(name)){
              return null;
          }
          return super.visitField(access, name, descriptor, signature, value);
      }

      @Override
      public MethodVisitor visitMethod((2)
          int access, String name, String descriptor,
          String signature, String[] exceptions)
      {
          //删除 String getAddress(){} 方法
          if("getAddress".equals(name)
             && "()Ljava/lang/String;".equals(descriptor))
          {
              return null;
          }
          return super.visitMethod(access, name, descriptor,
                                   signature, exceptions);
      }

      @Override
      public void visitSource(String source, String debug) {(3)
          //删除 SourceFile 属性
          //super.visitSource(source, debug);
      }

      @Override
      public AnnotationVisitor visitAnnotation(
          String descriptor, boolean visible)
      {
```

```
            return super.visitAnnotation(descriptor, visible);
        }
    };
    cr.accept(cv, 0);
    byte newData[] = cw.toByteArray();
```

上述代码的逻辑介绍如下。

- 在位置（1）处的visitField()方法中判断类字段名称是否是“address”，如果满足，visitField()方法就返回null，这样就达到了删除字段的目的。
- 在位置（2）处的visitMethod()方法中判断方法的名称以及方法描述符信息，如果满足，就返回null，这样就达到了删除getAddress()方法的目的。这里判断方法描述符的目的是可能有同名的重载方法。
- 在位置（3）处的visitSource()方法是用来访问类中的SourceFile属性，它没有返回值，就直接注释掉删除SourceFile属性的方式是重写并删除visitSource() 中的方法体。

5.6.3 增加Class成员

在介绍 ClassVisitor API 时提到，它的 API 方法需要按照一定的顺序来调用。例如，为 Class 添加一个字段，就不能在 visit()、visitAnnotation()、visitSource() 方法中调用 visitField() 方法。这在 ASM 中是不允许的，不过可以在 visitField() 方法中或者 visitEnd() 方法中调用。

首先介绍如何给 Class 添加字段。通常是在 visitEnd() 方法中，另外也可以根据具体的业务，在满足某种条件都是会后添加。举例如下。

```
public class AddingClassVisitor  extends ClassVisitor {

        public AddingClassVisitor( ClassVisitor classVisitor) {
            super(Opcodes.ASM9, classVisitor);
        }

        @Override
        public FieldVisitor visitField((1)
            int access, String name, String descriptor,
            String signature, Object value)
        {
            // 添加 address 字段
            if("address".equals(name)){
                cv.visitField(
                    Opcodes.ACC_PUBLIC,
                    "email","Ljava/lang/String;", null, null)
                    .visitEnd(); // 不要忘记调用 visitEnd()
            }
            return super.visitField(access, name, descriptor, signature, value);
        }

        @Override
        public void visitEnd() {(2)
            // 添加 TEL 字段

            cv.visitField(
                Opcodes.ACC_PUBLIC + Opcodes.ACC_STATIC + Opcodes.ACC_FINAL,
```

```
                "TEL", "Ljava/lang/String;", null, "133xxxxxxxx")
                .visitEnd(); // 不要忘记调用 visitEnd()
            super.visitEnd();
        }
    }
```

上述代码展示了添加 Field 的方式，在位置（1）处的 visitField() 方法中判断当满足字段名是 address 的时候就添加一个 email 字段。在 visitField() 中添加字段一定要注意判断，因为类中有多个 Field，该方法会多次触发。在位置（2）处的 visitEnd() 方法中添加了一个 TEL 字段，该方法总是会被调用，比较推荐这种做法。

该方法的添加方式与添加字段类似，举例如下。

```
public class AddingClassVisitor  extends ClassVisitor {

      public AddingClassVisitor( ClassVisitor classVisitor) {
          super(Opcodes.ASM9, classVisitor);
      }

      ..

      @Override
      public MethodVisitor visitMethod(
          int access, String name, String descriptor,
          String signature, String[] exceptions)
      {
          // 添加 setAddress() 方法的方式一
          if("getAddress".equals(name) && "
             ()Ljava/lang/String;".equals(descriptor))
          {
              MethodVisitor mv =
                  cv.visitMethod(Opcodes.ACC_PUBLIC,
                                 "setAddress",
                                 "(Ljava/lang/String;)V", null,null);

              mv.visitCode();
              mv.visitVarInsn(Opcodes.ALOAD, 0);
              mv.visitFieldInsn(Opcodes.GETFIELD,
                               "cn/sensorsdata/asm/AddingClassVisitor",
                               "address",
                               "Ljava/lang/String;");
              mv.visitInsn(Opcodes.ARETURN);
              mv.visitEnd();
              return mv;
          }
          return super.visitMethod(access, name, descriptor,
                                  signature, exceptions);
      }

      @Override
      public void visitEnd() {
          super.visitEnd();
          // 添加 setAddress() 方法的方式二
          MethodVisitor mv =  cv.visitMethod(Opcodes.ACC_PUBLIC,
                                             "setAddress",
                                             "(Ljava/lang/String;)V",
```

```
                                                    null, null);
            mv.visitCode();
            mv.visitVarInsn(Opcodes.ALOAD, 0);
            mv.visitFieldInsn(Opcodes.GETFIELD,
                        "cn/sensorsdata/asm/AddingClassVisitor",
                        "address",
                        "Ljava/lang/String;");
            mv.visitInsn(Opcodes.ARETURN);
            mv.visitEnd();
        }
    }
```

在上述代码位置（1）处的 visitMethod() 方法中，如果方法名和方法描述符满足特定规则时添加一个 setAddress() 方法，不满足就不添加。位置（2）处的 visitEnd() 方法也展示了添加 setAddress() 方法。关于 MethodVisitor 的用法，将在 5.7 节详细介绍。

5.6.4 修改 Class 成员

在 5.6.1 小节介绍转换类的方式中的例子时提到了如何修改 ClassFile 的版本号，那样的修改比较简单，现在思考如何修改下面例子中 FruitFactory 类中字段 price 和 DEFAULT_CHOOSE 的值。

```
public class FruitFactory {
      private static final String DEFAULT_CHOOSE = "Apple";
      private int price = 99;

      public String getFruitByType(int type) {
          if (type == 0) {
              return DEFAULT_CHOOSE;
          }
          return null;
      }
}
```

读者首先想到的方式可能是这样的。

```
public class ModifyingClassVisitor extends ClassVisitor{
      public ModifyingClassVisitor( ClassVisitor classVisitor) {
          super(Opcodes.ASM9, classVisitor);
      }
      @Override
      public FieldVisitor visitField(
          int access, String name, String descriptor,
          String signature, Object value)
      {
          // 判断字段名称，修改对应的值
          if("price".equals(name)){
              System.out.println("1111");
              value = 100;
          }
          if("DEFAULT_CHOOSE".equals(name)){
              value = "Raspberry";
          }
          return super.visitField(access, name, descriptor, signature, value);
      }
```

```
    ...
}
```

按照上述方式修改后的代码反编译的结果如图 5-5 所示。

```
//
// Source code recreated from a .class file by IntelliJ IDEA
// (powered by FernFlower decompiler)
//

package cn.sensorsdata.asm;

public class FruitFactory {
    private static final String DEFAULT_CHOOSE = "Raspberry";
    private int price = 99;

    public FruitFactory() {
    }

    public String getFruitByType(int var1) {
        return var1 == 0 ? "Apple" : null;
    }
}
```

图 5-5　FruitFactory2.class 反编译的结果

通过上述结果可以发现两个问题。

（1）对 price 字段的修改没有起作用。

（2）对 DEFAULT_CHOOSE 字段的修改起作用了，但是预期是 getFruitByType() 方法中对 DEFAULT_CHOOSE 引用的部分也一并被修改，事实却是没有变化。

解释如下。

（1）为什么对 price 字段的修改没有起作用？这里要涉及两个知识点，第一是 ClassVisitor.visitField() 方法中的 value 参数只针对常量才起作用，这也是 DEFAULT_CHOOSE 可以修改成功的原因；第二是对于类中字段的初始化时机，类中的非 static 常量的初始化实际上是在构造方法中初始化的。关于这两个知识点，可以参考 4.8 节的内容，也可以通过字节码来查看其初始化。

```
public class cn.sensorsdata.asm.FruitFactory
  ...
{
   private static final java.lang.String DEFAULT_CHOOSE;
     descriptor: Ljava/lang/String;
     flags: ACC_PRIVATE, ACC_STATIC, ACC_FINAL
     ConstantValue: String Apple

   private int price;
     descriptor: I
     flags: ACC_PRIVATE

   public cn.sensorsdata.asm.FruitFactory();
     descriptor: ()V
     flags: ACC_PUBLIC
     Code:
       stack=2, locals=1, args_size=1
```

```
         0: aload_0
         1: invokespecial #1  // Method java/lang/Object."<init>":()V
         4: aload_0
         5: bipush         99  // 初始化 price 值
         7: putfield      #2  // Field price:I
        10: return
      LineNumberTable:
        line 3: 0
        line 5: 4

SourceFile: "FruitFactory.java"
```

因此若要对 price 的初始值进行修改，实际上要修改构造方法中对应的指令，示例代码如下。

```
public class ModifyingClassVisitor extends ClassVisitor{
...

      @Override
      public MethodVisitor visitMethod(int access, String name,
             String descriptor, String signature, String[] exceptions)
      {
          MethodVisitor methodVisitor =
              super.visitMethod(access, name,
                                 descriptor, signature, exceptions);

          //判断方法是否是构造方法
          if("<init>".equals(name) && "()V".equals(descriptor)){
              methodVisitor = new MethodVisitor(Opcodes.ASM9, methodVisitor)
              {

                  @Override
                  public void visitFieldInsn(int opcode, String owner,
                                             String name, String descriptor)
                  {
//判断指令是否是 PUTFIELD、字段名是否是 price 等信息
                      if(opcode == Opcodes.PUTFIELD
                            && "cn/sensorsdata/asm/FruitFactory".equals(owner)
                            && "price".equals(name) && "I".equals(descriptor))
                      {
//添加 POP 指令，删除通过 5: bipush 99 添加到操作数栈中的 99
                        this.mv.visitInsn(Opcodes.POP);
//将 100 压入栈中
                        this.mv.visitVarInsn(Opcodes.BIPUSH,100);
                      }
                      super.visitFieldInsn(opcode, owner, name, descriptor);
                  }
              };
          }
          return methodVisitor;
      }
}
```

上述代码在 visitMethod() 方法中判断方法是不是构造方法，然后对其指令做处理。visitFieldInsn() 方法中指令的大致逻辑是判断指令是否是 PUTFIELD 以及是否满足字段 price 的定义，然后添加一个 POP 指令到操作数栈中，将原本压入操作数中 99 的值出栈，最后将 100 压入栈。

（2）为什么 getFruitByType() 方法中的 DEFAULT_CHOOSE 没有变化？先来观察字节码。

```
public class cn.sensorsdata.asm.combine.FruitFactory
Constant pool:
     ...
     #2 = Class    #18 // cn/sensorsdata/asm/combine/FruitFactory
     #3 = String   #19 // Apple
     ...
    #19 = Utf8         Apple
     ...
{
    private static final java.lang.String DEFAULT_CHOOSE;
      descriptor: Ljava/lang/String;
      flags: ACC_PRIVATE, ACC_STATIC, ACC_FINAL
      ConstantValue: String Apple

    public java.lang.String getFruitByType(int);
      descriptor: (I)Ljava/lang/String;
      flags: ACC_PUBLIC
      Code:
        stack=1, locals=2, args_size=2
           0: iload_1
           1: ifne     7
           4: ldc      #3    // String Apple(1)
           6: areturn
           7: aconst_null
           8: areturn
}
```

从 getFruitByType() 方法中位置（1）处的字节码中可以看到，使用 ldc 指令是加载常量池中的“Apple”字符串，并不是对类中定义的 DEFAULT_CHOOSE 常量进行引用。因此更改 DEFAULT_CHOOSE 的值并不会同时修改 getFruitByType() 方法中的值。想要修改 DEFAULT_CHOOSE 的值，就需要修改所有 ldc 指令指向“Apple”的值，使其加载“Raspberry”常量。

5.7 MethodVisitor API 介绍

MethodVisitor 是 Core API 中另一个比较重要的类，它用于生成和修改方法内容。本书第 3 章主要介绍了字节码基础，包括方法表、属性表等内容，知道方法体中的字节码信息是存储在 Code 属性中的；在第 4 章详细地介绍了字节码指令相关的知识，其中涉及了栈帧、局部变量表、操作数栈以及字节码指令的详细用法。MethodVisitor 中的方法就是围绕这些内容设计的。

MethodVistor 类的定义如下。

```
public abstract class MethodVisitor {

    // 表示此 API 的实现版本
    protected final int api;
    // 本 visitor 委托的对象
    protected MethodVisitor mv;

    public MethodVisitor(final int api)

    public MethodVisitor(final int api, final MethodVisitor methodVisitor)
```

```
public void visitParameter(final String name, final int access)

public AnnotationVisitor visitAnnotationDefault()

public AnnotationVisitor visitAnnotation( final String descriptor,
    final boolean visible)

public AnnotationVisitor visitTypeAnnotation(final int typeRef,
    final TypePath typePath,  final String descriptor,
    final boolean visible)

public void visitAnnotableParameterCount(final int parameterCount,
    final boolean visible)

public AnnotationVisitor visitParameterAnnotation( final int parameter,
    final String descriptor, final boolean visible)

public void visitAttribute(final Attribute attribute)

public void visitCode()

public void visitFrame( final int type, final int numLocal,
    final Object[] local, final int numStack, final Object[] stack)

public void visitInsn(final int opcode)

public void visitIntInsn(final int opcode, final int operand)

public void visitVarInsn(final int opcode, final int var)

public void visitTypeInsn(final int opcode, final String type)

public void visitFieldInsn( final int opcode,  final String owner,
    final String name,  final String descriptor)

@Deprecated
public void visitMethodInsn(final int opcode, final String owner,
          final String name, final String descriptor)

public void visitMethodInsn(final int opcode,final String owner,
  final String name, final String descriptor,final boolean isInterface)

public void visitInvokeDynamicInsn(final String name,
    final String descriptor,final Handle bootstrapMethodHandle,
    final Object... bootstrapMethodArguments)

public void visitJumpInsn(final int opcode, final Label label)

public void visitLabel(final Label label)

public void visitLdcInsn(final Object value)

public void visitIincInsn(final int var, final int increment)

public void visitTableSwitchInsn(
```

```
        final int min, final int max, final Label dflt, final Label... labels)

    public void visitLookupSwitchInsn(final Label dflt,
          final int[] keys,  final Label[] labels)

    public void visitMultiANewArrayInsn(final String descriptor,
          final int numDimensions)

    public AnnotationVisitor visitInsnAnnotation(
        final int typeRef, final TypePath typePath,
        final String descriptor, final boolean visible)

    public void visitTryCatchBlock(
        final Label start, final Label end,
        final Label handler, final String type)

    public AnnotationVisitor visitTryCatchAnnotation(
        final int typeRef, final TypePath typePath,
        final String descriptor, final boolean visible)

    public void visitLocalVariable(final String name,
        final String descriptor,final String signature,
        final Label start,final Label end,final int index)

    public AnnotationVisitor visitLocalVariableAnnotation(
        final int typeRef,final TypePath typePath,final Label[] start,
        final Label[] end,final int[] index,final String descriptor,
        final boolean visible)

    public void visitLineNumber(final int line, final Label start)

    public void visitMaxs(final int maxStack, final int maxLocals)

    public void visitEnd()
}
```

与 ClassVisitor 类似，MethodVisitor 也对方法的调用顺序进行了规定，使用时需要按照这个顺序来调用（为方便阅读，每一行中间添加了空行）。

```
( visitParameter )*

[ visitAnnotationDefault ]

( visitAnnotation | visitAnnotableParameterCount | visitParameterAnnotation
visitTypeAnnotation | visitAttribute )*

[ visitCode ( visitFrame | visitXInsn | visitLabel | visitInsnAnnotation | visitTryC
atchBlock | visitTryCatchAnnotation | visitLocalVariable | visitLocalVariableAnnotation |
visitLineNumber )* visitMaxs ] (1)

visitEnd.
```

因为 Java 中的方法并不一定有方法体，所以第四行［位置（1）］的方法是可选的；如果有方法体，那么必须先调用 visitCode()，在方法体的最后调用 visitMaxs() 方法，整体结束后不要忘了调用 visitEnd() 方法，用于通知 MethodVisitor 所有工作已处理完毕。

下面使用 MethodVisitor 来为 5.5.2 小节例子中的 SensorsDataAPI 类添加一个 getAddress() 方法，代码如下。

```
ClassWriter cw = new ClassWriter(0);
...

MethodVisitor methodVisitor =
   cw.visitMethod(ACC_PUBLIC,"getServerUrl","()Ljava/lang/String;",null,null);(1)
methodVisitor.visitCode();(2)
methodVisitor.visitVarInsn(ALOAD, 0);(3)
methodVisitor.visitFieldInsn(GETFIELD,
                        "cn/sensorsdata/asm/cw/SensorsDataAPI",
                        "serverUrl","Ljava/lang/String;");(4)
methodVisitor.visitInsn(ARETURN);(5)
methodVisitor.visitMaxs(1,1);(6)
methodVisitor.visitEnd();(7)
...
```

上述代码位置（1）通过调用 ClassWriter 的 visitMethod() 方法定义了 getServerUrl() 方法的基本信息，同时返回了 MethodVisitor 对象。位置（2）开始进行方法体内容的编写，首先需要调用 visitCode() 方法。位置（3）表示通过 ALOAD 指令将局部变量表索引位置 0 的值记载到操作数栈中。位置（4）表示通过 GETFIELD 指令获取类中的 serverUrl 字段的值到操作数栈中。位置（5）表示添加 ARETURN 指令，每一个方法都需要有 ARETURN 指令，这个不能少。位置（6）表示设置方法的操作数栈的最大深度以及局部变量表的最大长度，这里都是 1。最后不要忘记调用 visitEnd() 方法，如位置（7）所示。

以上就是 MethodVisitor 的简单应用，接下来详细介绍 MethodVisitor 中各方法的使用。

5.7.1 visitParameter()

此方法用于访问方法的参数信息，方法定义如下。

```
public void visitParameter(final String name, final int access)
```

其中 name 是参数名称，access 是参数的访问修饰符。该方法对应的是 ClassFile 中 method_info 结构的 MethodParameters 属性。若想获取 ClassFile 中方法的参数信息，需要在使用 javac 工具编译的时候添加 -parameters 选项，将 MethodParameters 属性一起编译到 ClassFile 中，命令格式如下。

```
javac -parameters XXX.java
```

例如：

```
//Java 代码
public void foo(final int p1, String p2){}

// 对应的字节码
public void foo(int, java.lang.String);
      ...
      MethodParameters:
        Name(1)                          Flags(2)
        p1                               final
        p2
```

上述字节码中位置（1）处的列表示参数的名称，位置（2）处的列表示参数的访问标识符。

5.7.2 visitAnnotationDefault()

此方法用于访问注解的默认值信息，方法定义如下。

```
public AnnotationVisitor visitAnnotationDefault()
```

举例如下。

```
@Target(ElementType.PARAMETER)
@Retention(RetentionPolicy.RUNTIME)
public @interface ParamCheck {
      String value() default "sensorsdata";
}
```

上述代码定义了一个 ParamCheck 注解，并且设置其 value 的默认值是 sensorsdata。其实 Java 源码中的注解继承自 java.lang.annotation.Annotation 接口，反编译 ParamCheck 类可以看到这一点。

```
/ParamCheck 继承自 Annotation
public interface cn.sensorsdata.asm.mv.ParamCheck extends java.lang.annotation.
Annotation

Constant pool:
    ...
     #7 = Utf8               sensorsdata

public abstract java.lang.String value();
      descriptor: ()Ljava/lang/String;
      flags: ACC_PUBLIC, ACC_ABSTRACT
      AnnotationDefault:
        default_value: s#7
```

从上述字节码也可以看到，value 方法的默认值是 sensorsdata。此内容对应的是 ClassFile 中 method_info 结构的 AnnotationDefault 属性，该属性是专门为注解设定的。

5.7.3 visitAnnotation()

此方法用于访问方法上的注解信息，方法定义如下。

```
public AnnotationVisitor visitAnnotation(
      final String descriptor, final boolean visible)
```

其中 descriptor 是注解的描述符，visible 表示注解是否可见。关于可见性，在 ClassFile 中是使用 RuntimeVisibleAnnotations 属性来记录运行时可见性的。

举例如下。

```
@Target(ElementType.METHOD)
@Retention(RetentionPolicy.RUNTIME)
public @interface VisibleAnnotationTest {
}


@Target(ElementType.METHOD)
@Retention(RetentionPolicy.SOURCE)
public @interface InVisibleAnnotationTest {
```

```
}

@VisibleAnnotationTest
@InVisibleAnnotationTest
public void foo(){
}
```

上述代码中定义了两个注解 Annotation，其中 VisibleAnnotationTest 运行时可见，即 Retention(RetentionPolicy.RUNTIME)；而 InVisibleAnnotationTest 是不可见的。foo() 方法上添加了这两个注解，foo() 方法对应的字节码如下。

```
Constant pool:
    ...
    #13 = Utf8          Lcn/sensorsdata/asm/mv/VisibleAnnotationTest;
    ...

public void foo();
      descriptor: ()V
      flags: ACC_PUBLIC
      Code:
        stack=0, locals=1, args_size=1
           0: return
        LineNumberTable:
          line 9: 0
        LocalVariableTable:
          Start  Length  Slot  Name   Signature
              0       1     0  this   Lcn/sensorsdata/asm/mv/Test;
      RuntimeVisibleAnnotations:
        0: #13()    //  Lcn/sensorsdata/asm/mv/VisibleAnnotationTest;
```

从字节码中可以看到，RuntimeVisibleAnnotations 属性列出了所有的“可见”注解。

5.7.4 visitTypeAnnotation()

在 5.4.12 节介绍了 visitTypeAnnotation，其作用是访问在方法上定义的类型注解，方法定义如下。

```
public AnnotationVisitor visitTypeAnnotation(
      final int typeRef,
      final TypePath typePath,
      final String descriptor,
      final boolean visible)
```

举例如下。

```
// 定义的注解
@Target(ElementType.TYPE_USE)
@Retention(RetentionPolicy.RUNTIME)
public @interface ParamTypeCheck {
}

// 在方法上使用
public void bar(@ParamTypeCheck int data){
}
```

上述代码定义了 Target 为 ElementType.TYPE_USE 的注解 ParamTypeCheck，foo() 方法的参

数使用了该注解。bar() 方法的字节码信息如下。

```
public void bar(int);
    ...
    RuntimeVisibleTypeAnnotations:
      0: #20(): METHOD_FORMAL_PARAMETER, param_index=0    // Lcn/sensorsdata/asm/mv/
ParamTypeCheck;
```

从字节码中可以看到 ParamTypeCheck 被定义为运行时可见，ClassFile 使用 RuntimeVisibleTypeAnnotations 属性来保存对应的值。

5.7.5 visitAnnotableParameterCount() 和 visitParameterAnnotation()

这两个方法用于访问方法中参数上的注解信息：visitAnnotableParameterCount() 方法用于表示方法中可添加注解的参数数目；visitParameterAnnotation() 方法用于访问方法参数上的注解内容，这一部分信息是存储在 RuntimeVisibleParameterAnnotations 属性中的。方法定义如下。

```
public void visitAnnotableParameterCount(
      final int parameterCount, final boolean visible)

public AnnotationVisitor visitParameterAnnotation(
      final int parameter, final String descriptor, final boolean visible)
```

举例如下。

```
// 定义的注解
@Target(ElementType.PARAMETER)
@Retention(RetentionPolicy.RUNTIME)
public @interface ParamCheck {
      String value() default "sensorsdata";
}

// 在方法中使用
public void foo(@ParamCheck int p1, @ParamTypeCheck int p2, int p3){
}
```

上述代码中 ParamCheck 注解的 Target 值为 ElementType.PARAMETER；foo() 方法中定义了 3 个参数，其中 p1 使用 ParamCheck 注解进行修饰。使用如下代码来测试上述 foo() 方法对应的输出。

```
byte[] bytes = ...
ClassReader cr = new ClassReader(bytes);
ClassWriter cw = new ClassWriter(0);
ClassVisitor cv = new ClassVisitor(Opcodes.ASM6, cw) {
      @Override
      public MethodVisitor visitMethod(
          int access, String name,
          String descriptor, String signature, String[] exceptions)
      {
          MethodVisitor methodVisitor =
              super.visitMethod(access, name, descriptor, signature, exceptions);
          methodVisitor = new MethodVisitor(Opcodes.ASM6, methodVisitor)
          {
              @Override
              public void visitAnnotableParameterCount(
```

```
                    int parameterCount, boolean visible)
                {
                    // 输出可添加注解的参数数目
                    System.out.println("visitAnnotableParameterCount: "
                                    + parameterCount + "===" + visible);
                    super.visitAnnotableParameterCount(parameterCount, visible);
                }
                @Override
                public AnnotationVisitor visitParameterAnnotation(
                    int parameter, String descriptor, boolean visible)
                {
                    // 输出 ElementType 为 PARAMETER 的注解信息
                    System.out.println(
                        "visitParameterAnnotation: " + parameter + "==="
                        + descriptor + "===" + visible);
                    return super.visitParameterAnnotation(
                        parameter, descriptor, visible);
                }
            };
            return methodVisitor;
        }
};
cr.accept(cv, 0);
```

输出结果如下。

```
visitAnnotableParameterCount: 3===true
visitParameterAnnotation: 0===Lcn/sensorsdata/asm/mv/ParamCheck;===true
```

5.7.6 visitAttribute()

此方法用于访问方法中定义的非标准属性，方法定义如下。

```
public void visitAttribute(final Attribute attribute)
```

所谓非标准是指非 JVM 中官方定义的属性，例如 JVM 规范中定义的 Code、ConstantValue、RuntimeVisibleParameterAnnotations 等这些属性都是标准属性。ClassFile 中允许用户定义自己的属性，用于扩展一些功能，在 6.7.3 小节会介绍如何在 ASM 中使用 Attribute。

5.7.7 visitCode()

此方法表示开始访问方法中的内容，如果方法是抽象方法，即没有方法体内容，就不需调用此方法，方法定义如下。

```
public void visitCode()
```

5.7.8 visitInsn()

此方法用于访问没有操作数的指令，方法定义如下。

```
public void visitInsn(final int opcode)
```

其中opcode表示操作码，这类指令有NOP、ACONST_NULL、ICONST_M1、ICONST_0、ICONST_1、ICONST_2、ICONST_3、ICONST_4、ICONST_5、LCONST_0、LCONST_1、FCONST_0、FCONST_1、FCONST_2、DCONST_0、DCONST_1、IALOAD、LALOAD、FALOAD、DALOAD、AALOAD、BALOAD、CALOAD、SALOAD、IASTORE、LASTORE、FASTORE、DASTORE、AASTORE、BASTORE、CASTORE、SASTORE、POP、POP2、DUP、DUP_X1、DUP_X2、DUP2、DUP2_X1、DUP2_X2、SWAP、IADD、LADD、FADD、DADD、ISUB、LSUB、FSUB、DSUB、IMUL、LMUL、FMUL、DMUL、IDIV、LDIV、FDIV、DDIV、IREM、LREM、FREM、DREM、INEG、LNEG、FNEG、DNEG、ISHL、LSHL、ISHR、LSHR、IUSHR、LUSHR、IAND、LAND、IOR、LOR、IXOR、LXOR、I2L、I2F、I2D、L2I、L2F、L2D、F2I、F2L、F2D、D2I、D2L、D2F、I2B、I2C、I2S、LCMP、FCMPL、FCMPG、DCMPL、DCMPG、IRETURN、LRETURN、FRETURN、DRETURN、ARETURN、RETURN、ARRAYLENGTH、ATHROW、MONITORENTER、MONITOREXIT。

例如将POP指令入栈，使用方式如下。

```
methodVisitor.visitInsn(Opcodes.POP)
```

5.7.9 visitIntInsc()

此方法用于访问有操作数为int类型的指令，方法定义如下。

```
public void visitIntInsn(final int opcode, final int operand)
```

其中opcode表示操作码，operand表示操作数。这类指令有：BIPUSH、SIPUSH、NEWARRAY。

例如将整数100压入操作数栈中，使用方式如下。

```
methodVisitor.visitIntInsc(Opcodes.BIPUSH, 100)
```

5.7.10 visitVarInsn()

此方法用于访问操作局部变量表的指令，方法定义如下。

```
public void visitVarInsn(final int opcode, final int var)
```

这类指令有ILOAD、ILOAD、LLOAD、FLOAD、DLOAD、ALOAD、ISTORE、LSTORE、FSTORE、DSTORE、ASTORE、RET。

例如加载局部变量表索引1位置的引用类型数据可以使用如下代码。

```
methodVisitor.visitVarInsn(Opcodes.ALOAD, 1)
```

5.7.11 visitTypeInsn()

此方法用于访问类型相关的指令，方法定义如下。

```
public void visitTypeInsn(final int opcode, final String type)
```

这类指令有 NEW、ANEWARRAY、CHECKCAST、INSTANCEOF。

例如向操作数栈插入 CHECKCAST 指令用于将栈顶元素强制类型转换成 string 类型，使用方式如下。

```
methodVisitor.visitTypeInsn(Opcodes.CHECKCAST, "java/lang/String")
```

5.7.12 visitFieldInsn()

此方法用于访问与操作字段相关的指令，方法定义如下。

```
public void visitFieldInsn(
      final int opcode, final String owner,
      final String name, final String descriptor)
```

说明如下。

- opcode对应操作字段的指令，包括GETSTATIC、PUTSTATIC、GETFIELD、PUTFIELD。
- owner表示字段所在的类的全限定名，例如“java/util/List”。
- name表示字段名称。
- descriptor表示字段描述符信息，例如“Ljava/lang/String;”。

以调用 java.lang.System 类中的静态字段 out 为例。

```
methodVisitor.visitFieldInsn(
      Opcodes.GETSTATIC, "java/lang/System", "out", "java/io/PrintStream");
```

5.7.13 visitMethodInsn()

此方法用于访问与操作方法相关的指令，方法定义如下。

```
public void visitMethodInsn(
      final int opcode, final String owner,
      final String name, final String descriptor,
      final boolean isInterface)
```

说明如下。

- opcode对应操作方法的指令，包括INVOKEVIRTUAL、INVOKESPECIAL、INVOKESTATIC、INVOKEINTERFACE。
- owner表示方法所在的类，例如“java/util/List”。
- name表示方法名称。
- descriptor表示方法的描述符信息，例如“(I)V”。
- isInterface用于标识此方法所在的类是否是接口类型。

以调用 java.lang.String 类的静态方法 valueOf(int) 为例。

```
methodVisitor.visitMethodInsn(
      Opcodes.INVOKESTATIC,
      "java/lang/String", "valueOf", "(I)Ljava/lang/String;",false);
```

5.7.14 visitInvokeDynamicInsn()

此方法用于访问 invokedynamic 指令，方法定义如下。

```
public void visitInvokeDynamicInsn(
      final String name, final String descriptor,
      final Handle bootstrapMethodHandle,
      final Object... bootstrapMethodArguments)
```

说明如下。

- name表示方法名。
- descriptor表示方法描述符。
- bootstrapMethodHandle引导方法对应的结构，此结构中定义了引导方法需要的数据，这里是Handle类型，该类型是ASM中定义的类型。
- bootstrapMethodArguments对应引导方法中的参数值，是一个数据结构。

Handle 类的定义如下。

```
public Handle(
      final int tag, final String owner,
      final String name, final String descriptor, final boolean isInterface)
```

构造方法中的参数说明如下。

- tag对应CONSTANT_MethodHandle_info属性结构中对应的reference_kind，即方法句柄的类型。
- owner表示方法所在类对应的全限定名。
- name表示方法名。
- descriptor表示方法描述符。
- isInterface用于标识此方法所在的类是否是接口类型。

举例如下。

```
// Java 代码
Runnable run = () ->{};

// ASM 代码
methodVisitor.visitInvokeDynamicInsn(
      "run",
      "()Ljava/lang/Runnable;",
      new Handle(
          Opcodes.H_INVOKESTATIC,
          "java/lang/invoke/LambdaMetafactory",
          "metafactory"
          "(Ljava/lang/invoke/MethodHandles$Lookup; \
          Ljava/lang/String;Ljava/lang/invoke/MethodType; \
          Ljava/lang/invoke/MethodType; \
          Ljava/lang/invoke/MethodHandle; \
          Ljava/lang/invoke/MethodType;)Ljava/lang/invoke/CallSite;"
          , false),
      new Object[]{
          Type.getType("()V"),
          new Handle(
              Opcodes.H_INVOKESTATIC,
              "cn/sensorsdata/asm/mv/Test",
              "lambda$bar$0",
              "()V",
              false),
```

```
        Type.getType("()V")}
);
```

上面例子中展示了 Java 8 中的 Lambda 的生成方式，读者可以感受一下，很少需要读者自己使用 ASM 来生成 Lambda 表达式，更多的是如何理解它。关于 invokedynamic 指令，在 8.4 节再详细介绍。

5.7.15 visitLabel()

用于在指定位置生成锚记点（也叫标记位置），方法定义如下。

```
public void visitLabel(final Label label)
```

关于此方法的用法会结合 5.7.19 小节的内容一起介绍。

5.7.16 visitJumpInsn()

用于访问或创建跳转指令，方法定义如下。

```
public void visitJumpInsn(final int opcode, final Label label)
```

其中 opcode 表示操作码，这类跳转指令有 IFEQ、IFNE、IFLT、IFGE、IFGT、IFLE、IF_ICMPEQ、IF_ICMPNE、IF_ICMPLT、IF_ICMPGE、IF_ICMPGT、IF_ICMPLE、IF_ACMPEQ、IF_ACMPNE、GOTO、JSR、IFNULL、IFNONNULL。label 表示对应指令执行满足条件时字节码指令偏移的位置。

举例如下。

```
public void bar(int a) {
      if (a == 0) {
          a++;
      }
}
```

上述方法对应的字节码如下。

```
public void bar(int);
     descriptor: (I)V
     flags: ACC_PUBLIC
     Code:
       stack=1, locals=2, args_size=2
          0: iload_1
          1: ifne          7
          4: iinc          1, 1
          7: return
```

其中 1: ifne 7 指令表示将操作数栈栈顶的元素与 0 比较，如果不等于 0 就跳转到字节码偏移“7: return”的位置。在 ASM 中，使用者肯定不需要关注跳转的具体字节码指令偏移位置，这些 ASM 会帮忙计算好，而实现这一步的关键就是通过 Label 来确定这个偏移位置。对应的 ASM 实现方式如下。

```
...
// 创建一个 Label 对象
Label label = new Label();(1)
```

```
// 此处的 label 表示当前跳转指令需要跳转到的目标位置，此时具体的目标位置还未确定
methodVisitor.visitJumpInsn(Opcodes.IFNE, label);(2)
// 执行 if 语句块中的 iinc 指令
methodVisitor.visitIincInsn(1, 1);
// 使用 visitLabel() 方法设置一个目标位置或者锚记点，这样 IFNE 指令满足条件就会跳到这个位置
methodVisitor.visitLabel(label);(3)
...
```

上述代码在位置（1）创建了一个 Label 对象。在位置（2）使用 visitJumpInsn() 方法将 IFNE 指令压入操作数栈，该指令判断栈顶的值是否满足不等于 0，如果满足就跳转到 label 指定的位置。此时 label 的具体位置还没有确定，在位置（3）调用 visitLabel() 方法，正式确定了 label 的具体位置。ASM 会负责计算好该 label 位置的具体字节码偏移值。visitLabel() 设定的目标位置，可以被多个跳转指令使用，它就像页面中的锚点一样，当单击的时候就会跳转到标记位置。

5.7.17 visitLdcInsn()

此方法用于访问或者将常量压入操作数栈中，方法定义如下。

```
public void visitLdcInsn(final Object value)
```

LDC 指令可以将 integer、float、long、double、string 等类型的常量压入操作数栈中。这一点与第 4 章介绍的 LDC 指令有些不同，LDC 指令的操作数是指向常量池的有效索引，visitLdcInsn(value) 方法是直接使用值，具体的常量池管理交给 ASM 内部来完成。

示例如下。

```
public void foo(){
      String a = "hello";
      String b = "hello";
      int c = 9;
      float d = 1.2f;
      double e = 1.5;
      long f = 100L;
}
```

对于上述 Java 方法中的变量，假如使用 visitLdcInsn() 方法添加指令，对应的代码如下。

```
methodVisitor.visitLdcInsn("hello");
methodVisitor.visitLdcInsn("hello");
methodVisitor.visitLdcInsn(9);
methodVisitor.visitLdcInsn(1.2f);
methodVisitor.visitLdcInsn(1.5d);
methodVisitor.visitLdcInsn(100L);
```

读者可能会有疑问：visitLdcInsn(9) 与 bipush 9 都是将 9 压入栈顶，那它们有什么区别呢？

bipush 9 是将数字压入栈顶，而 visitLdcInsn(9) 是将常量 9 压入栈顶，既然是常量，就会在常量池中记录下来，字节码如下所示。

```
Constant pool:
    ...
    #15 = Integer            9
    ...
    public void foo();
```

```
descriptor: ()V
flags: ACC_PUBLIC
Code:
  stack=2, locals=9, args_size=1
     ...
     6: ldc    #15      // int 9, 指向常量池中的结果
     ...
```

两种方式的效果是一样的，但是效率不一样。

5.7.18　visitIincInsn()

此方法用于访问或将 IINC 指令压入栈顶，方法定义如下。

```
public void visitIincInsn(final int var, final int increment)
```

其中 var 表示局部变量表中的位置，increment 表示增加的数量。IINC 是唯一一个对局部变量表中数据进行计算的指令。

例如实现 i += 9 这样的功能就可以这么做。

```
methodVisitor.visitIincInsn(var, 9)
```

5.7.19　visitTableSwitchInsn()

此方法用于访问或者将 tableswitch 指令压入栈顶，方法定义如下。

```
public void visitTableSwitchInsn(
        final int min, final int max, final Label dflt, final Label... labels)
```

回顾一下 tableswitch 指令，它是通过索引访问跳转表并跳转到指定的位置继续执行的，适用于分支比较集中连续的情况。跳转表的索引是连续的，每个索引对应一个跳转位置。根据以上知识，visitTableSwitchInsn() 方法中的参数说明如下。

- min 表示索引表中的最小值。
- max 表示索引表的最大值。
- dflt 表示索引表中默认跳转位置。
- labels 表示其他各索引的跳转位置。

结合实际例子来说明。

```
public void bar(int a) {
     switch (a){
         case 1:
             a=a+1;
             break;
         case 2:
             a=a+2;
             break;
         case 4:
             a=a+4;
             break;
     }
}
```

上面代码对应的字节码如下。

```
public void bar(int);
     descriptor: (I)V
     flags: ACC_PUBLIC
     Code:
       stack=2, locals=2, args_size=2
          0: iload_1
          1: tableswitch   { // 1 to 4
                       1: 32
                       2: 39
                       3: 50
                       4: 46
                 default: 50
            }
         32: iload_1
         33: iconst_1
         34: iadd
         35: istore_1
         36: goto          50
         39: iload_1
         40: iconst_2
         41: iadd
         42: istore_1
         43: goto          50
         46: iload_1
         47: iconst_4
         48: iadd
         49: istore_1
         50: return
```

参照上述字节码，其中 tableswitch() 对应的 ASM 实现如下。

```
Label label1 = new Label();
Label label2 = new Label();
Label label3 = new Label();
Label label4 = new Label();
methodVisitor.visitTableSwitchInsn(
      1, 4, label3, new Label[]{label1, label2, label3, label4});
...
methodVisitor.visitLabel(label3);
methodVisitor.visitInsn(RETURN);
```

其中label3为默认索引的位置，从上述字节码中可以看到此位置是50: return，其他label的用法类似。这里要注意，visitLabel() 方法要在目标指令之前调用，例如 36: goto 50，字节码指令偏移位置的指令是return，使用 ASM 时就要在 visitInsn(RETURN) 方法之前调用 visitLabel()。

5.7.20 visitLookupSwitchInsn()

此方法用于访问或者将 lookupswitch 指令压入栈顶，方法定义如下。

```
public void visitLookupSwitchInsn(
      final Label dflt, final int[] keys, final Label[] labels)
```

根据第 3 章我们知道 lookupswitch 适用于分支比较松散的情况，它的使用方式与 visitTableSwitchInsn()

比较相似。visitLookupSwitchInsn() 方法中的参数说明如下。

- dflt 表示跳转表中默认跳转位置。
- keys 表示跳转表中的 key 值。
- labels 表示指定的 key 值跳转的位置。

结合实际例子来理解。

```
public void bar(int a) {
    switch (a){
        case 1:
            a=a+1;
            break;
        case 10:
            a=a+2;
            break;
        case 20:
            a=a+4;
            break;
    }
}
```

上面代码对应的字节码如下。

```
public void bar(int);
    descriptor: (I)V
    flags: ACC_PUBLIC
    Code:
      stack=2, locals=2, args_size=2
         0: iload_1
         1: lookupswitch  { // 3
                       1: 36
                      10: 43
                      20: 50
                 default: 54
            }
        36: iload_1
        37: iconst_1
        38: iadd
        39: istore_1
        40: goto          54
        43: iload_1
        44: iconst_2
        45: iadd
        46: istore_1
        47: goto          54
        50: iload_1
        51: iconst_4
        52: iadd
        53: istore_1
        54: return
```

参照上述字节码，其中 lookupswitch() 对应的 ASM 实现如下。

```
Label label1 = new Label();
Label label2 = new Label();
Label label3 = new Label();
Label label4 = new Label();
```

```
methodVisitor.visitLookupSwitchInsn(
        label4, new int[]{1, 10, 20}, new Label[]{label1, label2, label3});
```

上述代码结合使用方式和 API 定义很容易理解，在此不多做介绍。

5.7.21 visitTryCatchBlock()

此方法用于实现 try-catch 语句块，方法定义如下。

```
public void visitTryCatchBlock(
        final Label start, final Label end, final Label handler, final String type)
```

在 4.7.4 小节中详细地介绍了 try-catch-finally 的原理，从中我们知道 Java 中 try-catch 语句块使用的是异常表（exception table）来保存。通过异常表，读者能够很容易理解 visitTryCatchBlock() 中各个参数的意义，在此就不多做表述了。

5.7.22 visitLocalVariable和visitLineNumber()

```
public void visitLocalVariable(
        final String name, final String descriptor,
        final String signature, final Label start,
        final Label end, final int index)
public void visitLineNumber(final int line, final Label start)
```

visitLocalVariable() 方法用于设定 Code 属性中的 LocalVariableTable 表中的某一项的值，调试器可以使用它来确定方法执行期间给定局部变量的值。visitLineNumber() 则用于设置行号信息，调试器可以使用它来确定代码的哪一部分对应于源文件中的给定行号。

5.7.23 visitFrame()

此方法用于访问栈中信息，方法定义如下。

```
public void visitFrame(
        final int type, final int numLocal,
        final Object[] local, final int numStack, final Object[] stack)
```

在第 3 章中详细地介绍了什么是栈映射帧以及 StackMapTable 属性的作用，我们知道从 Java 6 开始，在编译时会生成一系列的栈映射帧，以此来加快 JVM 校验的速度。栈映射帧记录了某些指令执行之前的局部变量和操作数栈的状态，而为了节省空间，并不会为每条指令都生成栈映射帧，而是在跳转目标（jump target）、异常处理（exception handlers）位置或者在无条件跳转指令（unconditional jump instructions）后面生成对应的栈映射帧，这种方式我们又称为压缩形式（compressed form），与之相对应的方式称为展开形式（expanded form）。

visitFrame() 方法中各参数的意义如下。

- type 表示栈映射帧的类型，必须是 Opcodes.F_NEW、Opcodes.F_SAME、Opcodes.F_

SAME1、Opcodes.F_APPEND、Opcodes.F_CHOP、Opcodes.F_FULL之一，其中如果Opcodes.F_NEW只用于展开形式，其他形式可以对比第3章介绍的栈映射帧的类型描述。

- numLocal表示局部变量的数量。
- local表示局部变量的类型，基本类型使用Opcodes.TOP、Opcodes.INTEGER、Opcodes.FLOAT、Opcodes.LONG、Opcodes.DOUBLE、Opcodes.NULL、Opcodes.UNINITIALIZED_THIS；引用类型使用内部名字的字符串，例如java/lang/Runnable。如果numLocal = 0，此处传入null即可。
- numStack表示操作数栈中的元素数量。
- stack表示操作数栈中的元素类型，参考local中的类型定义。

举例如下。

```
int data = 0;
public void checkAndSetData(int data) {
      if (data >= 0) {
          this.data = data;
      } else {
          throw new IllegalArgumentException();
      }
}
```

对应的字节码如下。

```
public void checkAndSetData(int);
    descriptor: (I)V
    flags: ACC_PUBLIC
    Code:
      stack=2, locals=2, args_size=2
     0: iload_1
     1: iflt          12
     4: aload_0
     5: iload_1
     6: putfield      #2  // Field data:I
     9: goto           20
    12: new            #3  // class java/lang/IllegalArgumentException
    15: dup
    16: invokespecial #4 // Method java/lang/IllegalArgumentException."<init>":()V
    19: athrow
    20: return

    StackMapTable: number_of_entries = 2
      frame_type = 12 /* same */
      frame_type = 7 /* same */
```

根据上面的字节码，能够推断出来需要两个栈映射帧：一个是iflt指令的跳转目标12: new，同时它也是无条件跳转指令9: goto指令的下一条指令；另一条是20: return，它是9: goto指令的跳转目标，同时它也是无条件跳转指令19: athrow的下一条指令。根据前面对StackMapTable的了解可知，每个方法默认有一个隐式的栈映射帧，可以从方法的描述符中推断出来。将这两个栈映射帧与隐式的栈映射帧对比，可以知道它们压缩形式的帧类型（frame_type）为same_frame，即局部变量和操作数栈没有变化。

使用ASM的实现方式如下。

```
...
Label label1 = new Label();
methodVisitor.visitJumpInsn(IFLT, label1);
```

```
methodVisitor.visitVarInsn(ALOAD, 0);
methodVisitor.visitVarInsn(ILOAD, 1);
methodVisitor.visitFieldInsn(
      PUTFIELD, "cn/sensorsdata/asm/mv/Test", "data", "I");
Label label2 = new Label();
methodVisitor.visitJumpInsn(GOTO, label2);
methodVisitor.visitLabel(label1);
methodVisitor.visitFrame(Opcodes.F_SAME, 0, null, 0, null);
methodVisitor.visitTypeInsn(NEW, "java/lang/IllegalArgumentException");
methodVisitor.visitInsn(DUP);
methodVisitor.visitMethodInsn(
      INVOKESPECIAL, "java/lang/IllegalArgumentException", "<init>", "()V", false);
methodVisitor.visitInsn(ATHROW);
methodVisitor.visitLabel(label2);
methodVisitor.visitFrame(Opcodes.F_SAME, 0, null, 0, null);
methodVisitor.visitInsn(RETURN);
...
```

其中 Opcodes.F_SAME 表示与前一帧相比，局部变量和操作数栈没有变化。

从上面的介绍可以看出，计算一个方法所有的栈映射帧并不是一件容易的事情，包括计算局部变量表和操作数栈的最大值，好在可以在创建 ClassWriter 的时候设置 flags 让 ASM 帮我们自动计算这些值，具体的使用方式如下。

- 当为 new ClassWriter(0) 时，表示 ASM 什么都不做，用户需要自己去计算栈映射帧、局部变量表和操作数栈的最大值。
- 当为 new ClassWriter(ClassWriter.COMPUTE_MAXS) 时，ASM 将自动计算局部变量表和操作数栈的最大值，但是用户仍然需要调用 MethodVisitor 的 visitMaxs() 方法，不过 visitMaxs() 方法中传的参数将被忽略。此种方式还是需要用户自己去计算栈映射帧。
- 当为 new ClassWriter(ClassWriter.COMPUTE_FRAMES) 时，ASM 将自动计算栈映射帧、局部变量表和操作数栈的最大值，并且不需要 MethodVisitor 的 visitFrame() 方法，不过 visitMaxs() 方法仍然需要被调用，其值也会被忽略，换句话说，COMPUTE_FRAMES 包括 COMPUTE_MAXS。

ClassWriter 提供了自动计算帧的方式，其代价就是需要消耗更多的资源，其中 COMPUTE_MAXS 性能降低约 10%，而 COMPUTE_FRAMES 则降低约 20%。如果用户选择自己来计算栈映射帧，用户可以使用 Opcodes.F_NEW 这个类型，即 visitFrame(F_NEW, nLocals, locals, nStacks, stacks)，这样 ClassWriter 将会为用户压缩这些帧。

5.7.24 visitMaxs()

此方法用于访问或设置方法的操作数栈和局部变量表的最大值，方法定义如下。

```
public void visitMaxs(final int maxStack, final int maxLocals)
```

如果使用了 COMPUTE_MAXS 选项，则只需要调用即可，不需要考虑其中参数的值。

举例如下。

```
ClassWriter cw = new ClassWriter(COMPUTE_MAXS);
...
MethodVisitor methodVisitor =
```

```
        cw.visitMethod(ACC_PUBLIC, "add", "(II)I", null, null);
    methodVisitor.visitCode();
    Label label0 = new Label();
    methodVisitor.visitLabel(label0);
    methodVisitor.visitLineNumber(6, label0);
    methodVisitor.visitVarInsn(ILOAD, 1);
    methodVisitor.visitVarInsn(ILOAD, 2);
    methodVisitor.visitInsn(IADD);
    methodVisitor.visitInsn(IRETURN);
    methodVisitor.visitMaxs(0, 0);
    methodVisitor.visitEnd();
```

上述代码是生成 add(int a, int b) 方法的 ASM 片段，并且 ClassWriter 中添加了 COMPUTE_MAXS 的参数，这样 ASM 就不会考虑 visitMaxs(0, 0) 中的参数值，ASM 会帮我们计算正确的结果。

图 5-6 所示为生成后的结果，可以看到其 maxStack 和 maxLocals 并不是上述代码中设置的值。

```
public int add(int, int);
  descriptor: (II)I
  flags: ACC_PUBLIC
  Code:
    stack=2, locals=3, args_size=3
       0: iload_1
       1: iload_2
       2: iadd
       3: ireturn
```

图 5-6　maxStack 和 maxLocals 的结果展示

5.7.25　visitEnd()

此方法是 MethodVisitor 最后需要调用的方法，用于通知 visitor 所有的操作已处理完毕。

```
public void visitEnd()
```

至此我们介绍了 MethodVisitor 中绝大部分的 API 并简单地介绍了用法，接下来看看更多关于 MethodVisitor 的使用方法。

5.8　方法的转换和修改

5.8.1　方法生成

假设需要生成如下代码中的类。

```
package cn.sensorsdata.asm.mv;

public final class Utils {
     public static int add(int a, int b) {
         return a + b;
```

```
        }
}
```

Utils 类中的 add() 方法功能比较简单，可以根据源码在脑海中构建其对应的字节码。其实现思路是：加载两个局部变量表中的值到操作数栈中，再调用 IRETURN 指令即可。假如方法体中的内容很复杂，只通过源码在脑海中构建对应的指令就变得无法操作了，这种情况先使用 javap 工具输出对应的字节码内容，参照字节码做相应的处理即可。

例如，这个例子中 add() 方法的字节码如下。

```
public static int add(int, int);
     descriptor: (II)I
     flags: ACC_PUBLIC, ACC_STATIC
     Code:
       stack=2, locals=2, args_size=2
          0: iload_0
          1: iload_1
          2: iadd
          3: ireturn
```

使用 MethodVisitor 的实现方式如下。

```
// 参数 0 表示手动计算操作数栈和局部变量表的最大值以及栈映射帧
ClassWriter classWriter = new ClassWriter(0);
// 创建对应的类的基本信息
classWriter.visit(
      Opcodes.V1_8,
      Opcodes.ACC_PUBLIC + Opcodes.ACC_FINAL,
      "cn/sensorsdata/asm/mv/Utils", null, "java/lang/Object",null);
// 创建 add() 方法
MethodVisitor mv = classWriter.visitMethod(
      Opcodes.ACC_PUBLIC + Opcodes.ACC_STATIC, "add", "(II)I", null, null);
// 调用 visitCode() 方法，表示接下来处理方法体中的内容
mv.visitCode();
// 因为是 static 类型，所以局部变量表的第一个位置就是方法第一个参数的值
mv.visitVarInsn(Opcodes.ALOAD, 0);
mv.visitVarInsn(Opcodes.ALOAD, 1);
mv.visitInsn(Opcodes.IADD);
// 调用返回指令
mv.visitInsn(Opcodes.IRETURN);
// 设置计算操作数栈和局部变量表的最大值，从字节码可以看出最大值都是 2
mv.visitMaxs(2,2);
// 调用 visitEnd() 方法表示方法处理完毕
mv.visitEnd();
// 同样不要忘了调用 ClassVisitor 的 visitEnd 方法
classWriter.visitEnd();
// 将 data 数据保存到 Utils.class 文件中即可
byte[] data = classWriter.toByteArray();
```

上述代码中已添加了详细的注释，读者对照着理解即可。对于刚接触 ASM 的读者，要注意如下几点。

- ClassVisitor 和 MethodVisitor 中的方法调用顺序要明确，必须按照官方给定的顺序调用。
- 不要忘记调用 xRETURN 指令。如果方法返回的是 void，读者很可能会漏掉 RETURN 指令，字节码校验会报错，请不要漏掉。

图 5-7 所示为反编译上面代码生成的结果。

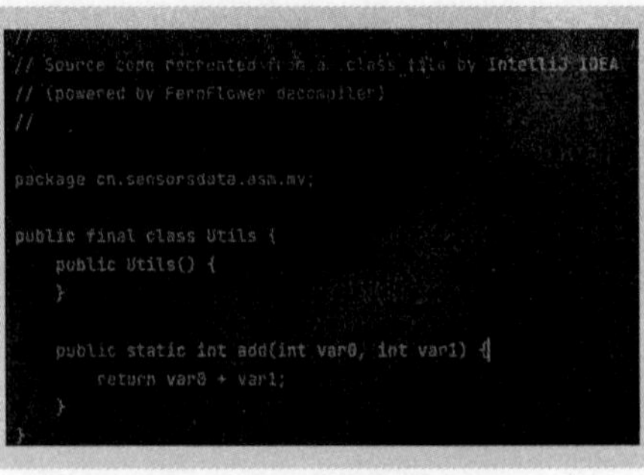

```
//
// Source code recreated from a .class file by IntelliJ IDEA
// (powered by FernFlower decompiler)
//

package cn.sensorsdata.asm.mv;

public final class Utils {
    public Utils() {
    }

    public static int add(int var0, int var1) {
        return var0 + var1;
    }
}
```

图 5-7　反编译结果展示

细心的读者可能发现了，上述反编译的结果中 add() 方法的参数名并不是 a 和 b，这是因为没有添加 MethodParameters 属性。这个属性不是必需的，没有也不影响执行结果。

5.8.2　删除方法和方法体内容

5.6.2 小节已介绍了删除方法的方式，不过这种方式是直接将方法删除掉，其风险还是比较大的，因为该方法可能在其他地方被使用。若要让方法不起作用，还有一种方法，就是删除方法体中的内容，而保留方法的定义。

一个实际场景：假如下面代码中的 getMacAddress() 方法是第三方 JAR 包提供的用于获取 MAC 地址信息的方法，MAC 地址属于敏感信息，那么该 JAR 包将无法通过合规检测。此时就可以借助 ASM 将方法体内容删除，让其返回一个空字符串或者 null，这样就不影响其他代码对这个方法的调用。

```
public String getMacAddress(Object device){
      return device.toString();
}
```

定义 RemovingVisitor，用来删除方法体的内容，代码如下。

```
public static class RemovingVisitor extends ClassVisitor {
      public RemovingVisitor(ClassVisitor classVisitor) {
          super(Opcodes.ASM6, classVisitor);
      }
      public MethodVisitor visitMethod(
          int access, String name, String descriptor,
          String signature, String[] exceptions)
      {
          if("getMacAddress".equals(name) &&
             "(Ljava/lang/Object;)Ljava/lang/String;".equals(descriptor) )
          {
              MethodVisitor mv =
                  cv.visitMethod(access,name,descriptor,signature,exceptions);
              mv.visitCode();
              mv.visitLdcInsn("");
              mv.visitInsn(Opcodes.ARETURN);
              mv.visitMaxs(1,2);
              mv.visitEnd();
              return mv;
          }
          return super.visitMethod(access, name, descriptor,
```

```
                                    signature, exceptions);
        }
    }
```

上面代码首先判断了方法名和方法描述符（建议读者也这么做，防止有重载的方法被误操作），当满足条件时重新生成方法体中的内容，并且返回这个方法用于替代原有的方法，这样就达到了清空方法的目的。

5.8.3 优化方法中的指令

方法中的指令有必要优化吗？当然，非常有必要。有很多工具提供这样的能力，例如 ProGuard 可以删除没有使用的方法，还能对方法中的指令进行优化，达到减小文件体积的目的。

举例如下。

```
ICONST_0
IADD
```

上面指令表示将一个整数与 0 进行相加并将结果压入操作数栈中，这显然是无意义的操作，这种情况就可以对其进行优化。针对这种情况，只要判断连续两条指令如果是 ICONST_0 和 IADD，那么将这两条指令删除即可。

具体思路是：不同的指令需要调用 MethodVisitor 类中的不同方法，ICONST_0、IADD 指令都需要调用 visitInsn() 方法，可以在此方法中判断指令是否是 ICONST_0，如果是 ICONST_0 指令就直接返回，并判断下一条指令是否是 IADD，如果是也直接返回，这样就达到了删除两条指令的目的。需要注意的是 ICONST_0 后面的指令可能不是 IADD，也就是说下一条指令是调用了 MethodVisitor 的其他 visit*Insn() 方法，此时需要在这些方法中将 ICONST_0 添加回去，避免 ICONST_0 指令的丢失。

具体的实现代码如下。

```
public static class RemovingUselessCode extends MethodVisitor {

        private final static int SEEN_NOTHING = 0;
        private final static int SEEN_ICONST_0 = 1;
        // 定义 ICONST_0 指令的状态，SEEN_NOTHING 表示未发现 ICONST_0 指令
        protected int state = SEEN_NOTHING;

        public RemovingUselessCode(MethodVisitor methodVisitor) {
            super(Opcodes.ASM6, methodVisitor);
        }

        @Override
        public void visitCode() {
            super.visitCode();
        }

        @Override
        public void visitInsn(int opcode) {
            // 如果上一条指令是 ICONST_0，并且当前指令是 IADD，就直接返回，
            // 以达到删除两条指令的目的
            if (state == SEEN_ICONST_0) {
                if (opcode == Opcodes.IADD) {
                    state = SEEN_NOTHING;
                    return;
```

```
                }
            }
            visitInsn();
            // 如果指令是 ICONST_0，就直接返回，并记录状态
            if (opcode == Opcodes.ICONST_0) {
                state = SEEN_ICONST_0;
                return;
            }
            super.visitInsn(opcode);
        }

        @Override
        public void visitIntInsn(int opcode, int operand) {
            // 判断是否需要补上 ICONST_0
            visitInsn();
            super.visitIntInsn(opcode, operand);
        }

        @Override
        public void visitFieldInsn(
            int opcode, String owner, String name, String descriptor)
        {
            // 判断是否需要补上 ICONST_0
            visitInsn();
            super.visitFieldInsn(opcode, owner, name, descriptor);
        }

        ...

        void visitInsn() {
            // 如果上一条指令是 ICONST_0，因为这一条指令不是 IADD，
            // 就需要补上 ICONST_0，并修改状态
            if (state == SEEN_ICONST_0) {
                mv.visitInsn(Opcodes.ICONST_0);
            }
            state = SEEN_NOTHING;
        }
}
```

注意上面只重写了两个 visit*Insn() 方法，其他方法按照此方式去实现即可。

下面是一个能够产生无意义指令的代码，读者将其编译后可以先观察其构造方法和 foo() 方法中的与 ICONST_0 相关的字节码，然后执行 RemovingUselessCode 以后再查看结果。

```
public class C {
      int c = 0;
      public void foo() {
          int a = 10;
          int b = a + 0;
      }
}
```

通过上面的方式去操作字节码比较烦琐、复杂，使用 Tree API 处理这种情况将会简单很多，我们将在第 6 章介绍这些内容。

5.9 ASM工具包介绍

前文我们介绍了 ASM Core API 中比较重要的 3 个类，除此之外，ASM 还提供了一些非常有用的工具类，这些类位于 org.objectweb.asm.util 包中。这些工具类中有的使用在开发过程中，有的则提供一些通用功能，例如 Type 类就是这样的工具类，下面先来看看 Type 的用法。

5.9.1 Type

在使用 ASM 的时候，经常需要处理各种类型。比如，获取类的全限定名、获取方法的描述符，通过方法描述符获得其中的参数类型、参数数量、返回值类型等操作，这些操作都可以通过 Type 类处理。另外 Type 还对常用的基本类型都进行了封装，可以轻松地做到类型转换。

（1）首先可以使用 Type 获取方法描述符中的参数数量和返回值，代码如下。

```
// 根据方法描述符构建一个 Type 对象
Type type = Type.getMethodType("(Ljava/lang/String;JZ)I");(1)
//getArgumentTypes() 获取方法描述符中参数类型 Type[]
int argumentsCount = type.getArgumentTypes().length;(2)
//getReturnType() 获取方法描述符中返回值类型 Type，根据方法描述符可知它是一个 int 类型
String returnDescriptor = type.getReturnType().getDescriptor();(3)
System.out.println("args count: " + argumentsCount);
System.out.println("return descriptor: " + returnDescriptor);
```

输出结果如下。

```
args count: 3
return descriptor: I
```

上面代码位置（1）通过 getMethodType() 方法，将方法描述符转换成 Type 类型。再通过位置（2）的 getArgumentTypes() 获取方法中的所有参数数组，通过数组的长度即可计算出参数数量。最后通过位置（3）的 getReturnType() 方法获取返回值 Type，通过 Type 的 getDescriptor() 方法可以获取到对应的字段描述符。

（2）Type 还支持从 Class、Method、Constructor 等对象中获取到对应的 Type，相关方法如下。

```
public static Type getReturnType(final Method method)
public static String getDescriptor(final Class<?> clazz)
public static String getConstructorDescriptor(final Constructor<?> constructor)
public static Type getType(final Class<?> clazz)
public static Type getType(final Constructor<?> constructor)
public static Type getType(final Method method)
```

例如，下面这段代码通过 String.class 对象获取到类的描述符、类名和全限定名（也叫内部名，ClassFile 中使用的名）。

```
Type type = Type.getType(String.class);
System.out.println("Descriptor: " + type.getDescriptor());
System.out.println("ClassName: " + type.getClassName());
System.out.println("InternalName(全限定名): " + type.getInternalName());
```

输出结果如下。

```
Descriptor: Ljava/lang/String;
```

```
ClassName: java.lang.String
InternalName(全限定名): java/lang/String
```

（3）对于常用的基本类型，Type 也提供了封装，可以很方便地使用它们，如 Type.INT_TYPE、Type.FLOAT_TYPE、Type.LONG_TYPE、Type.DOUBLE_TYPE、Type.BYTE_TYPE、Type.SHORT_TYPE、Type.VOID_TYPE、Type.CHAR_TYPE 等。

例如，输出 TYPE.INT_TYPE 的描述符。

```
System.out.println("Int 类型的描述符: " + Type.INT_TYPE.getDescriptor());
```

输出结果如下。

```
Int 类型的描述符：I
```

另外，在开发中经常需要计算某个类型所占的“槽”大小。例如在局部变量表和操作数栈中，long、double 类型占用两个槽、int 类型占一个槽，这个时候可以通过 Type 的 getSize() 方法获取结果。

```
public int getSize()
```

这个方法对我们动态计算类型的大小很有帮助，实际开发中也会经常使用到。

这里留给大家一个简单的练习题：对于下面的方法描述符，如何使用 Type 中的方法使其返回值类型为 VOID 以及在方法参数的头尾各追加一个 Object 类型，即修改后的结果是“(IJZLjava/lang/Object;)V”。

```
"(IJZ)B"
```

希望大家能够实际动手操作一下，熟悉 Type 这个工具的使用。

5.9.2 TraceClassVisitor

在使用 ASM 生成和转换类的时候，常常需要确认结果是否是我们想要的，通常的做法是将 ClassWriter.toByteArray() 的结果写入 ClassFile 中，然后反编译或者通过 javap 命令查看其字节码是否符合预期来确认结果。ASM 提供了类似的功能，使用 TraceClassVisitor 这个类就可以输出类的字节码信息，便于开发时调试。TraceClassVisitor 继承自 ClassVisitor，这就意味着我们可以像使用其他 ClassVisitor 一样使用它。

例如，用其解析下面的代码。

```
public final class Utils {
      public static int add(int a, int b) {
          return a + b;
      }
}
//获取 Utils.class 的内容
byte[] bytes = FileUtils.readFileToByteArray(file);
ClassReader classReader = new ClassReader(bytes);
ClassWriter classWriter = new ClassWriter(0);
//把 TraceClassVisitor 当作普通的 ClassVisitor 使用
TraceClassVisitor traceClassVisitor =
      new TraceClassVisitor(classWriter, new PrintWriter(System.out));
classReader.accept(traceClassVisitor,0);
```

上面代码中 TraceClassVisitor 接收了一个 PrintWriter 对象，用于输出结果。PrintWriter 有多个构造方法，不同的构造方法可以指明输出的目标位置。上面代码是将结果输出到控制台中，运行结果如下。

```
// class version 52.0 (52)
// access flags 0x31
public final class cn/sensorsdata/asm/mv/Utils {

    // compiled from: Utils.java

    // access flags 0x1
    public <init>()V
     L0
      LINENUMBER 3 L0
      ALOAD 0
      INVOKESPECIAL java/lang/Object.<init> ()V
      RETURN
      MAXSTACK = 1
      MAXLOCALS = 1

    // access flags 0x9
    public static add(II)I
     L0
      LINENUMBER 5 L0
      ILOAD 0
      ILOAD 1
      IADD
      IRETURN
      MAXSTACK = 2
      MAXLOCALS = 2
}
```

可以看到输出结果与 javap 的结果非常像。关于其背后的原理会在 5.9.4 小节进行介绍。

5.9.3 CheckClassAdapter

ClassWriter 只负责把结果写入到 .class 中，并不会检测方法的指令是否正确或方法的调用顺序是否正确。这就使得开发者必须很小心，否则生成的 .class 很可能存在问题导致不能被 JVM 正常加载。CheckClassAdapter 可以判断生成或转换的字节码是否正确，这在实际开发中很有用。

其构造方法如下。

```
public class CheckClassAdapter extends ClassVisitor {
    /**
     * Constructs a new {@link CheckClassAdapter}.
     * <i>Subclasses must not use this constructor</i>.
     * Instead, they must use the {@link #CheckClassAdapter(int, ClassVisitor,
     * boolean)} version.
     *
     * @param classVisitor the class visitor to which this adapter must
     * delegate calls.
     */
    public CheckClassAdapter(final ClassVisitor classVisitor) {
      this(classVisitor, true);
    }
    ...
}
```

CheckClassAdapter 跟 TraceClassVisitor 类似，都是 ClassVisitor 的子类。根据使用时机不同，CheckClassAdapter 有如下两种使用方式。

（1）在生成或转换类过程中使用。

（2）在生成或转换类完成后使用。

接下来介绍具体使用方式。

（1）在生成或转换类过程中使用。下面是这种方式对应的代码模板。

```
//1. 类生成过程中的模板
ClassWriter classWriter = new ClassWriter(0);
// 此处使用 CheckClassAdapter
CheckClassAdapter checkClassAdapter = new CheckClassAdapter(classWriter);
checkClassAdapter.visit();
...
checkClassAdapter.visitEnd();


//2. 类转换过程中的模板
byte[] bytes = null;
ClassReader classReader = new ClassReader(bytes);
ClassWriter classWriter = new ClassWriter(0);
ClassVisitor cv = new MyClassVisitor(classWriter);

CheckClassAdapter cca = new CheckClassAdapter(cv);
classReader.accept(cca, 0);
```

上面是生成和转换过程中 CheckClassAdapter 的两种用法，下面举例“类的生成过程”中的用法。

```
// 生成 LogUtils 类
ClassWriter classWriter = new ClassWriter(0);
// 使用 CheckClassAdapter
CheckClassAdapter checkClassAdapter = new CheckClassAdapter(classWriter);
checkClassAdapter.visit(
      Opcodes.V1_8, Opcodes.ACC_PUBLIC+Opcodes.ACC_FINAL,
      "cn/sensorsdata/asm/mv/LogUtils",
      null, "java/lang/Object", null);
MethodVisitor mv = checkClassAdapter.visitMethod(
                   Opcodes.ACC_PUBLIC+Opcodes.ACC_STATIC,
                   "log", "(Ljava/lang/String;)V", null, null);
mv.visitCode();
mv.visitFieldInsn(Opcodes.GETSTATIC,
                    "java/lang/System",
                    "out",
                    "Ljava/io/PrintStream;");
mv.visitVarInsn(Opcodes.ALOAD, 0);
mv.visitMethodInsn(Opcodes.INVOKEVIRTUAL,
                     "java/io/PrintStream",
                     "println", "(Ljava/lang/String;)V", false);// (1)
mv.visitInsn(Opcodes.RETURN);// (2)
mv.visitMaxs(2, 1); // (3)
mv.visitEnd();
checkClassAdapter.visitEnd();
byte[] data = classWriter.toByteArray();
String filePath = "src/main/java/cn/sensorsdata/asm/mv/";
FileUtils.writeByteArrayToFile(new File(filePath + "LogUtils.class"), data);
```

上述代码会生成如下的类。

```
package cn.sensorsdata.asm.mv;

public final class LogUtils {
      public static void log(String var0) {
          System.out.println(var0);
      }
}
```

假如把位置（1）的“(Ljava/lang/String;)V”改成“(I)V”，使其方法参数不正确，运行时就会报如下错误。

```
java.lang.IllegalArgumentException: Error at instruction 2: Argument 1: expected I,
but found R log(Ljava/lang/String;)V
```

假如把位置（2）的 return 指令注释掉，使其没有 return 语句，运行时会报如下错误。

```
java.lang.IllegalArgumentException: Execution can fall off the end of the code log
(Ljava/lang/String;)V
```

假如把位置（3）的 visitMax() 方法注释掉，运行时会报如下错误。

```
java.lang.IllegalArgumentException: Data flow checking option requires valid, non zero
maxLocals and maxStack.
```

以上就是一些示例代码，上述有的异常信息并不是那么容易理解，例如 return 语句的异常描述，这也是为什么我们多次提示不要忘了 return 指令的原因。

（2）在生成或转换类完成后使用。CheckClassAdapter 还提供了一个静态 verify() 方法，用于对生成或转换后的结果进行验证，其定义如下。

```
public static void verify(
      final ClassReader classReader, final boolean printResults,
      final PrintWriter printWriter)
public static void verify(
      final ClassReader classReader, final ClassLoader loader,
      final boolean printResults, final PrintWriter printWriter)
```

其中参数 classReader 为需要被校验的数据来源、printResults 表示是否输出校验结果、printWriter 表示输出结果的位置（例如控制台、文件中），需要注意的是 printWriter 总会输出异常栈信息。

举例如下。

```
ClassWriter classWriter = new ClassWriter(0);
classWriter.visit(
      Opcodes.V1_8,
      Opcodes.ACC_PUBLIC + Opcodes.ACC_FINAL,
      "cn/sensorsdata/asm/mv/LogUtils", null, "java/lang/Object", null);
MethodVisitor mv = classWriter.visitMethod(
      Opcodes.ACC_PUBLIC + Opcodes.ACC_STATIC,
      "log", "(Ljava/lang/String;)V", null, null);
mv.visitCode();
mv.visitFieldInsn(
      Opcodes.GETSTATIC,
      "java/lang/System",
      "out", "Ljava/io/PrintStream;");
mv.visitVarInsn(Opcodes.ALOAD, 0);
```

```
mv.visitMethodInsn(
      Opcodes.INVOKEVIRTUAL,
      "java/io/PrintStream",
      "println", "(Ljava/lang/String;)V", false);
mv.visitInsn(Opcodes.RETURN);
mv.visitMaxs(2,1);
mv.visitEnd();
classWriter.visitEnd();
byte[] data = classWriter.toByteArray();

//verify 校验
ClassReader classReader = new ClassReader(data);
CheckClassAdapter.verify(classReader, false, new PrintWriter(System.out));
String filePath = "src/main/java/cn/sensorsdata/asm/mv/";
FileUtils.writeByteArrayToFile(new File(filePath + "LogUtils.class"), data);
```

上述代码先生成了一个 LogUtils 类，通过 toByteArray() 获取到结果，根据结构构建 ClassReader 对象，最后使用 CheckClassAdapter 的 verify() 方法校验。CheckClassAdapter 非常有用，在开发中使用它对字节码进行校验可以提前发现问题，节省很多的排查时间。

5.9.4 ASMifier

5.9.2 小节介绍了 TraceClassVisitor 类的使用，现在观察该类的部分源码。

```
public final Printer p;
public TraceClassVisitor(final PrintWriter printWriter) {
    this(null, printWriter);
}
public TraceClassVisitor(final ClassVisitor classVisitor,
                         final PrintWriter printWriter)
{
    this(classVisitor, new Textifier(), printWriter);
}
public TraceClassVisitor(
      final ClassVisitor classVisitor,
      final Printer printer, final PrintWriter printWriter)
{
    super(/* latest api = */ Opcodes.ASM10_EXPERIMENTAL, classVisitor);
    this.printWriter = printWriter;
    this.p = printer;
}
```

它的第三个构造方法中，参数 Printer 的作用是将 ClassVisitor 中的 visit() 方法的结果字符化并输出。Printer 是个抽象类，该类中定义了很多的 visit*() 方法，几乎与 ClassVisitor 中的 visit*() 方法一一对应。

例如，观察 TraceClassVisitor 中部分 visit() 方法的源码。

```
@Override
public void visit(
      final int version, final int access, f
      inal String name, final String signature, f
      inal String superName, final String[] interfaces)
{
    // 变量 p 就是 Printer 对象
```

```
    p.visit(version, access, name, signature, superName, interfaces);
    super.visit(version, access, name, signature, superName, interfaces);
}
@Override
public void visitSource(final String file, final String debug)
{
    p.visitSource(file, debug);
    super.visitSource(file, debug);
}
@Override
public ModuleVisitor visitModule(
      final String name, final int flags, final String version)
{
    Printer modulePrinter = p.visitModule(name, flags, version);
    return new TraceModuleVisitor(
        super.visitModule(name, flags, version), modulePrinter);
}
```

从上述代码片段可以看到，TraceClassVisitor 中 visit() 方法中都有 Printer 的身影。

Printer 有两个子类，分别是 Textifier 和 ASMifier，其中 Textifier 的作用是输出 ClassFile 的字节码文本内容，TraceClassVisitor 中默认使用的 Printer 实现就是 Textifier，这就是 TraceClassVisitor 背后的原理。ASMifier 的作用则是输出 ClassFile 内容对应的 ASM 代码。

例如，使用 ASMifier 来输出在 5.9.2 小节介绍的 Utils 类的内容。

```
// 先读取 Utils 类的内容
String filePath = "src/main/java/cn/sensorsdata/asm/mv/";
File file = new File(filePath + "Utils.class");
byte[] bytes = FileUtils.readFileToByteArray(file);
ClassReader classReader = new ClassReader(bytes);
ClassWriter classWriter = new ClassWriter(0);
//TraceClassVisitor 中使用 ASMifier 类
TraceClassVisitor traceClassVisitor = new TraceClassVisitor(
      classWriter, new ASMifier(), new PrintWriter(System.out));
classReader.accept(traceClassVisitor, 0);
```

其输出结果如下。

```
package asm.cn.sensorsdata.asm.mv;
import org.objectweb.asm.AnnotationVisitor;
import org.objectweb.asm.Attribute;
import org.objectweb.asm.ClassReader;
import org.objectweb.asm.ClassWriter;
import org.objectweb.asm.ConstantDynamic;
import org.objectweb.asm.FieldVisitor;
import org.objectweb.asm.Handle;
import org.objectweb.asm.Label;
import org.objectweb.asm.MethodVisitor;
import org.objectweb.asm.Opcodes;
import org.objectweb.asm.RecordComponentVisitor;
import org.objectweb.asm.Type;
import org.objectweb.asm.TypePath;
public class UtilsDump implements Opcodes {

      public static byte[] dump () throws Exception {

          ClassWriter classWriter = new ClassWriter(0);
```

```
            FieldVisitor fieldVisitor;
            RecordComponentVisitor recordComponentVisitor;
            MethodVisitor methodVisitor;
            AnnotationVisitor annotationVisitor0;

            classWriter.visit(
                V1_8, ACC_PUBLIC | ACC_FINAL | ACC_SUPER,
                "cn/sensorsdata/asm/mv/Utils",
                null, "java/lang/Object", null);

            classWriter.visitSource("Utils.java", null);

            {
                methodVisitor = classWriter.visitMethod(
                    ACC_PUBLIC, "<init>", "()V", null, null);
                methodVisitor.visitCode();
                Label label0 = new Label();
                methodVisitor.visitLabel(label0);
                methodVisitor.visitLineNumber(3, label0);
                methodVisitor.visitVarInsn(ALOAD, 0);
                methodVisitor.visitMethodInsn(
                    INVOKESPECIAL, "java/lang/Object", "<init>", "()V", false);
                methodVisitor.visitInsn(RETURN);
                methodVisitor.visitMaxs(1, 1);
                methodVisitor.visitEnd();
            }
            {
                methodVisitor = classWriter.visitMethod(
                    ACC_PUBLIC | ACC_STATIC, "add", "(II)I", null, null);
                methodVisitor.visitCode();
                Label label0 = new Label();
                methodVisitor.visitLabel(label0);
                methodVisitor.visitLineNumber(5, label0);
                methodVisitor.visitVarInsn(ILOAD, 0);
                methodVisitor.visitVarInsn(ILOAD, 1);
                methodVisitor.visitInsn(IADD);
                methodVisitor.visitInsn(IRETURN);
                methodVisitor.visitMaxs(2, 2);
                methodVisitor.visitEnd();
            }
            classWriter.visitEnd();

            return classWriter.toByteArray();
        }
}
```

看到这里读者是不是想到了什么？没错，可以通过这种方式来让 ASM 帮我们生成 ASM 代码。这非常有用，当我们不知道如何使用 ASM 去实现一个方法时，可以先将这个方法的源码编译成 .class，再通过 TraceClassVisitor 这种操作来查看方法对应的 ASM 代码。

不过这种方式还是有一点儿麻烦，现在为大家介绍一款插件——IntelliJ IDEA，用该插件可以很方便地实现这一步。IntelliJ IDEA 的 plugins 搜索关键词“asm”会显示图 5-8 所示的结果。

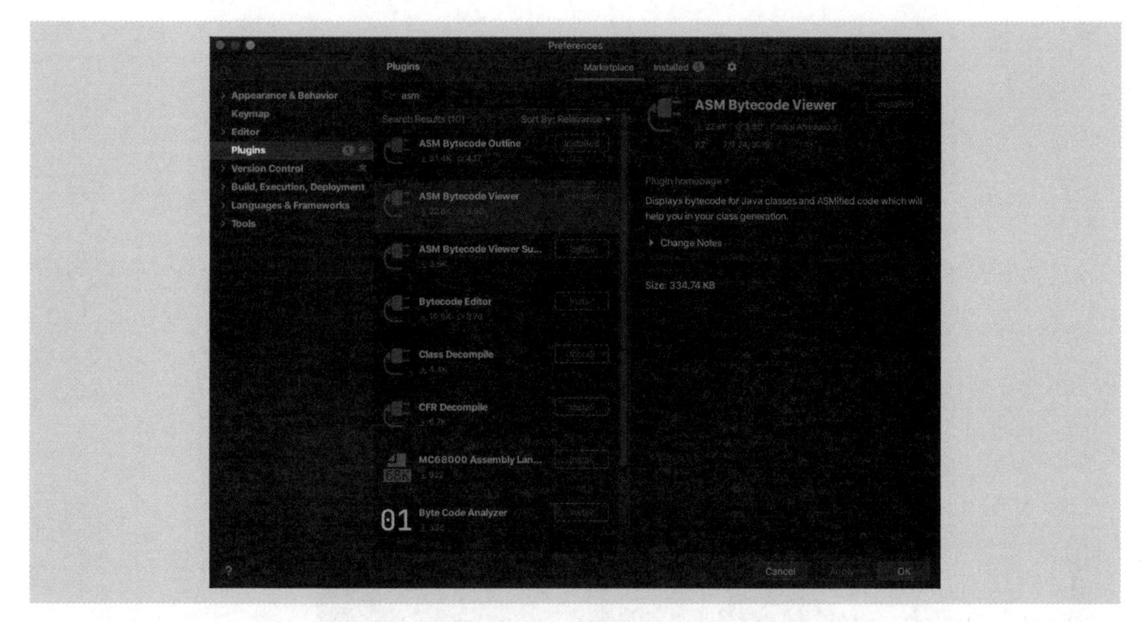

图 5-8 ASM 插件

其中前两个插件选择安装一个即可，安装好后在 Java 源码中单击鼠标右键，会弹出图 5-9 所示的菜单。

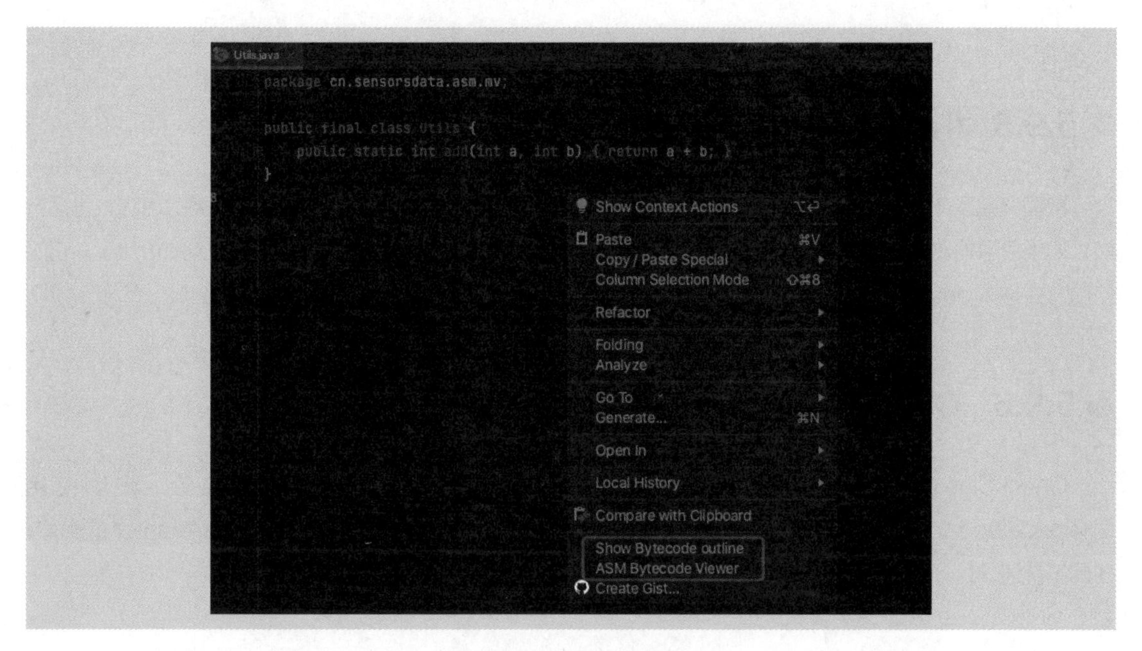

图 5-9 使用 ASM 插件

选择其中的一项，会在 IDEA 的右侧显示图 5-10 所示的窗口。

注意观察顶部的 3 个选项卡，其中“Bytecode”显示了 Java 类对应的字节码信息，“ASMified”显示了 ASM 生成该类的 Java 代码，“Groovified”则显示了 ASM 生成该类的 Groovy 代码。

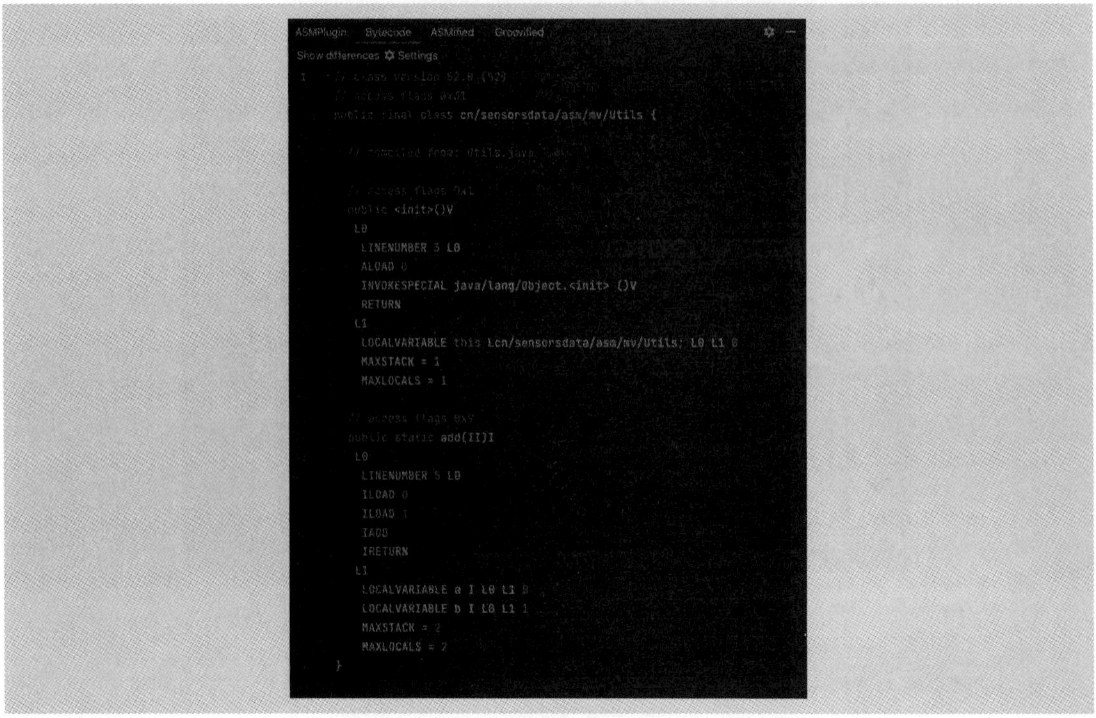

```
ASMPlugin  Bytecode  ASMified  Groovified
Show differences  Settings
1  // class version 52.0 (52)
   // access flags 0x31
   public final class cn/sensorsdata/asm/mv/Utils {

     // compiled from: Utils.java

     // access flags 0x1
     public <init>()V
      L0
       LINENUMBER 3 L0
       ALOAD 0
       INVOKESPECIAL java/lang/Object.<init> ()V
       RETURN
      L1
       LOCALVARIABLE this Lcn/sensorsdata/asm/mv/Utils; L0 L1 0
       MAXSTACK = 1
       MAXLOCALS = 1

     // access flags 0x9
     public static add(II)I
      L0
       LINENUMBER 5 L0
       ILOAD 0
       ILOAD 1
       IADD
       IRETURN
      L1
       LOCALVARIABLE a I L0 L1 0
       LOCALVARIABLE b I L0 L1 1
       MAXSTACK = 2
       MAXLOCALS = 2
   }
```

图 5-10　ASM 插件结果窗口

5.9.5　TraceMethodVisitor

前面介绍了 TraceClassVisitor，通过示例可以看到 TraceClassVisitor 输出了整个类的内容，包括方法。如果只想输出方法中的结果，则可以使用 TraceMethodVisitor 类，该类继承自 MethodVisitor。它的使用方式与 TrackClassVisitor 的类似，在此不赘述。

5.9.6　CheckMethodAdapter

CheckClassAdapter 会对整个类进行校验，如果只想对某些方法做校验，可以使用 CheckMethodAdapter 类，该类继承自 MethodVisitor。它的使用方式与 CheckClassAdapter 的类似，在此不赘述。

5.9.7　LocalVariableSorter

思考一个问题：如何在方法中添加局部变量？

这里提供一个思路：首先得确定当前方法中有哪些局部变量，并且得知道这些局部变量在局部变量表中的对应位置，然后将新创建的局部变量放在局部变量表中合理的位置上。最简单的方式是确定 maxlocals 的大小，然后放在 maxlocals 后面未使用的位置上，不过前提是得事先知道 maxlocals 的值。

这些工作显然是比较麻烦的，而 LocalsVariableSorter 则可以帮助我们解决这些问题。

LocalsVariableSorter 是 MethodVisitor 的子类，通过 LocalsVariableSorter 的 newLocal() 方法可以很方便地在方法中创建变量，开发者不需要去关注如何管理局部变量表。newLocal() 方法的定义如下。

```
/**
  * Constructs a new local variable of the given type.
  *
  * @param type the type of the local variable to be created.
  * @return the identifier of the newly created local variable.
  */
public int newLocal(final Type type) {
```

该方法的作用是根据 Type 计算类型的大小，并返回该类型在局部变量表中的对应位置。

举例如下。

```
package cn.sensorsdata.asm.mv;

public class C {

      public int foo(int value) throws Exception {
          try {
              value = 10000 / value;
              Thread.sleep(100);
          } catch (Exception e) {
              throw e;
          }
          return value;
      }
}
```

上述代码中定义了一个 foo() 方法，现在需要统计该方法的运行耗时情况，若要统计方法的耗时，首先需要在方法开始的位置创建一个变量，用于保存当前时间戳，然后在方法退出的时候计算时间差并输出结果。foo() 方法有两个退出指令，一个是 return 指令（正常退出），一个是 throw 指令（异常退出）。

插入后的逻辑类似如下。

```
public class C {

      public int foo(int value) throws Exception {
          long time = System.currentTimeMillis();
          try {
              value = 10000 / value;
              Thread.sleep(100);
          } catch (Exception e) {
              time = System.currentTimeMillis() - time;
              System.out.println("cost time: " + time);
              throw e;
          }
          time = System.currentTimeMillis() - time;
          System.out.println("cost time: " + time);
          return value;
      }
}
```

对应的 ASM 的实现结果如下。

```
public class TestLocalVariableSorter {
```

```
//创建ClassVisitor对象，并重写visitMethod()方法，用于获取类的方法信息
public static class AddTimerCostVisitor extends ClassVisitor{

    public AddTimerCostVisitor(int api, ClassVisitor classVisitor) {
        super(api, classVisitor);
    }

    @Override
    public MethodVisitor visitMethod(
        int access, String name, String descriptor,
        String signature, String[] exceptions)
    {
        MethodVisitor mv = super.visitMethod(
            access, name, descriptor, signature, exceptions);

        //不处理抽象方法和构造方法
        if (mv != null && !"<init>".equals(name)
            && !"<clinit>".equals(name)) //(1)
        {
            boolean isAbstractMethod =
                (access & Opcodes.ACC_ABSTRACT) != 0;
            boolean isNativeMethod = (access & Opcodes.ACC_NATIVE) != 0;
            if (!isAbstractMethod && !isNativeMethod) {
                mv = new AddTimerCostMethodVisitor(
                    api, access, descriptor, mv);
            }
        }
        return mv;
    }
}
//创建AddTimerCostMethodVisitor，继承自LocalVariablesSorter
public static class AddTimerCostMethodVisitor
    extends LocalVariablesSorter implements Opcodes //(2)
{

    private int timeSlotIndex = 0;

    protected AddTimerCostMethodVisitor( int api, int access,
        String descriptor, MethodVisitor methodVisitor)
    {
        super(api, access, descriptor, methodVisitor);
    }

    @Override
    public void visitCode()
    {
        super.visitCode();
        //调用newLocal()方法，创建long类型的局部变量
        //newLocal()方法会返回long在局部变量表中的位置
        //这里将结果保存在timeSlotIndex中
        timeSlotIndex = newLocal(Type.LONG_TYPE); //(3)
        mv.visitMethodInsn(
            INVOKESTATIC,
            "java/lang/System",
            "currentTimeMillis", "()J", false);
        mv.visitVarInsn(LSTORE, timeSlotIndex);
```

```
        }

        @Override
        public void visitInsn(int opcode) {
            // 在退出方法之前插入代码，这里判断如果指令是 return 或 athrow
            // 就代表退出方法了
            if ((opcode >= IRETURN && opcode <= RETURN) || opcode == ATHROW) {//(4)
                // 调用 System.currentTimeMillis() 方法
                mv.visitMethodInsn(
                    INVOKESTATIC,
                    "java/lang/System", "currentTimeMillis", "()J", false);
                // 加载局部变量到操作数栈中
                mv.visitVarInsn(LLOAD, timeSlotIndex);
                // 将栈顶两个值相减
                mv.visitInsn(LSUB);
                // 并将结果保存到 timeSlotIndex 位置处
                mv.visitVarInsn(LSTORE, timeSlotIndex);
                mv.visitFieldInsn(
                    GETSTATIC,
                    "java/lang/System",
                    "out", "Ljava/io/PrintStream;");
                // 使用 StringBuilder 来构造 "cost time: " + time 的值
                mv.visitTypeInsn(NEW, "java/lang/StringBuilder");
                mv.visitInsn(DUP);
                mv.visitMethodInsn(
                    INVOKESPECIAL,
                    "java/lang/StringBuilder", "<init>", "()V", false);
                mv.visitLdcInsn( "cost time: ");
                mv.visitMethodInsn(
                    INVOKEVIRTUAL,
                    "java/lang/StringBuilder",
                    "append",
                    "(Ljava/lang/String;)Ljava/lang/StringBuilder;", false);
                mv.visitVarInsn(LLOAD, timeSlotIndex);
                mv.visitMethodInsn(
                    INVOKEVIRTUAL,
                    "java/lang/StringBuilder",
                    "append",
                    "(J)Ljava/lang/StringBuilder;", false);
                mv.visitMethodInsn(
                    INVOKEVIRTUAL,
                    "java/lang/StringBuilder",
                    "toString", "()Ljava/lang/String;", false);
                mv.visitMethodInsn(
                    INVOKEVIRTUAL,
                    "java/io/PrintStream",
                    "println", "(Ljava/lang/String;)V", false);
            }
            super.visitInsn(opcode);
        }

    }
}
```

上述代码在位置(1)处判断方法名是不是构造方法和静态代码块，如果不是就对其使用 AddTimerCostMethodVisitor [位置(2)] 中的逻辑处理。AddTimerCostMethodVisitor 继承自

LocalVariables Sorter，并重写 visitCode() 方法，添加一个 long 类型的局部变量 [位置（3）]。在 visitCode() 方法中这么做相当于在进入方法的时候添加一个局部变量。然后重写 visitInsn() 方法，在位置（4）处判断是否是 return 或 throw 指令，如果是，则表示方法将退出，在此位置计算出最终的结果，并输出。注意上面例子并未重写 visitMax() 方法，这里让 ClassWriter 完成即可。

5.9.8 GeneratorAdapter

GeneratorAdatper 继承自 LocalVariableSorter，它对 MethodVisitor 中的方法进行了重新封装，使得我们可以采用更加简单的方式去处理方法。例如，可以使用 push() 方法将数值 129 或者字符串“123”压入操作数栈中，而不需要我们具体使用哪个指令。通过 push() 方法的源码更加容易理解。

```
public void push(final int value) {
    if (value >= -1 && value <= 5) {
      mv.visitInsn(Opcodes.ICONST_0 + value);
    } else if (value >= Byte.MIN_VALUE && value <= Byte.MAX_VALUE) {
      mv.visitIntInsn(Opcodes.BIPUSH, value);
    } else if (value >= Short.MIN_VALUE && value <= Short.MAX_VALUE) {
      mv.visitIntInsn(Opcodes.SIPUSH, value);
    } else {
      mv.visitLdcInsn(value);
    }
}

public void push(final String value) {
    if (value == null) {
      mv.visitInsn(Opcodes.ACONST_NULL);
    } else {
      mv.visitLdcInsn(value);
    }
}
```

可以看到 GeneratorAdatper 的 push() 方法根据 value 的类型以及值的大小来匹配不同的指令。这样的方式可以大大减少对指令的记忆，类似的方法还有很多，建议读者通过阅读源码来理解这些方法的用途。

例如，通过 GeneratorAdatper 来生成如下代码。

```
public class Example {
      public static void main(String[] args) {
          System.out.println("Hello world!");
      }
}
```

对应的代码如下。

```
ClassWriter cw = new ClassWriter(0);
cw.visit(
      Opcodes.V1_8, Opcodes.ACC_PUBLIC,
      "Example", null, "java/lang/Object", null);

//1. 构建一个方法，根据方法的字符串定义以获取对应的封装结果
Method m = Method.getMethod("void <init> ()");
```

```
// 使用 GeneratorAdapter 创建方法
GeneratorAdapter mg = new GeneratorAdapter(Opcodes.ACC_PUBLIC, m, null, null, cw);
// 加载非静态方法的第 0 个参数到操作数栈中，即 this
mg.loadThis();
// 调用父类 Object 的构造方法
mg.invokeConstructor(Type.getType(Object.class), m);
// 返回操作数栈栈顶的值
mg.returnValue();
mg.endMethod();

//2. 构建 main() 方法
m = Method.getMethod("void main (String[])");
mg = new GeneratorAdapter(
      Opcodes.ACC_PUBLIC + Opcodes.ACC_STATIC, m, null, null, cw);
// 获取 System 类的 out 属性，其中 owner 和描述符信息都是通过 Type 来指定的
mg.getStatic(Type.getType(System.class), "out", Type.getType(PrintStream.class));
// 将字符串常量压入栈顶中
mg.push("Hello world!");
mg.invokeVirtual(
      Type.getType(PrintStream.class),
      Method.getMethod("void println (String)"));
mg.returnValue();
mg.endMethod();
cw.visitEnd();
```

可以看到，通过上述代码来生成类和方法，代码量会小很多。

5.9.9 AdviceAdapter

在实际的开发过程中，经常需要在方法开始或者结束的时候做一些操作，例如前面统计方法耗时的例子中，需要在进入方法时和退出方法时插入不同的代码。AdviceAdapter 对此进行了封装，提供了如下两个方法。

```
public abstract class AdviceAdapter extends GeneratorAdapter implements Opcodes
{
    protected void onMethodEnter() {}
    protected void onMethodExit(final int opcode) {}
}
```

可以看到 AdviceAdapter 是一个抽象类，并且继承自 GeneratorAdapter，其中 onMethodEnter() 方法表示进入方法的回调，onMethodExit() 方法表示退出方法的回调。下面介绍使用 AdviceAdapter 来修改统计方法耗时的例子。

```
public static class MethodTimeCostAdapter extends AdviceAdapter {
      private int timeSlotIndex = 0;
      protected MethodTimeCostAdapter(
          int api, MethodVisitor methodVisitor,
          int access, String name, String descriptor)
      {
          super(api, methodVisitor, access, name, descriptor);
      }
      @Override
      protected void onMethodEnter() {
```

```
            super.onMethodEnter();
            timeSlotIndex = newLocal(Type.LONG_TYPE);
            invokeStatic(
                Type.getType(System.class),
                Method.getMethod("long currentTimeMillis()"));
            storeLocal(timeSlotIndex);
        }
        @Override
        protected void onMethodExit(int opcode) {
            super.onMethodExit(opcode);
            invokeStatic(
                Type.getType(System.class),
                Method.getMethod("long currentTimeMillis()"));
            loadLocal(timeSlotIndex);
            math(SUB, Type.LONG_TYPE);
            storeLocal(timeSlotIndex);
            getStatic(
                Type.getType(System.class),
                "out", Type.getType(PrintStream.class));
            newInstance(Type.getType(StringBuilder.class));
            dup();
            invokeConstructor(
                Type.getType(StringBuilder.class),
                Method.getMethod("void <init> ()"));
            push("cost time: ");
            invokeVirtual(
                Type.getType(StringBuilder.class),
                Method.getMethod("StringBuilder append(String str)"));
            loadLocal(timeSlotIndex);
            invokeVirtual(
                Type.getType(StringBuilder.class),
                Method.getMethod("StringBuilder append(String str)"));
            invokeVirtual(
                Type.getType(StringBuilder.class),
                Method.getMethod("String toString()"));
            invokeVirtual(
                Type.getType(PrintStream.class),
                Method.getMethod("void println(String)"));
        }
}
```

上述的代码中综合了 AdviceAdapter、GeneratorAdapter、LocalVariableSorter 中的知识，熟练掌握后，可以大大提升开发效率。

5.10 其他实例

在实际的开发中，有一些场景是我们经常遇到的，具有一定的代表性，本节为大家介绍几种常见的场景。

5.10.1 方法替换

所谓方法替换，就是使用自己库中定义的方法替换方法中已有的方法，达到“偷梁换柱”的效果。例如在 Android 中，我们无法修改 android.jar 包，也就无法对包中的 webView.loadUrl(String) 方法进行 Hook，但可以在其使用的位置进行方法替换。

考虑如下代码，将 foo() 方法中的 webView.loadUrl() 使用后面注释位置的代码替换。

```
public class MethodReplace {

      public static class WebView{
          public void loadUrl(String url){
              System.out.println("start loading: " + url);
          }
      }

      public static class TrackUtils{
          public static void loadUrl(WebView webView, String url){
              //do something
              webView.loadUrl(url);
              //do something
          }
      }

      public static void foo() {
          WebView webView = new WebView();
          webView.loadUrl("https://www.sensorsdata.cn");
          // TrackUtils.loadUrl(webView, "https://www.sensorsdata.cn");
      }
}
```

如果要替换 webView.loadUrl()，需要确认调用 loadUrl() 时操作数栈的状态。根据 loadUrl() 方法的 access 和方法描述符可知，调用 loadUrl() 方法时栈顶的前两个数据为 webview、https://www.sensorsdata.cn，它们为调用 loadUrl() 方法的必要信息。

具体操作如下。

```
public class MethodReplaceAdapter extends MethodVisitor {

      public MethodReplaceAdapter(int api, MethodVisitor methodVisitor) {
          super(api, methodVisitor);
      }

      @Override
      public void visitMethodInsn(
          int opcode, String owner, String name,
          String descriptor, boolean isInterface)
      {
          // 判断方法的签名信息，以及判断方法的 owner
          if ("loadUrl".equals(name)
              && "(Ljava/lang/String;)V".equals(descriptor)
              && checkOwnerIsWebView(owner))
          {
              opcode = Opcodes.INVOKESTATIC;
              owner = "cn/sensorsdata/asm/demo/MethodReplace$TraceUtils";
                descriptor = "(Lcn/sensorsdata/asm/demo/MethodReplace$WebView;Ljava/
lang/String;)V";
          }
          super.visitMethodInsn(opcode, owner, name, descriptor, isInterface);
      }
```

```
    private boolean checkOwnerIsWebView(String owner) {
        if ("cn/sensorsdata/asm/demo/MethodReplace$WebView".equals(owner)) {
            return true;
        }
        // 使用其他方式验证，例如 Class
        return false;
    }
}
```

上述代码在位置（1）处判断方法的名称、方法描述符信息以及判断 owner 是否是 WebView 的子类，然后替换其中的操作码、owner、方法基本信息。需要注意 TraceUtils.loadUrl(WebView,String) 方法的参数是 WebView 加上 webView.loadUrl(String) 方法的参数。这是因为在触发 webView.loadUrl() 方法时，即 invokevirtual 指令执行之前，操作数栈栈顶的内容就是 owner + 方法参数。TraceUtils.loadUrl() 方法中只关注 WebView 参数，但也请设计成与 webView.loadUrl() 一致。这样字节码操作会简单一些，把逻辑从这里转移到 TraceUtils.loadUrl() 方法中，后期操作会更加灵活。另外 checkOwnerIsWebView() 方法用于确认是否是 WebView 以及 WebView 的子类，判断子类的方式我们会在后面 9.4 节介绍。

5.10.2 方法参数复用

方法替换是将原有的代码使用其他方法替换，一般比较推荐这种方式，还有一种方法替换的方式，就是在目标代码后面直接追加。以 MethodReplace 的 foo() 方法为例，最终效果如下。

```
public class MethodReplace {
    ...
    public static void foo() {
        WebView webView = new WebView();
        webView.loadUrl("https://www.sensorsdata.cn");
        // 在上面 loadUrl() 后面插入此段代码
        TrackUtils.loadUrl(webView, "https://www.sensorsdata.cn");
    }
}
```

实现此种功能的关键点就是如何构建 TrackUtils.loadUrl() 方法中的参数，其实这个问题是要解决在调用 webView.loadUrl() 时操作数栈中的值如何进行保存，解决方式大致可以分为如下两种。

- 第一种：当方法参数较少时，可以考虑使用 dup* 相关的指令对操作数栈栈顶的元素进行复制，例如上例中的 loadUrl() 方法，有一个参数，加上 owner，总共有两个，这时候可以使用 dup2 指令。
- 第二种：当方法参数较多时，可以考虑将操作数栈中的值保存到局部变量表中。

我们来看看第二种方式，继续以 MethodReplace 的 foo() 方法为例，当执行 webView.loadUrl() 之前，其栈顶元素是 webview、https://www.sensorsdata.cn，通过 *store 指令依次将栈顶元素出栈并保存到局部变量表中，出栈顺序是 https://www.sensorsdata.cn、webview，出栈之后执行 webView.loadUrl() 方法时需要的信息已不在栈中，所以需要重新将它们加载到栈中，具体的实现逻辑如下。

```
public class MethodReplaceAdapter extends AdviceAdapter {

    protected MethodReplaceAdapter(
```

```
            int api, MethodVisitor methodVisitor, int access,
            String name, String descriptor)
        {
            super(api, methodVisitor, access, name, descriptor);
        }

        @Override
        public void visitMethodInsn(
            int opcode, String owner, String name,
            String descriptor, boolean isInterface)
        {
            if ("loadUrl".equals(name)
                && "(Ljava/lang/String;)V".equals(descriptor)
                && checkOwnerIsWebView(owner))
            {
                    //(1)
                    // 从描述符中获取参数内容
                    Type[] argTypes = Type.getArgumentTypes(descriptor);
                    // 创建一个 List，用于保存栈中值在局部变量表中的位置
                    List<Integer> positionList = new ArrayList<>();
                    for (int index = argTypes.length - 1 ; index >= 0; index--) {
                        int position = newLocal(argTypes[index]);
                        storeLocal(position, argTypes[index]);
                        // 按照栈中的顺序进行存储
                        positionList.add(0, position);
                    }
                    //(2)
                    // 将 owner 的值也添加到局部变量表中
                    Type ownerType = Type.getType(
                        "Lcn/sensorsdata/asm/demo/MethodReplace$WebView;");
                    int ownerPosition = newLocal(ownerType);
                    storeLocal(ownerPosition, ownerType);
                    positionList.add(0, ownerPosition);

                    // 将方法调用需要的数据加载到栈中
                    positionList.forEach(this::loadLocal);//(3)
                    // 调用原始方法
                    super.visitMethodInsn(
                        opcode, owner, name, descriptor, isInterface);
                    // 再次将方法调用需要的数据加载到栈中
                    positionList.forEach(this::loadLocal);
                    mv.visitMethodInsn(
                        Opcodes.INVOKESTATIC,
                        "cn/sensorsdata/asm/demo/MethodReplace$TraceUtils",
                        "loadUrl",
                        "(Lcn/sensorsdata/asm/demo/MethodReplace$WebView;\
                            Ljava/lang/String;)V",
                        false);
                    return;
            }
            super.visitMethodInsn(opcode, owner, name, descriptor, isInterface);
        }

        ...
}
```

上述代码位置（1）到位置（2）处表示根据 webview.loadUrl() 方法的方法描述符信息获得参数类型 Type，然后根据 Type 使用 newLocal() 创建对应的局部变量，并依次将操作数栈栈顶的值存到局部变量表中，同时使用 positionList 记录具体的位置。位置（2）表示将 owner 也存放在局部变量表中，此时 positionList 中保存了操作数栈栈顶中到局部变量表的位置。位置（3）依次将 positionList 记录的局部变量表的值压入操作数栈中，执行 super.visitMethodInsn()，也就是调用原 webView.loadUrl()，然后加载局部变量表中的值，执行 TraceUtils.loadUrl() 方法，这样就实现了追加方法的效果。

5.11 小结

本章详细地介绍了 ASM Core API 中的 3 个核心类——ClassReader、ClassVisitor、ClassWriter，并且介绍了 utils 和 commons 包下一些开发中比较有用的工具类，最后举了一些例子来演示实际开发中可能遇到的问题（这里只介绍了一小部分，后文中还会有更多的介绍），希望读者在学习本章的过程中能够动手操作，多多练习才会理解得更深刻。

6. ASM基础之Tree API

在 5.8.3 小节介绍了如何优化 ICONST_0 IADD 指令，可以看到它的实现细节较复杂。如果能将方法中的所有指令存放在一个列表中，那么只需要判断 IADD 指令的前一个指令是不是 ICONST_0，然后决定是否删除这两个指令即可。幸运的是使用 Tree API 的 MethodNode 很容易做到这一点。

本章将详细介绍 Tree API 的相关知识，并且在章末会对 ASM 框架的兼容性、原理进行讨论。

6.1 Tree API简介

在介绍 Core API 时，介绍了一些非常核心的类，如 ClassVisitor、MethodVisitor、FieldVisitor 等，同样的，在 Tree API 中也有与之相对应的类：ClassNode、MethodNode、FieldNode。Tree API 是基于 Core API 上实现的，接下来依次对这些类进行介绍。

6.2 ClassNode API介绍

先来看看 ClassNode 的部分源码。

```
public class ClassNode extends ClassVisitor {
    public int version;
    public int access;
    public String name;
    public String signature;
    public String superName;
    public List<String> interfaces;
    public String sourceFile;
    public String sourceDebug;
    public ModuleNode module;
    public String outerClass;
    public String outerMethod;
    public String outerMethodDesc;
    public List<AnnotationNode> visibleAnnotations;
    public List<AnnotationNode> invisibleAnnotations;
    public List<TypeAnnotationNode> visibleTypeAnnotations;
    public List<TypeAnnotationNode> invisibleTypeAnnotations;
    public List<Attribute> attrs;
    public List<InnerClassNode> innerClasses;
    public String nestHostClass;
    public List<String> nestMembers;
    public List<String> permittedSubclasses;
    public List<RecordComponentNode> recordComponents;
    public List<FieldNode> fields;
    public List<MethodNode> methods;

    public ClassNode()
    public ClassNode(final int api)

    @Override
    public void visit(
```

```
    final int version,
    final int access,
    final String name,
    final String signature,
    final String superName,
    final String[] interfaces) {
  this.version = version;
  this.access = access;
  this.name = name;
  this.signature = signature;
  this.superName = superName;
  this.interfaces = Util.asArrayList(interfaces);
}
// 省略其他 visit() 方法

@Override
public FieldVisitor visitField(
    final int access,
    final String name,
    final String descriptor,
    final String signature,
    final Object value) {
  FieldNode field = new FieldNode(access, name, descriptor, signature, value);
  fields.add(field);
  return field;
}

@Override
public MethodVisitor visitMethod(
    final int access,
    final String name,
    final String descriptor,
    final String signature,
    final String[] exceptions) {
  MethodNode method =
      new MethodNode(access, name, descriptor, signature, exceptions);
  methods.add(method);
  return method;
}

@Override
public void visitEnd() {
  // Nothing to do
}

// 关注 check() 方法
public void check(final int api) {
  if (api < Opcodes.ASM9 && permittedSubclasses != null) {
    throw new UnsupportedClassVersionException();
  }
  if (api < Opcodes.ASM8
      && ((access & Opcodes.ACC_RECORD) != 0 || recordComponents != null))
  {
    throw new UnsupportedClassVersionException();
  }
  if (api < Opcodes.ASM7
```

```
            && (nestHostClass != null || nestMembers != null))
        {
          throw new UnsupportedClassVersionException();
        }
        ...
        for (int i = fields.size() - 1; i >= 0; --i) {
          fields.get(i).check(api);
        }
        for (int i = methods.size() - 1; i >= 0; --i) {
          methods.get(i).check(api);
        }
      }

      //关注 accept() 方法
      public void accept(final ClassVisitor classVisitor) {
        // Visit the header
        String[] interfacesArray = new String[this.interfaces.size()];
        this.interfaces.toArray(interfacesArray);
        classVisitor.visit(
            version, access, name, signature, superName, interfacesArray);
        // Visit the source
        if (sourceFile != null || sourceDebug != null) {
          classVisitor.visitSource(sourceFile, sourceDebug);
        }
        // Visit the module
        if (module != null) {
          module.accept(classVisitor);
        }
        ...
        // Visit the fields
        for (int i = 0, n = fields.size(); i < n; ++i) {
          fields.get(i).accept(classVisitor);
        }
        // Visit the methods
        for (int i = 0, n = methods.size(); i < n; ++i) {
          methods.get(i).accept(classVisitor);
        }
        classVisitor.visitEnd();
      }
}
```

上述 ClassNode 源码中保留了一些核心逻辑，观察源码可以得出如下结论。

（1）ClassNode 继承自 ClassVisitor 类。

（2）ClassNode 重写了 ClassVisitor 中的 visit() 方法，并将结果保存在类的成员变量中。例如 visitMethod() 方法中创建了一个 MethodNode 对象，该对象保存在 ClassNode 的 methods 列表中。

（3）ClassNode 重写了 visitEnd() 方法，但方法体是空的。

（4）ClassNode 新增了 check() 方法，从此方法的实现可以看出它做了一些兼容性相关的操作。

（5）ClassNode 新增了 accept(ClassVisitor) 方法，该方法的作用是将 ClassNode 中保存的内容分发给 ClassVisitor，方式是调用 ClassVisitor 中的对应方法。该方法是 ClassVisitor 和 ClassNode 相互转换的关键，也就是 Core API 和 Tree API 相互转换的关键。

（6）对比 ClassVisitor 知识，ClassNode 类中的字段意思很容易理解，例如 name 表示类名、

signature 表示类签名。这里可以确定 ClassNode 使用成员变量保存 ClassFile 中的结构信息。

接下来介绍 ClassNode 的相关使用方式。

6.2.1 类的生成

类的生成的代码如下。

```
//1.创建 ClassNode 对象 (1)
//注意在初始化时 fields、methods、interfaces 等 List 类型会初始化为空列表
ClassNode classNode = new ClassNode(Opcodes.ASM6);

//2.设置类的相关属性 part1(2)
classNode.version = Opcodes.V1_8;
classNode.access = Opcodes.ACC_PUBLIC + Opcodes.ACC_ABSTRACT;

// 设置字段 (3)
FieldNode versionField = new FieldNode(
      Opcodes.ASM6,
      Opcodes.ACC_PUBLIC + Opcodes.ACC_STATIC + Opcodes.ACC_FINAL,
      "VERSION", "Ljava/lang/String;", null, "9.0.0");
classNode.fields.add(versionField);
FieldNode serverUrlField = new FieldNode(
      Opcodes.ASM6,
      Opcodes.ACC_PROTECTED ,
      "serverUrl", "Ljava/lang/String;", null, "http://www.sensorsdata.cn");
classNode.fields.add(serverUrlField);

// 设置类的相关属性 part2(4)
classNode.name = "cn/sensorsdata/asm/cw/SenorsDataAPI";
classNode.superName = "java/lang/Object";

//3.设置方法 (5)
//constructorMethod 并未实现构造方法体中的内容
MethodNode constructorMethod = new MethodNode(
      Opcodes.ASM6, Opcodes.ACC_PUBLIC,
      "<init>","(Ljava/lang/String;)V",null, null);
classNode.methods.add(constructorMethod);
MethodNode trackMethod = new MethodNode(
      Opcodes.ASM6,
      Opcodes.ACC_PUBLIC + Opcodes.ACC_ABSTRACT,
      "track", "(Ljava/lang/String;)V",null, null);
classNode.methods.add(trackMethod);

//4.保存 (6)
// 本章后面内容介绍
```

上述代码位置（1）初始化 ClassNode 对象，ClassNode 构造方法中会将其成员变量 fields、methods、interfaces 等 List 类型初始化为空列表。位置（2）部分的代码表示设置 ClassFile 的版本和访问标识符信息。位置（3）部分的代码创建了两个 FieldNode 对象，分别表示 SensorsDataAPI 类中的 VERSION 和 serverUrl 字段，并将这两个 FieldNode 对象添加到 ClassNode 的 fields 列表中。位置（4）部分的代码也是设置 Class 的基本信息，这里设置了名称和父类信息。位置（5）部分的代码创

建了两个 MethodNode 对象，分表表示 SensorsDataAPI 类中的构造方法和 track() 方法，并将这两个 MethodNode 对象添加到 ClassNode 的 methods 列表中。位置（5）部分的代码表示将 ClassNode 生成的结果保存起来，不过这部分会结合 6.3 节详细介绍。

相比 ClassVisitor，使用 ClassNode 生成类这种方式在速度上要降低约 30%，不过好处是不用关注 ClassNode 中方法和属性的设置顺序，例如位置（2）和（4）设置类的基本信息，不用考虑像 ClassVisitor 那样有明确的调用顺序。

6.2.2 类的转换和修改

Tree API 使用成员变量保存 ClassFile 内容解析后的数据，因此对于类的转换就是需要搞清楚如何操作 ClassNode 的成员变量。通常情况我们可以直接操作 ClassNode 对象，不过为了保持良好的代码结构，一般会定义如下 Transfomer。

```
public class ClassNodeTransformer {
      private ClassNodeTransformer transformer;

      public ClassNodeTransformer(ClassNodeTransformer transformer) {
          this.transformer = transformer;
      }

      public void transform(ClassNode classNode) {
          if (this.transformer != null) {
              this.transformer.transform(classNode);
          }
      }
}
```

上述代码中定义了一个 ClassNodeTransformer，它的构造方法也接收一个 ClassNodeTransformer 类，并且提供 transform(ClassNode) 类，利用这种结构可以构建成一个 Transformer 链。

首先，使用此结构来为类添加一个字段，其代码如下。

```
public class AddFieldTransformer extends ClassNodeTransformer {
      // 设置标志用于判断是否找到目标字段
      private boolean isFoundUserNameField = false;

      public AddFieldTransformer(ClassNodeTransformer transformer) {
          super(transformer);
      }

      @Override
      public void transform(ClassNode classNode) {
          // 遍历所有的 fields，查看对应的字段
          for(FieldNode fieldNode:classNode.fields){
              if("userName".equals(fieldNode.name)){
                  isFoundUserNameField = true;
                  break;
              }
          }
          // 如果没有找到 userName 字段，就添加一个
          if(!isFoundUserNameField){
```

```
            FieldNode userNameField = new FieldNode(
                Opcodes.ACC_PRIVATE, "userName",
                "Ljava/lang/String;", null, null);
            classNode.fields.add(userNameField);
        }

        super.transform(classNode);
    }
}
```

上述代码在 transform() 判断 ClassNode 中是否有 userName 字段，如果没有就添加。同样，也可以通过 Transformer 删除类中的成员。例如：

```
public class DeleteMethodTransformer extends ClassNodeTransformer {

    private String methodNameAndDesc;

    public DeleteMethodTransformer(
        ClassNodeTransformer transformer, String methodNameAndDesc)
    {
        super(transformer);
        this.methodNameAndDesc = methodNameAndDesc;
    }

    @Override
    public void transform(ClassNode classNode) {
        Iterator<MethodNode> i = classNode.methods.iterator();
        while (i.hasNext()) {
            MethodNode mn = i.next();
            if (methodNameAndDesc.equals(mn.name + mn.desc) ) {
                i.remove();
            }
        }
        super.transform(classNode);
    }
}
```

上述代码用于删除符合方法名 + 方法描述的方法。可以将上面不同功能的 Transformer 组合在一起，形成一个 Transfomer 链，例如：

```
ClassNode classNode = ...;
ClassNodeTransformer transformer =
    new AddFieldTransformer(new DeleteMethodTransformer(null, "getUserName()V"));
transformer.transform(classNode);
```

6.3 ClassNode 与 Core API 相互转换

6.2.1 小节遗留了一个问题：如何将 ClassNode 转换为 ClassFile 字节数据？以及如何将 ClassFile 字节数据转换成 ClassNode？

我们知道 Tree API 是基于 Core API 的，也就是说只要将 Tree API 转换到 Core API 上就可以了。再直白点就是如何与 ClassReader、ClassWriter、ClassVisitor 这些 Core API 一起使用且相互转换。

6.3.1 ClassNode的特性

根据 6.2 节的介绍，ClassNode 具有如下两个特性。

（1）ClassNode 是 ClassVisitor 的子类，因此它具有 ClassVisitor 的所有特性。

（2）ClassNode 提供了一个 accept(ClassVisitor) 方法，该方法的逻辑是根据 ClassNode 中定义的字段调用参数 ClassVisitor 中对应的 visit() 方法。

首先根据 ClassNode 是 ClassVisitor 子类的特性，可以将其作为参数传递给 ClassReader 的 accept() 方法。使用方式如下。

```
byte[] bytes = ... // ClassFile data
ClassReader classReader = new ClassReader(bytes);
ClassNode classNode = new ClassNode(Opcodes.ASM6){

};
classReader.accept(classNode, 0);
```

按照这种使用方式，ClassFile 中的数据都会保存在 ClassNode 对象中。

再根据 ClassNode 的 accept() 方法实现，accept() 可以接收一个 ClassWriter 对象，再通过 ClassWriter 对象的 toByteArray() 方法就能获取到 ClassFile 的数据。使用方式形式如下。

```
ClassNode classNode = ...
ClassWriter classWriter = new ClassWriter(0);
classNode.accept(classWriter);
byte[] fileData = classWriter.toByteArray();
```

按照这种方式就能获取到 ClassNode 中的数据。

6.3.2 与Core API相互转换

综合 ClassNode 的两种特性和对应的使用方式，在转换类时，可以通过如下代码将其与 ClassReader、ClassWriter、ClassVisitor 组合在一起。

```
byte[] bytes = FileUtils.readFileToByteArray(file);
ClassReader classReader = new ClassReader(bytes);
ClassWriter classWriter = new ClassWriter(0);
// 使用匿名类的方式实现 ClassVisitor 子类
ClassVisitor classVisitor = new ClassVisitor(Opcodes.ASM6, classWriter) {
      // 重写相关方法
};
// 使用匿名类的方式实现 ClassNode 子类
ClassNode classNode = new ClassNode(Opcodes.ASM6){
      @Override
      public void visitEnd() {
          // 关键点是此处的 classVisitor
          accept(classVisitor);
      }
};
classReader.accept(classNode, 0);
byte[] fileData = classWriter.toByteArray();
```

注意，为方便阅读，ClassVisitor 和 ClassNode 使用了匿名类的实现方式。其中关键逻辑是在 ClassNode 的 visitEnd() 实现中调用 accept() 方法，这步操作表示将 ClassNode 的内容分发给 ClassVisitor 对象，这就是与其他 ClassVisitor 组合使用的关键所在。ClassReader 的 accept() 方法不一定非要接收 ClassNode 参数，也可以将 ClassNode 当作普通的 ClassVisitor 使用。下面的例子展示了这种用法。

```
// 定义 ClassNode，用于删除类中的 track(String) 方法
public class MyClassAdapter extends ClassNode {
      public MyClassAdapter(int api, ClassVisitor classVisitor) {
           super(api);
           this.cv = classVisitor;
      }

      @Override
      public void visitEnd() {
           super.visitEnd();
           Iterator<MethodNode> i = methods.iterator();
           while (i.hasNext()) {
               MethodNode mn = i.next();
               if ("track(Ljava/lang/String;)V".equals(mn.name + mn.desc) )
               {
                    i.remove();
               }
           }
           accept(this.cv);
      }
}
```

上述代码定义了的类 MyClassAdapter 继承自 ClassNode，MyClassAdapter 的构造方法中接收一个 ClassVisitor 参数，并将此参数赋值给 cv 字段。在 visitEnd() 方法中实现了删除 track() 方法的逻辑，最后调用 accept() 方法，将结果分发给 ClassVisitor 对象。完整的代码如下。

```
String filePath = "src/main/java/cn/sensorsdata/asm/treeapi/";
File file = new File(filePath + "SensorsDataAPI.class");
byte[] bytes = FileUtils.readFileToByteArray(file);
ClassReader classReader = new ClassReader(bytes);
ClassWriter classWriter = new ClassWriter(0);
//classVisitor1 用于删除 VERSION 字段
ClassVisitor classVisitor1 = new ClassVisitor(Opcodes.ASM6, classWriter) (1)
{
      @Override
      public FieldVisitor visitField(
           int access, String name, String descriptor,
           String signature, Object value)
      {
           if(name.equals("VERSION")){
               return null;
           }
           return super.visitField(access, name, descriptor, signature, value);
      }
};
ClassNode classNode = new MyClassAdapter(Opcodes.ASM6, classVisitor1); (2)
//classVisitor2 用于删除 serverUrl 字段
ClassVisitor classVisitor2 = new ClassVisitor(Opcodes.ASM6, classNode) (3)
{
      @Override
```

```
        public FieldVisitor visitField(
            int access, String name, String descriptor,
            String signature, Object value)
        {
            if(name.equals("serverUrl")){
                return null;
            }
            return super.visitField(access, name, descriptor, signature, value);
        }
};
classReader.accept(classVisitor2, 0);
byte[] fileData = classWriter.toByteArray();
File newFile = new File(filePath + "SensorsDataAPI3.class");
FileUtils.writeByteArrayToFile(newFile, fileData);
```

为演示MyClassAdapter的使用，上述代码在位置（1）和（2）处定义了两个ClassVisitor的匿名实现，用于删除类中的 VERSION 和 serverUrl 字段。这么做目的是演示位置（3）处声明的 MyClassAdapter 对象处理完以后，其他 ClassVisitor 能够继续消费。这就是 ClassNode 与其他 Core API 在一起使用的方式之一。这种方式是在 ClassNode 实现的构造方法中传递一个 ClassVisitor，那么思考能否在 ClassVisitor 的实现中使用 ClassNode 来做类转换呢?

这样也是可以的，举例如下。

```
public class DeleteFieldAdapter extends ClassVisitor {
      private ClassVisitor nextCV;
      public DeleteFieldAdapter(ClassVisitor classVisitor ) {
// 在 super() 方法中，创建一个 ClassNode 对象
          super(Opcodes.ASM6, new ClassNode(Opcodes.ASM6));(1)
// 保存 classVisitor 对象
          this.nextCV = classVisitor;
      }

      @Override
      public void visitEnd() {
// 获取 ClassNode 对象
          ClassNode classNode = (ClassNode) this.cv;
// 对 ClassNode 做转换
          Iterator<FieldNode> i = classNode.fields.iterator();
          while (i.hasNext()) {
              FieldNode mn = i.next();
              if ("serverUrl".equals(mn.name + mn.desc) ) {
                  i.remove();
              }
          }
// 将结果委托给 nextCV 继续执行
          classNode.accept(this.nextCV);
      }
}
```

上述代码在 DeleteFieldAdapter 类的构造方法调用父类构造方法时创建了一个 ClassNode，如位置（1）所示。然后在 DeleteFieldAdapter 的 visitEnd() 方法中获取 ClassNode 对象对 Class 做转换，转换后再调用 ClassNode 的 accept() 方法将结果分发给 DeleteFieldAdapter 类的构造方法中的 classVisitor 对象。

以上 MyClassAdapter 和 DeleteFieldAdapter 是比较典型的两种实现方式，这部分内容并非一定要

记住，在使用时参考一下即可。

6.4 MethodNode API 介绍

与 ClassNode 的实现方式类似，MethodNode 继承自 MethodVisitor，也使用成员变量来存储方法的结构信息，例如方法名、描述符、字节码指令等。为方便理解，下面列出 MethodNode 的部分源码。

```
public class MethodNode extends MethodVisitor {

    public int access;
    public String name;
    public String desc;
    public String signature;
    public List<String> exceptions;
    public List<ParameterNode> parameters;
    public List<AnnotationNode> visibleAnnotations;
    public List<AnnotationNode> invisibleAnnotations;
    public List<TypeAnnotationNode> visibleTypeAnnotations;
    public List<TypeAnnotationNode> invisibleTypeAnnotations;
    public List<Attribute> attrs;
    public Object annotationDefault;
    public int visibleAnnotableParameterCount;
    public List<AnnotationNode>[] visibleParameterAnnotations;
    public int invisibleAnnotableParameterCount;
    public List<AnnotationNode>[] invisibleParameterAnnotations;
    public InsnList instructions;
    public List<TryCatchBlockNode> tryCatchBlocks;
    public int maxStack;
    public int maxLocals;
    public List<LocalVariableNode> localVariables;
    public List<LocalVariableAnnotationNode> visibleLocalVariableAnnotations;
    public List<LocalVariableAnnotationNode> invisibleLocalVariableAnnotations;

    /** Whether the accept method has been called on this object. */
    private boolean visited;

    public MethodNode() {
      this(/* latest api = */ Opcodes.ASM9);
      if (getClass() != MethodNode.class) {
        throw new IllegalStateException();
      }
    }

    public MethodNode(final int api) {
      super(api);
      this.instructions = new InsnList();
    }

    public MethodNode(
        final int access,
        final String name,
        final String descriptor,
```

```
      final String signature,
      final String[] exceptions) {
    this(/* latest api = */ Opcodes.ASM9,
        access, name, descriptor, signature, exceptions);
    if (getClass() != MethodNode.class) {
      throw new IllegalStateException();
    }
  }

  public MethodNode(
      final int api,
      final int access,
      final String name,
      final String descriptor,
      final String signature,
      final String[] exceptions) {
    super(api);
    this.access = access;
    this.name = name;
    this.desc = descriptor;
    this.signature = signature;
    this.exceptions = Util.asArrayList(exceptions);
    if ((access & Opcodes.ACC_ABSTRACT) == 0) {
      this.localVariables = new ArrayList<>(5);
    }
    this.tryCatchBlocks = new ArrayList<>();
    this.instructions = new InsnList();
  }

  @Override
  public void visitParameter(final String name, final int access) {
    if (parameters == null) {
      parameters = new ArrayList<>(5);
    }
    parameters.add(new ParameterNode(name, access));
  }

  ...

  @Override
  public void visitCode() {
    // Nothing to do
  }

  @Override
  public void visitInsn(final int opcode) {
    instructions.add(new InsnNode(opcode));
  }

  @Override
  public void visitIntInsn(final int opcode, final int operand) {
    instructions.add(new IntInsnNode(opcode, operand));
  }

  @Override
  public void visitVarInsn(final int opcode, final int var) {(1)
```

```
    instructions.add(new VarInsnNode(opcode, var));
  }

  @Override
  public void visitMethodInsn(
      final int opcodeAndSource,
      final String owner,
      final String name,
      final String descriptor,
      final boolean isInterface) {
    if (api < Opcodes.ASM5
        && (opcodeAndSource & Opcodes.SOURCE_DEPRECATED) == 0)
    {
      // Redirect the call to the deprecated version of this method
      super.visitMethodInsn(
          opcodeAndSource, owner, name, descriptor, isInterface);
      return;
    }
    int opcode = opcodeAndSource & ~Opcodes.SOURCE_MASK;

    instructions.add(
        new MethodInsnNode(opcode, owner, name, descriptor, isInterface));
  }

  ...

  @Override
  public void visitMaxs(final int maxStack, final int maxLocals) {
    this.maxStack = maxStack;
    this.maxLocals = maxLocals;
  }

  @Override
  public void visitEnd() {
    // Nothing to do
  }

  public void check(final int api) {
    if (api == Opcodes.ASM4) {
      if (parameters != null && !parameters.isEmpty()) {
        throw new UnsupportedClassVersionException();
      }
      if (visibleTypeAnnotations != null
          && !visibleTypeAnnotations.isEmpty()) {
        throw new UnsupportedClassVersionException();
      }
      if (invisibleTypeAnnotations != null
          && !invisibleTypeAnnotations.isEmpty()) {
        throw new UnsupportedClassVersionException();
      }
      ...
    }
  }

  public void accept(final ClassVisitor classVisitor) {
    String[] exceptionsArray = exceptions == null ?
```

```
            null : exceptions.toArray(new String[0]);
        MethodVisitor methodVisitor =
            classVisitor.visitMethod(access, name, desc,
                                     signature, exceptionsArray);
        if (methodVisitor != null) {
          accept(methodVisitor);
        }
      }

      public void accept(final MethodVisitor methodVisitor) {
        // Visit the parameters
        if (parameters != null) {
          for (int i = 0, n = parameters.size(); i < n; i++) {
            parameters.get(i).accept(methodVisitor);
          }
        }
        ...

        methodVisitor.visitEnd();
      }
}
```

观察上述 MethodNode 的源码，可以看出其与 ClassNode 的实现原理是一样的，从中可以得出如下结论。

（1）MethodNode 继承自 MethodVisitor 类。

（2）MethodNode 重写了 MethodVisitor 中的 visit() 方法，并将结果保存在类的成员变量中。例如位置（1）处的 visitVarInsn() 方法中创建了一个 VarInsnNode 对象，该对象记录了对应的指令信息，并添加到 MethodNode 的 instructions 列表中。

（3）MethodNode 重写了 visitEnd() 方法，但方法体是空的。

（4）MethodNode 新增了 check() 方法，从此方法的实现可以看出它做了一些兼容性相关的操作。

（5）MethodNode 新增了两个 accept() 方法，其中 accept(ClassVisitor) 方法的作用是为 ClassVisitor 添加一个方法；accept(MethodVisitor) 的作用是根据 MethodNode 中的字段调用 MethodVisitor 中的对应的方法，与 ClassNode 中 accept() 的用法一致。该方法是 MethodVisitor 和 MethodNode 相互转换的关键，也就是 Core API 和 Tree API 相互转换的关键。

（6）对比 MethodVisitor 知识，MethodNode 类中的字段意思很容易理解，例如 name 表示方法名、desc 表示方法描述符。

在 MethodNode 中一个比较重要字段是 InsnList instructions，该字段中保存了方法中的指令信息，InsnList 的结构如下。

```
public class InsnList implements Iterable<AbstractInsnNode> {
        int size();
        AbstractInsnNode getFirst();
        AbstractInsnNode getLast();
        AbstractInsnNode get(int index);
        boolean contains(AbstractInsnNode insn);
        int indexOf(AbstractInsnNode insn);
        void accept(MethodVisitor mv);
        ListIterator iterator();
        ListIterator iterator(int index);
```

```
        AbstractInsnNode[] toArray();
        void set(AbstractInsnNode location, AbstractInsnNode insn);
        void add(AbstractInsnNode insn);
        void add(InsnList insns);
        void insert(AbstractInsnNode insn);
        void insert(InsnList insns);
        void insert(AbstractInsnNode location, AbstractInsnNode insn);
        void insert(AbstractInsnNode location, InsnList insns);
        void insertBefore(AbstractInsnNode location, AbstractInsnNode insn);
        void insertBefore(AbstractInsnNode location, InsnList insns);
        void remove(AbstractInsnNode insn);
        void clear();
}
```

InsnList 是一个双向链表结构，元素之间的指向定义在节点 AbstractInsnNode 的 previousInsn 和 nextInsn 字段中，InsnList 则提供了操作节点的方法。AbstractInsnNode 类的核心代码如下。

```
public abstract class AbstractInsnNode {
    AbstractInsnNode previousInsn;
    AbstractInsnNode nextInsn;
    public int getOpcode()
    public abstract int getType();
    public AbstractInsnNode getPrevious()
    public AbstractInsnNode getNext()
    public abstract AbstractInsnNode clone(Map<LabelNode, LabelNode> clonedLabels);
}
```

从中可以看出 AbstractInsnNode 是一个抽象类，它的实现类与 MethodVisitor 类中的 visitXxx 方法相关。例如，从上述代码可以看出 AbstractInsnNode 是一个抽象类，它的实现类命名是按照 MethodVisitor 类中的 visit() 方法来实现的。例如 visitVarInsn() 方法对应的实现是 TypeInsnNode、visitMethod() 方法对应的实现是 MethodInsnNode。Label、Frame 以及 LineNumber 虽然不是指令，但也是用 AbstractInsnNode 的实现来表示，这样做是为了在使用 MethodVisitor 操作方法时调用的 API 顺序与 InsnList 列表中元素的顺序保持一致。另外，根据 AbstractInsnNode 的结构定义，在使用时需要注意，每一个 AbstractInsnNode 对象，在 InsnList 中不能多次出现，如果想要构造同样的指令可使用 clone 方法。

接下来将结合这些知识，介绍 ClassNode 的相关使用方式。

6.4.1 方法的生成

使用 MethodNode 来生成下面这个类和其方法中的内容。

```
public final class CommonUtils {

      public static int add(int a, int b){
         return a + b;
      }
}
```

对应的实现如下。

```
public static void generateCommonUtils() throws IOException {
```

```
        //1. 创建 ClassNode，构建类需要的基本信息 (1)
        ClassNode cn = new ClassNode();
        cn.access = Opcodes.ACC_PUBLIC + Opcodes.ACC_FINAL;
        cn.name = "cn/sensorsdata/asm/treeapi/CommonUtils";
        cn.superName = "java/lang/Object";
        cn.version = Opcodes.V1_8;

        //2. 创建 MethodNode，构建方法需要的基本信息 (2)
        MethodNode mn = new MethodNode(

            Opcodes.ACC_PUBLIC + Opcodes.ACC_STATIC, "add", "(II)I", null, null);
        InsnList instructions = mn.instructions;
        instructions.add(new VarInsnNode(Opcodes.ILOAD, 0));
        instructions.add(new VarInsnNode(Opcodes.ILOAD, 1));
        instructions.add(new InsnNode(Opcodes.IADD));
        instructions.add(new InsnNode(Opcodes.IRETURN));
        mn.maxLocals = 2;
        mn.maxStack = 2;
        // 将 MethodNode 添加到 ClassNode 中
        cn.methods.add(mn);

        //3. 将数据保存到文件中 (3)
        ClassWriter classWriter = new ClassWriter(0);
        cn.accept(classWriter);
        byte[] data = classWriter.toByteArray();
        String filePath = "src/main/java/cn/sensorsdata/asm/treeapi/";
        File file = new File(filePath + "CommonUtils.class");
        FileUtils.writeByteArrayToFile(file, data);
}
```

上述代码位置（1）部分创建了一个 ClassNode 并设置了类的基本信息。位置（2）部分创建一个 MethodNode 对象，MethodNode 的构造方法中添加了方法的访问标识符、方法名等基本信息；然后获取 MethodNode 的 instructions 指令列表添加 add() 方法的指令；接着设置方法的操作数栈和局部变量表的最大深度；最后将 mn 对象添加到 ClassNode 的 methods 字段中，这样就创建了一个方法并添加到类中。位置（3）部分是将数据保存到文件中。

类似 ClassNode，使用 MethodNode 生成类这种方式速度也要比 MethodVisitor 低一些，不过好处是不用关注 MethodNode 中方法和属性的设置顺序，不用考虑像 MethodVisitor 那样有明确的调用顺序。

6.4.2 方法的转换和修改

对方法进行转换就是对 MethodNode 中的字段进行操作。例如清空方法中的内容，只需要将 instructions 列表清空，再追加一个 xreturn 指令即可。在 5.8.3 小节介绍了如何使用 Core API 来优化 ICONST_0 IADD 指令，为达到这个目的，在实现中需要定义重写很多 visit() 方法，并且还需要定义一些状态变量，可以看出其实现逻辑还是较复杂的，这种场景使用 MethodNode 来操作就比较合适。MethodNode 版的代码如下。

```
public void transform(MethodNode methodNode) {
        InsnList insnList = methodNode.instructions;
        Iterator<AbstractInsnNode> iterator = insnList.iterator();
```

```
        while (iterator.hasNext()) {
            AbstractInsnNode ain = iterator.next();
            if (ain instanceof InsnNode && ain.getOpcode() == Opcodes.IADD) {
                AbstractInsnNode previous = ain.getPrevious();// 获取上一条指令
                if (previous instanceof InsnNode
                    && previous.getOpcode() == Opcodes.ICONST_0)
                {
                    iterator.remove();// 删除 ICONST_0 指令
                    iterator.next();  // 指向 IADD 指令
                    iterator.remove();// 删除 IADD 指令
                } else {
                    iterator.next(); // 重新指向 IADD 指令
                }
            }
        }
}
```

上述代码通过遍历 MethodNode 中的指令，当遇到 IADD 指令时，就判断前一条指令是否是 ICONST_0，如果满足就将这两条指令删除。可以看到当需要对前后指令做判断时，MethodNode 的用法会简单很多。

6.5 MethodNode与Core API相互转换

类似ClassNode与Core API的转换内容，本节我们来看看MethodNode如何与Core API进行转换。

6.5.1 MethodNode的特性

根据 6.4 节的介绍，MethodNode 具有如下两个特性。

（1）MethodNode 是 MethodVisitor 的子类，因此它具有 MethodVisitor 的所有特性。

（2）MethodNode 提供了两个 accept() 方法，两个 accept() 方法的参数不一样，一个是 ClassVisitor，另一个是 MethodVisitor，可以通过 accept() 方法将 MethodNode 中的数据添加到 ClassVisitor 中或向 MethodVisitor 分发。

6.5.2 与Core API相互转换

根据以上特性，MethodNode 与 Core API 相互转换有下 3 种方式。

第一种是继承自 MethodNode，形式如下。

```
public class MyMethodAdapter extends MethodNode {
      public MyMethodAdapter(
          int access, String name, String descriptor,
          String signature, String[] exceptions, MethodVisitor methodVisitor)
      {
          super(access, name, descriptor, signature, exceptions);
          this.mv = mv;
```

```
        }

        @Override
        public void visitEnd() {
            // 对 MethodNode 进行转换操作
            //... 省略

            accept(mv);
        }
}
```

这种形式在构造方法接收一个 MethodVisitor 参数，在 visitEnd() 方法中对 MethodNode 转换后调用 accept(MethodVisitor) 方法将数据分发给 MethodVisitor。

第二种是在 MethodVisitor 中构建 MethodNode，其形式如下。

```
public class MyMethodAdapter extends MethodVisitor {

        private MethodVisitor next;
        public MyMethodAdapter(int api, MethodVisitor methodVisitor) {
            super(api, new MethodNode(api));
            next = methodVisitor;
        }

        @Override
        public void visitEnd() {
            super.visitEnd();
            MethodNode mn = (MethodNode) mv;
            // 对 MethodNode 进行转换操作
            ...

            mn.accept(next);
        }
}
```

此种形式是继承自 MethodVisitor，在构造方法中创建 MethodNode 对象，并调用父类的 super(api, methodvisitor) 构造方法，此时 MethodVisitor 中的 mv 字段的值为创建的 MethodNode 对象，这么操作后方法中的数据将保存在 MethodNode 对象中，所以在 visitEnd() 方法中调用 MethodNode 的 accept() 方法将 MethodNode 中的数据转换到 MethodVisitor 中。

以上两种方式都是利用 MethodNode 的 accept(MethodVisitor) 方法来完成数据的转换，MethodNode 还有一个 accept(ClassVisitor) 方法，它的用法示例如下。

```
public class MyClassVisitor extends ClassVisitor {
        public MyClassVisitor(int api, ClassVisitor classVisitor) {
            super(api, classVisitor);
        }

        @Override
        public MethodVisitor visitMethod(
            int access, String name, String descriptor,
            String signature, String[] exceptions)
        {
            return new MethodNode(
                this.api, access, name, descriptor, signature, exceptions)
            {
```

```
            @Override
            public void visitEnd() {
                // 对 MethodNode 进行转换操作
                ...

                accept(MyClassVisitor.this.cv);
            }
        };
    }
}
```

此种形式是创建了一个 MethodNode 匿名内部类，并在 visitEnd() 方法中调用 accept(ClassVisitor) 方法。

以上就是 MethodNode 与 Core API 相互转换的 3 种方式，在实际使用场景中常专门使用 MethodNode 来处理方法中的内容，而类的其他部分则继续使用 Core API。

6.6 Core API和Tree API如何选择

相比Core API，Tree API会占用更多的内存，但使用Tree API进行类转换更加容易一些。举个例子，假设类中可能有 a、b 两个方法，当存在 b 方法时就将 a 方法删除。使用 Core API 实现此功能可能需要遍历两遍类，第一遍是寻找 b 方法，如果 b 方法存在，就需要遍历第二遍，而使用 Tree API 就只需要遍历两遍 ClassNode 的 methods 列表。

对于一些复杂的场景，使用 Core API 很可能需要遍历多遍类，并且需要保存一些状态参数，这就使得代码也变得复杂。因此当使用 Core API 不能通过一次遍历即可完成对类的转换时，就需要考虑使用 Tree API 来操作。但这也并非绝对，例如混淆代码的操作是不能够通过一次遍历来解决的，因为它需要先获取到类与混淆名之间的映射关系，还需要知道类使用的位置，这就需要遍历所有的类。这种情况下使用 Tree API 就不是一个好的选择，因为这需要将所有需要混淆的类保存在内存中。而使用 Core API 将会是一个不错的选择，遍历两遍，第一遍是计算类和混淆名的映射关系，这个映射结构只需要很少的内存即可；第二遍根据映射关系再对类进行转换。

6.7 其他

6.7.1 方法分析

计算机科学中有一门关于程序分析的课程“静态程序设计”，它是编程语言应用层面下的一个细分领域，并且是一个非常重要的核心内容。静态程序分析是指在不运行代码的方式下，通过词法分析、语法分析、控制流分析、数据流分析等技术对程序代码进行扫描，验证代码是否满足规范性、安全性、可靠性、可维护性等指标的一种代码分析技术。现在很多 IDE 工具都能做到提示代码的风险，例如空指针问题、潜在的内存泄漏问题、死代码消除优化（在编译器的机器无关优化环节，将不会对程序执行结果产生影响的代码删除）等功能，这些都依赖于静态程序分析。

静态程序分析是一个非常深的话题，在此不展开讨论。ASM 在 org.objectweb.asm.tree.analysis 包中提供了基于 Tree API 的代码分析工具，这些工具的类图关系如图 6-1 所示。

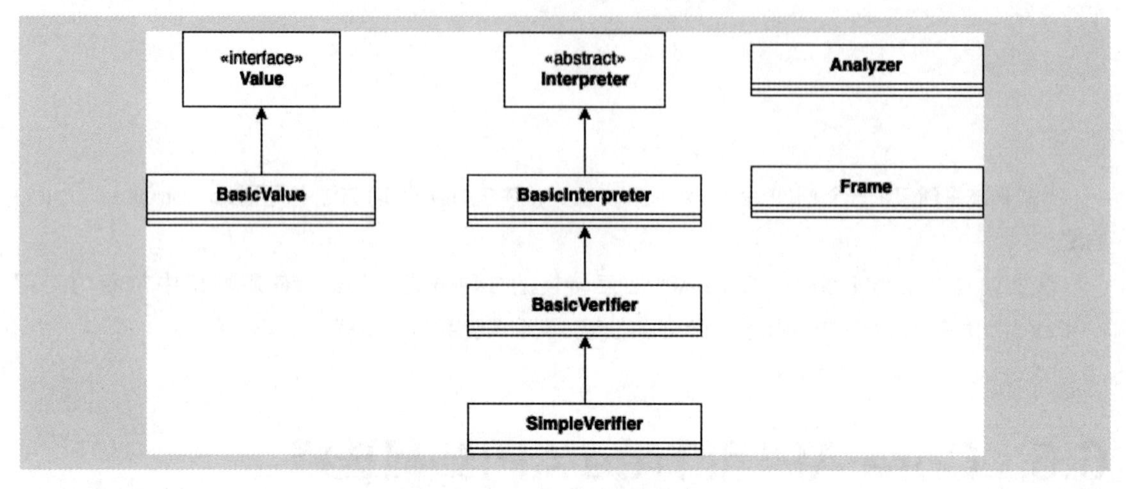

图 6-1　analysis 包的类图关系

具体的说明如下。

（1）Analyzer 是代码分析的主要工具类。

（2）BasicInterpreter 是基本的数据流分析器，是抽象类 Interpreter 的一个子类，用于构建 Analyzer 对象。它利用定义在 BasicValue 类中声明的 7 个值集来模拟字节码指令的运行效果，这 7 个值为：

- UNINITIALIZED_VALUE 表示所有“可能的值”；
- INT_VALUE 表示所有类型可能为 int、short、byte、boolean 或 char values 的值；
- FLOAT_VALUE 表示所有 float 类型的值；
- LONG_VALUE 表示所有 long 类型的值；
- DOUBLE_VALUE 表示所有 double 类型的值；
- REFERENCE_VALUE 表示所有对象或者数组类型的值；
- RETURNADDRESS_VALUE 所示所有协程的值。

（3）BasicVerifier 是基本数据流校验器，是 BasicInterpreter 的子类，用于实现对字节码指令是否正确的校验。

（4）SimpleVerifier 是简单数据流校验器，是 BasicVerifier 的子类。它使用更多的集合来模拟字节码指令的执行，所以它可以检测出更多的错误。

这些工具类的基本使用模板如下。

```
//1. 创建 Analyzer 对象，构造方法中传入 Interpreter 的子类，这里是 BasicInterpreter
Analyzer<BasicValue> analyzer = new Analyzer<BasicValue>(new BasicInterpreter());
//2. 分析给定的 MethodNode，其中 owner 表示此方法所在的类
analyzer.analyze(owner, methodNode);
//3. 获取上一次调用 analyze() 方法后的分析结果
//Frame 表示方法中每条指令对应的栈帧状态，它的大小与指令的长度保持一致（包括 labels）
// 如果指令对应的 Frame 为 null，则表示该指令是无法执行到的
Frame<BasicValue>[] frames = analyzer.getFrames();
```

如果方法中的指令无法执行，那么指令对应的 Frame 将为 null，可以利用这个特性来实现“死代码”功能的检测，代码如下。

```
public class RemoveDeadCodeAdapter extends MethodVisitor {
      private String owner;
      private MethodVisitor next;
      public RemoveDeadCodeAdapter(
          String owner, int access String name, String desc, MethodVisitor mv)
      {
          super(Opcodes.ASM6, new MethodNode(access, name, desc, null, null));
          this.owner = owner;
          next = mv;
      }

      @Override
      public void visitEnd() {
          MethodNode mn = (MethodNode) mv;
          Analyzer<BasicValue> a = new Analyzer<BasicValue>(new BasicInterpreter());
          try {
              a.analyze(owner, mn);
              Frame<BasicValue>[] frames = a.getFrames();
              AbstractInsnNode[] insns = mn.instructions.toArray();
              for (int i = 0; i < frames.length; ++i) {
                  if (frames[i] == null && !(insns[i] instanceof LabelNode)) {
                      mn.instructions.remove(insns[i]);
                  }
              }
          } catch (AnalyzerException ignored) {
          }
          mn.accept(next);
      }
}
```

另外，5.9.3 小节介绍的 ASM 工具类 CheckClassAdapter，此类可以校验字节码指令是否合法。例如可以检测 ISTORE 1 ALOAD 1 这样的指令序列不合法，其背后的原理就是利用 ASM 的代码分析工具来实现的。

下面是 CheckClassAdapter 的 verify() 方法的源码，可以看到 Analyzer 的使用。

```
public class CheckClassAdapter extends ClassVisitor {
    /**
     * Checks the given class.
     *
     * @param classReader the class to be checked.
     * @param loader a <code>ClassLoader</code> which will be used to
     * load referenced classes. May be  {@literal null}.
     * @param printResults whether to print the results of the bytecode verification.
     * @param printWriter where the results (or the stack trace in case of error)
     * must be printed.
     */
    public static void verify(
        final ClassReader classReader,
        final ClassLoader loader,
        final boolean printResults,
        final PrintWriter printWriter) {
      ClassNode classNode = new ClassNode();
      classReader.accept(
```

```
            new CheckClassAdapter(
              /*latest*/ Opcodes.ASM10_EXPERIMENTAL, classNode, false) {},
            ClassReader.SKIP_DEBUG);

        Type superType = classNode.superName ==
            null ? null : Type.getObjectType(classNode.superName);
        List<MethodNode> methods = classNode.methods;

        List<Type> interfaces = new ArrayList<>();
        for (String interfaceName : classNode.interfaces) {
          interfaces.add(Type.getObjectType(interfaceName));
        }
        // 对类中的方法进行分析
        for (MethodNode method : methods) {
          // 此处使用的是简单数据流校验器
          SimpleVerifier verifier =
              new SimpleVerifier(
                  Type.getObjectType(classNode.name),
                  superType,
                  interfaces,
                  (classNode.access & Opcodes.ACC_INTERFACE) != 0);
          Analyzer<BasicValue> analyzer = new Analyzer<>(verifier);
          if (loader != null) {
            verifier.setClassLoader(loader);
          }
          try {
            analyzer.analyze(classNode.name, method);
          } catch (AnalyzerException e) {
            e.printStackTrace(printWriter);
          }
          if (printResults) {
            printAnalyzerResult(method, analyzer, printWriter);
          }
        }
        printWriter.flush();
      }
      ...
}
```

由此，在使用 Tree API 时，可以使用 BasicVerifier 基本数据流校验工具来对字节码进行校验，具体用法如下。

```
ublic class BasicVerifierAdapter extends MethodVisitor {
      String owner;
      MethodVisitor next;

      public BasicVerifierAdapter(String owner, int access, String name,
                                  String desc, MethodVisitor mv) {
          super(Opcodes.ASM6, new MethodNode(access, name, desc, null, null));
          this.owner = owner;
          next = mv;
      }

      @Override
      public void visitEnd() {
          MethodNode mn = (MethodNode) mv;
          // 此处使用 BasicVerifier
```

```
            Analyzer<BasicValue> a = new Analyzer<BasicValue>(new BasicVerifier());
            try {
                a.analyze(owner, mn);
            } catch (AnalyzerException e) {
                throw new RuntimeException(e.getMessage());
            }
            mn.accept(next);
        }
    }
```

关于方法分析还有很多内容，例如可以用来分析潜在的空指针问题，如果读者对这些感兴趣，可以先从静态程序分析的理论知识学起。

6.7.2 兼容性探讨

你可能已经注意到了 ASM 提供的核心 API 类的构造方法中需要传入 api 参数，如 ClassVisitor 的构造方法。

```
public ClassVisitor(final int api) {
    this(api, null);
}
```

这里对应有几个问题。

- 这里api参数的作用是什么，又为什么要这样设计？
- 随着Java的不断发展，越来越多的特性也会被添加到Java中，例如Java 7引入的invokedynamic指令、Java 8的Type Annotation、Java 9的模块化以及Java 11的Nest Host等特性，ASM又是如何对此做兼容的呢？
- 你可能还注意到Opcodes类中定义的api参数的值是从ASM4开始的，那为什么没有ASM1、ASM2、ASM3呢？

实际上 ASM 4.0 之前的版本是没有向后兼容的，这显然是不友好的做法。因此从 ASM 4.0 开始引入了向后兼容机制，它的目标是确保所有未来的 ASM 版本能够向后一直兼容到 ASM 4.0（即使在类文件格式中引入了新功能），这就是 api 的值是从 ASM4 开始的原因。不过 ASM 的向后兼容做得并不彻底，例如用户使用 ASM 4.0，并且定义 api 为 ASM4，当用户升级到 ASM 6.0 时还需要考虑手动将 api 的值升级到 ASM6，即升级 ASM 版本后，还需要考虑升级 api 的问题，而这一步需要用户参与，具体原因稍后说明。

在兼容性方面还需要考虑如何兼容新的 Java 特性，例如基于 ASM 4.0 编写的代码并不能支持 invokeydynamic 指令，当用户升级到 ASM 5.0 后遇到 invokedynamic 指令，ASM 又如何处理呢？如果只是简单的忽略可能会产生意想不到的后果，ASM 处理此场景的原则是，尽可能早地抛出异常，让用户来针对异常进行分析和适配。例如 MethodVisitor 的 visitInvokeDynamicInsn() 方法所展示的处理逻辑。

```
public void visitInvokeDynamicInsn(
      final String name, final String descriptor,
      final Handle bootstrapMethodHandle,
      final Object... bootstrapMethodArguments)
{
    if (api < Opcodes.ASM5) { // 在此对 api 参数进行校验
```

```
        throw new UnsupportedOperationException(REQUIRES_ASM5);
    }
    ...
}
```

观察上述代码可以发现 visitInvokeDynamicInsn() 方法中判断了 api 的值，如果小于 ASM5 就抛出异常。

下面用一个实际的例子来描述 ASM 的兼容机制。假设有两个属性将被添加到 Java 8 的 ClassFile 中，其中一个属性用于存储作用（author）信息，另一个属性用于存储版权（license）信息。并且假设这两个属性是通过 ASM 5.0 的 ClassVisitor 类中的 visitSource() 和 visitLicense() 方法进行访问的。方法定义如下。

```
void visitLicense(String license);
void visitSource(String author, String source, String debug);
```

其中第一个方法 visitLicense() 方法用于访问版权信息；第二个方法是对 visitSource() 方法的重载，重载方法中多了一个 author 字段，原本的 visitSource() 方法仍然存在，只不过被标注为废弃方法。

```
@Deprecated
void visitSource(String source, String debug);
```

license 和 author 这两个属性是可选的，例如调用 visitSource(String,String,String) 方法时，author 字段可以为 null。在我们设定的这个例子中，ASM 4.0 对应的实现代码类似如下。

```
public class MyClassAdapter extends ClassVisitor {
      public MyClassAdapter(ClassVisitor cv) {
          super(ASM4, cv);
      }

      public void visitSource(String source, String debug) {
          super.visitSource(source, debug);
      }
}
```

当 ASM 版本升级到 ASM 5.0 时，我们应该移除 visitSource(String, String) 方法，调整后的实现代码如下。

```
class MyClassAdapter extends ClassVisitor {
      public MyClassAdapter(ClassVisitor cv) {
          super(ASM5, cv);
      }

      public void visitSource(String author, String source, String debug) {
          super.visitSource(author, source, debug);
      }

      public void visitLicense(String license) {
          super.visitLicense(license);
      }
}
```

那么调整后的代码又是如何兼容 ASM 4.0 的呢？对比 ASM 4.0 和 ASM 5.0 中 ClassVisitor 的内部实现，首先是 ASM 4.0 的实现如下。

```
public abstract class ClassVisitor {
      int api;
      ClassVisitor cv;
      public ClassVisitor(int api, ClassVisitor cv) {
          this.api = api;
          this.cv = cv;
      }
      ...
      public void visitSource(String source, String debug) {
          if (cv != null)
              cv.visitSource(source, debug);
      }
}
```

ASM 5.0 实现如下。

```
public abstract class ClassVisitor {
      // 该方法被标注为废弃方法，并且内部根据 api 的值会重定向到新的 visitSource() 方法
      @Deprecated
      public void visitSource(String source, String debug) {
          if (api < ASM5) {
              if (cv != null) cv.visitSource(source, debug);
          } else {
              visitSource(null, source, debug);
          }
      }
      public void visitSource(String author, String source, String debug) {
          if (api < ASM5) {
              if (author == null) {
                visitSource(source, debug);
              } else {
                throw new RuntimeException();
              }
          } else {
              if (cv != null) cv.visitSource(author, source, debug);
          }
      }
      public void visitLicense(String license) {
          if (api < ASM5) throw new RuntimeException();
          if (cv != null) cv.visitSource(source, debug);
      }
}
```

首先 ASM 4.0 的 MyClassAdapter 代码不存在任何问题，当升级到 ASM 5.0 后，此时 MyClassAdapter 将继承 ASM 5.0 的 ClassVisitor，不过代码中的 api 值仍然为 ASM4（api < ASM5）。这种情况下对比 ASM 5.0 和 ASM 4.0 中 ClassVisitor 的 visitSource(String, String) 方法的实现，ASM 5.0 中根据 api 的值，继续调用 visitSource(source, debug) 方法，可以看出这样也没有什么问题。

另外如果 author 值为 null，调用 visitSource(String, String, String) 方法，其内部会重定向到老版本的 visitSource(String,String) 方法。可是当 author 不为 null 或者在 ClassFile 中获取到了 license 信息时就会抛出 RuntimeException() 异常。

如果升级到 ASM 5.0 时，将代码也一并升级，此时 MyClassAdapter 5.0 继承自 ClassVisitor 5.0，api 的值为 ASM5，visitSource() 和 visitLicense() 的逻辑仅仅是分发给下一个 cv，老版本的 visitSource(String,String) 会将结果重定向到新的 visitSource(String,String,String) 方法，这样就不会存

在问题，这就是 ASM 处理兼容性的逻辑。

那是不是就意味着只要我们升级到新版本的 ASM，只需将 api 的值升级到最新就可以了呢？是，但也不是，因为一旦 api 的值每次都使用最大版本的 ASMX，老的代码很可能会漏掉一些新版本添加的实现。例如，前面提到的 visitInvokeDynamicInsn() 方法，首先可以确定的是 ASM 4.0 中没有此方法。假如代码逻辑 MethodAdapter 中需要对所有的 visit() 做处理，此时升级到 ASM 5.0 并且同时修改 api 的值为 ASM5，MethodAdapter 不会报错，但产生的后果就是漏掉了对 visitInvokeDynamicInsn() 方法的适配。这种情况将 api 设置到最大值就显得不合适，通常的做法是保留一个尽量小的 api 值，这样当升级 ASM 或者处理高版本的 ClassFile 时，让 ASM 抛出异常，以便我们能够发现问题。

6.7.3 Attribute

3.4.10 小节我们详细地介绍了 ClassFile 中的属性，可以看出属性是 ClassFile 中扩展性最强的一个点，开发者可以定义自己的属性。因为这个机制，读者可能会想到通过自定义属性来实现一些“骚操作”，不过不建议这么做。自定义属性相比使用注解会麻烦很多，除非因为某些原因必须要自定义属性或者解析其他人提供的属性，这时可以使用 ASM 提供的 Attribute API。

默认情况下，ClassReader 在解析到非标准属性时会创建 Attribute 对象，并调用 visitAttribute(Attribute) 方法（ClassVisitor、MethodVisitor、FieldVisitor 都有此方法）。Attribute 对象只保存原始数据对应的 byte[] 数组，用户获取到 byte[] 数组后需要自行解析其内容。

下面使用一个例子来介绍如何使用 Attribute。假设我们想要为 ClassFile 添加一个 Comment 属性，其对应的代码如下。

```
public final class CommentAttribute extends Attribute {
    // 定义属性的名称，这个名称是 JVM 中的名称，例如标准属性 SourceFile
    private static final String TYPE = "Comment";
    // 评论的具体内容
    private final String content;

    protected CommentAttribute(String content) {
        super(TYPE);
        this.content = content;
    }

    @Override
    public boolean isUnknown() {
        return false;
    }

    @Override
    protected Attribute read(
        ClassReader classReader, int offset, int length,
        char[] charBuffer, int codeAttributeOffset, Label[] labels)
    {
        //classReader.readUTF8 用于从常量池中获取 UTF-8 类型的常量
        return new CommentAttribute(classReader.readUTF8(offset, charBuffer));
    }

    @Override
```

```
        protected ByteVector write(
            ClassWriter classWriter, byte[] code, int codeLength,
            int maxStack, int maxLocals)
        {
            //classWriter.newUTF8 用于在常量池中创建 UTF-8 类型常量的索引
            return new ByteVector().putShort(classWriter.newUTF8(content));
        }
}
```

其中 read() 方法用于解码数据，write() 方法则用于编码数据，read() 方法必须返回一个新的 Attribute 对象。在生成类时可以按照如下方式使用。

```
public static void generateSensorsDataAPI() throws IOException {
        ClassWriter cw = new ClassWriter(0);
        cw.visit(
            Opcodes.V1_8,
            Opcodes.ACC_PUBLIC + Opcodes.ACC_ABSTRACT,
            "cn/sneosrsdata/asm/attribute/SensorsDataAPI",
            null, "Ljava/lang/Object;", null);
        ...
        // 此处使用 CommentAttribute
        cw.visitAttribute(new CommentAttribute("Here is SensorsData."));
        cw.visitEnd();
        String filePath = "src/main/java/cn/sensorsdata/asm/attribute/";
        File file = new File(filePath + "SensorsDataAPI.class");
        byte[] data = cw.toByteArray();
        FileUtils.writeByteArrayToFile(file, data);
}
```

使用 javap 命令查看上述代码生成的 .class 文件内容。

```
public abstract class cn.sensorsdata.asm.attribute.SensorsDataAPI extends Ljava.
lang.Object;
...
Constant pool:
    ...
    #16 = Utf8              Comment
    #17 = Utf8              Here is SensorsData.
{
    ...
}
    Comment: length = 0x2 (unknown attribute)
     00 11
```

可以看到 Comment 属性的内容对应在常量池 #17 的位置。同样，为了能够解析自定义的属性，还需要按照如下方式去使用它。

```
ClassReader cr = ...;
ClassVisitor cv = ...;
cr.accept(cv, new Attribute[] { new CommentAttribute("") }, 0);
```

即在 ClassReader 的 accept() 方法中声明 CommentAttribute。

6.7.4　ASM 框架分析

到目前为止，ASM 的绝大多数用法已经介绍完毕，如果想要继续深入研究 ASM，就需要去研究它的源码。本小节与大家一起简单看看 ASM 内部的工作流程是怎样的，我们以处理 Field 作为切入点来介绍。首先看下面这个例子。

```
public class Foo {
      public static final String TMP = "start";
}

public class Main {
      public static class FirstAdapter extends ClassVisitor {
          public FirstAdapter(ClassVisitor classVisitor) {
              super(Opcodes.ASM6, classVisitor);
          }

          @Override
          public FieldVisitor visitField(
              int access, String name, String descriptor,
              String signature, Object value)
          {
              //追加 _First 字符串
              return super.visitField(
                  access, name, descriptor, signature, value + "_First");
          }
      }
      public static class SecondAdapter extends ClassVisitor {
          public SecondAdapter(ClassVisitor classVisitor) {
              super(Opcodes.ASM6, classVisitor);
          }

          @Override
          public FieldVisitor visitField(
              int access, String name, String descriptor,
              String signature, Object value)
          {
              //追加 _Second 字符串
              return super.visitField(
                  access, name, descriptor, signature, value + "_Second");
          }
      }

      public static void transform() throws IOException {
          ClassReader cr = new ClassReader("cn.sensorsdata.asm.design.Foo");
          ClassWriter cw = new ClassWriter(0);

          FirstAdapter firstAdapter = new FirstAdapter(cw);
          SecondAdapter secondAdapter = new SecondAdapter(firstAdapter);

          cr.accept(secondAdapter, 0);
          File newFile =
              new File("src/main/java/cn/sensorsdata/asm/design/Foo2.class");
          FileUtils.writeByteArrayToFile(newFile, cw.toByteArray());
      }

      public static void main(String[] args) throws IOException {
          transform();
      }
}
```

上述代码中定义了一个 Foo 类，Foo 类中定义了 TMP 字符串常量。FirstAdapter 和 SecondAdapter 用于对 Foo 类中的 TMP 字段进行转换。上述 Main() 方法运行以后，Foo 类的 TMP 最终结果是 start_Second_First。可以看到结果与 Adapter 的声明顺序似乎有关系，上述结果对应的时序图如图 6-2 所示。

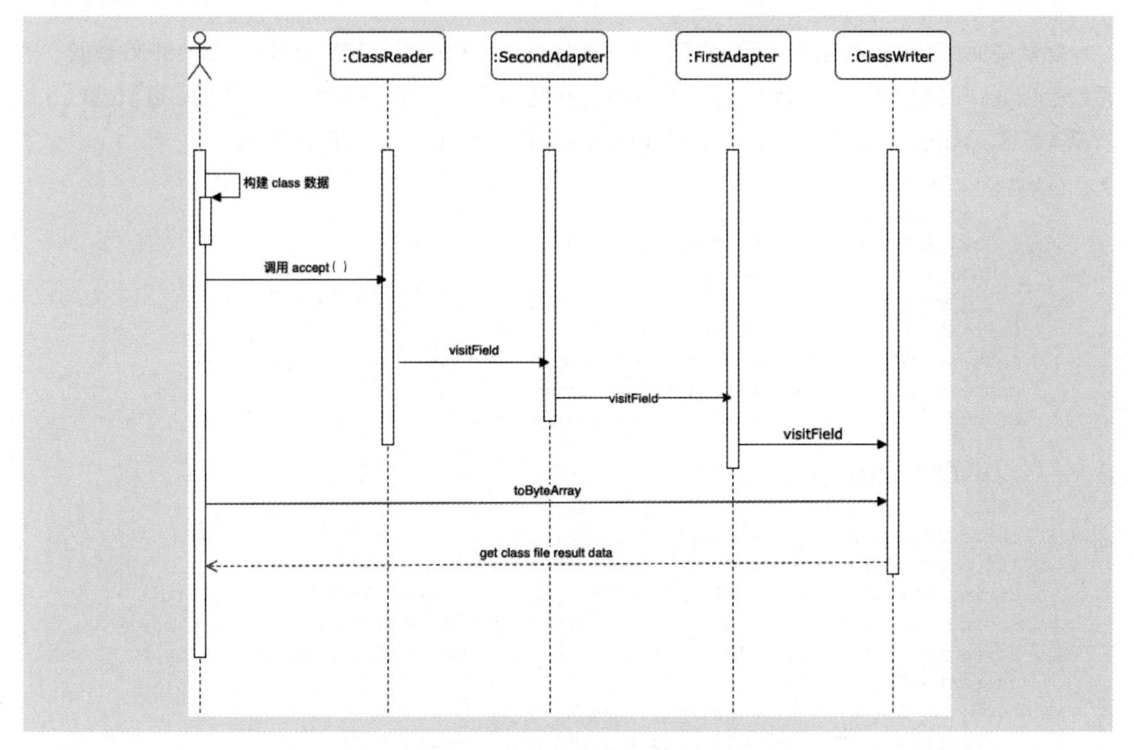

图 6-2 ASM 运行的时序图

对照着图 6-2 所示的时序图，先来看看 ClassReader 构造方法中的逻辑。

```
ClassReader(
      final byte[] classFileBuffer,
      final int classFileOffset, final boolean checkClassVersion)
{
    this.classFileBuffer = classFileBuffer;
    this.b = classFileBuffer;

    if (checkClassVersion && readShort(classFileOffset + 6) > Opcodes.V18) (1)
    {
      throw new IllegalArgumentException(
          "Unsupported class file major version "
          + readShort(classFileOffset + 6));
    }

    int constantPoolCount = readUnsignedShort(classFileOffset + 8);(2)
    ...
    maxStringLength = currentMaxStringLength;
    // The Classfile's access_flags field is just after the last constant pool entry
    header = currentCpInfoOffset;

    // Allocate the cache of ConstantDynamic values, if there is at least one
    constantDynamicValues =
```

```
        hasConstantDynamic ? new ConstantDynamic[constantPoolCount] : null;

    bootstrapMethodOffsets = hasBootstrapMethods ? (3)
        readBootstrapMethodsAttribute(currentMaxStringLength) : null;
}
```

观察 ClassReader 构造方法可以看到，它主要从 classFileBuffer 字节数组获取 ClassFile 数据，然后校验 ClassFile 版本信息 [位置（1）]、获取常量池长度 [位置（2）]、计算引导方法的偏移量 [位置（3）] 等基本信息。真正对 ClassFile 数据进行解析是在 ClassReader 的 accept() 方法中。

```
public void accept(
      final ClassVisitor classVisitor,
      final Attribute[] attributePrototypes,
      final int parsingOptions)
{
    Context context = new Context();
    context.attributePrototypes = attributePrototypes;
    context.parsingOptions = parsingOptions;
    context.charBuffer = new char[maxStringLength];

    // 计算类的头信息 (1)
    char[] charBuffer = context.charBuffer;
    int currentOffset = header;
    int accessFlags = readUnsignedShort(currentOffset);
    String thisClass = readClass(currentOffset + 2, charBuffer);
    String superClass = readClass(currentOffset + 4, charBuffer);
    String[] interfaces = new String[readUnsignedShort(currentOffset + 6)];
    currentOffset += 8;
    for (int i = 0; i < interfaces.length; ++i) {
      interfaces[i] = readClass(currentOffset, charBuffer);
      currentOffset += 2;
    }
    ...
    // 调用 classVisitor 的 visit() 方法 (2)
    classVisitor.visit(
        readInt(cpInfoOffsets[1] - 7),
        accessFlags, thisClass, signature, superClass, interfaces);

    // 调用 classVisitor 的 visitSource() 方法
    if ((parsingOptions & SKIP_DEBUG) == 0
        && (sourceFile != null || sourceDebugExtension != null)) {
      classVisitor.visitSource(sourceFile, sourceDebugExtension);
    }
    ...

    // 获取 class 中的字段信息，最终在 readField() 方法中调用 visitField() 方法 (3)
    int fieldsCount = readUnsignedShort(currentOffset);
    currentOffset += 2;
    while (fieldsCount-- > 0) {
      currentOffset = readField(classVisitor, context, currentOffset); (4)
    }
    ...

    // Visit the end of the class
    classVisitor.visitEnd();
}
```

```
private int readField((5)
      final ClassVisitor classVisitor, final Context context, final int fieldInfoOffset) {
    char[] charBuffer = context.charBuffer;

    // Read the access_flags, name_index and descriptor_index fields
    int currentOffset = fieldInfoOffset;
    int accessFlags = readUnsignedShort(currentOffset);
    String name = readUTF8(currentOffset + 2, charBuffer);
    String descriptor = readUTF8(currentOffset + 4, charBuffer);
    currentOffset += 6;
    ...

    // Visit the field declaration
    // 此处调用 visitField() 方法
    FieldVisitor fieldVisitor = classVisitor.visitField((6)
        accessFlags, name, descriptor, signature, constantValue);

    if (fieldVisitor == null) {
      return currentOffset;
    }
    ...

    // Visit the end of the field
    fieldVisitor.visitEnd();
    return currentOffset;
}
```

可以看到 accept() 方法中解析 ClassFile 文件中的内容：位置（1）处解析类的基本信息（包括版本、父类、接口）、位置（2）处根据 ClassFile 内容依次调用 ClassVisitor 的 visitXxx() 方法、位置（3）、（4）处解析 Class 中的 Field 信息。位置（5）处的 readField() 方法中获取 Field 的具体信息，并且调用 ClassVisitor 的 visitField() 方法 [位置（6）]。以本小节为例，ClassRead 的 accept() 方法接收参数 secondAdapter，所以这里调用的是 secondAdapter 的 visitField() 方法。接下来我们再看看 ClassVisitor 的 visitField() 中的代码。

```
public abstract class ClassVisitor {
    protected final int api;
    /**
     * The class visitor to which this visitor must delegate method calls.
     * May be {@literal null}.
     */
    protected ClassVisitor cv;

    public ClassVisitor(final int api, final ClassVisitor classVisitor) {
      if (api != Opcodes.ASM9
          && api != Opcodes.ASM8
          && api != Opcodes.ASM7
          && api != Opcodes.ASM6
          && api != Opcodes.ASM5
          && api != Opcodes.ASM4
          && api != Opcodes.ASM10_EXPERIMENTAL) {
        throw new IllegalArgumentException("Unsupported api " + api);
      }
      if (api == Opcodes.ASM10_EXPERIMENTAL) {
```

```
            Constants.checkAsmExperimental(this);
        }
        this.api = api;
        this.cv = classVisitor;
    }

    ...
    public FieldVisitor visitField(
            final int access,
            final String name,
            final String descriptor,
            final String signature,
            final Object value) {
        if (cv != null) {
            return cv.visitField(access, name, descriptor, signature, value);
        }
        return null;
    }

    public void visitEnd() {
        if (cv != null) {
            cv.visitEnd();
        }
    }
}
```

从上述 ClassVisitor 的代码可以看到，在 visitField() 方法中，最终还是调用了 cv 对象的 visitField() 方法，所以 secondAdapter 中的 cv 对象是 firstAdapter，而 firstAdapter 的 cv 对象是 cw（ClassWriter）对象。因此不考虑 ClassWriter 中 visitField() 逻辑的情况下，根据各 visitor 的处理顺序是

SecondAdapter → FirstAdapter → ClassWriter，因此最终的结果是 start_Second_First。ClassWriter 这个类继承自 ClassVisitor，但它有一些特殊，它是将各 visitor 处理结果进行汇总的地方，其 visitField() 源码如下。

```
public class ClassWriter extends ClassVisitor {
    private FieldWriter firstField;
    private FieldWriter lastField;

    ...

    @Override
    public final FieldVisitor visitField(
            final int access,
            final String name,
            final String descriptor,
            final String signature,
            final Object value) {
        FieldWriter fieldWriter =
                new FieldWriter(symbolTable, access, name, descriptor, signature, value);
        if (firstField == null) {
            firstField = fieldWriter;
        } else {
            lastField.fv = fieldWriter;
        }
        return lastField = fieldWriter;
    }
```

```
    public byte[] toByteArray() {
      // 1.First step: compute the size in bytes of the ClassFile structure
      // The magic field uses 4 bytes
      // 10 mandatory fields (minor_version, major_version
      // constant_pool_count, access_flags, this_class
      // super_class, interfaces_count, fields_count
      // methods_count and attributes_count) use 2 bytes each
      //  and each interface uses 2 bytes too
      int size = 24 + 2 * interfaceCount;
      int fieldsCount = 0;
      FieldWriter fieldWriter = firstField;
      while (fieldWriter != null) {
        ++fieldsCount;
        size += fieldWriter.computeFieldInfoSize();
        fieldWriter = (FieldWriter) fieldWriter.fv;
      }
      ...
      // 2.Second step: allocate a ByteVector of the
      // correct size (in order to avoid any array copy in
      // dynamic resizes) and fill it with the ClassFile content
      // 创建存放最终结果的对象
      ByteVector result = new ByteVector(size);
      result.putInt(0xCAFEBABE).putInt(version);
      symbolTable.putConstantPool(result);
      int mask = (version & 0xFFFF) < Opcodes.V1_5 ? Opcodes.ACC_SYNTHETIC : 0;
      result.putShort(accessFlags & ~mask).putShort(thisClass).putShort(superClass);
      result.putShort(interfaceCount);
      for (int i = 0; i < interfaceCount; ++i) {
        result.putShort(interfaces[i]);
      }
      result.putShort(fieldsCount);
      fieldWriter = firstField;
      while (fieldWriter != null) {
        fieldWriter.putFieldInfo(result);// 依次设置 Field 内容 (1)
        fieldWriter = (FieldWriter) fieldWriter.fv;
      }
      ...
      // 3.Third step: replace the ASM specific instructions, if any
      if (hasAsmInstructions) {
        return replaceAsmInstructions(result.data, hasFrames);
      } else {
        return result.data;
      }
    }
}
```

可以看到每次调用 ClassWriter 的 visitField() 方法时都会创建一个 FieldWriter 对象，FieldWriter 继承自 FieldVisitor，在 toByteArray() 中会通过链式的方式依次调用 FieldWriter 的 putFieldInfo(ByteVetor) 方法（位置（1））将结果设置到 ByteVetor 对象中。ByteVector 是一个大小可扩展的并且使用 byte[] 存储数据的结构，toByteArray() 利用此结构来存储 ClassFile 结构中的各个部分内容。

以上通过处理 Field 这一个点，带大家整体回顾一下 ASM 处理过程，可以看出在处理元素内容时采用的是 visitor 模式，而 visitor 之间采用的是链式模式，有兴趣的读者可以自己按照上面的处理流程写一个

Demo 来加深理解。本小节只是起到抛砖引玉的作用，ASM 还有很多值得学习的地方。

6.8 小结

本章介绍了 ASM 中 Tree API 相关的知识，并探讨了 ASM 处理兼容的策略以及 ASM 框架处理流程。至此本书关于 ASM 部分的内容已完结，接下来会通过一个小的项目将本书介绍的 Gradle、字节码、ASM 等内容串联起来。

7. ASM实现全埋点——基础部分

经过了前文内容的学习，读者已经了解了 Gradle 插件的知识以及 ASM 框架的使用。从本章开始，我们将通过一个具体的案例来把这些内容串联起来，以加深读者对这些知识的理解。

7.1 目标

本案例比较简单，目标是能够自动采集 Android 中按钮的单击事件。采集的事件中包括如下用于数据分析的元素内容。

- element_content：按钮上设置的文字。
- screen_name：按钮所在页面的名称，通常是Activity或者Fragment对应类的全限定名。
- title：按钮所在页面的标题，例如ActionBar中的title值。
- time：事件触发的时间戳。

虽然只是一个案例，但还是希望读者能够按照一个标准的项目来做，最终结果是将它分享并发布出去，所以此案例中将包含如下清单。

- Android Demo：用于测试全埋点的Android工程。
- SDK：提供全埋点业务逻辑的代码。
- Gradle插件：基于AGP实现了AOP功能的Gradle插件。

7.2 实现步骤

7.2.1 创建Demo工程和SDK模块

首先打开 Android Studio，选择菜单 File→NewProject，在弹出的窗口中选择“Empty Project” Android 工程，单击 Next 按钮，填写 Android 工程的信息。Android 工程名为 AutoTrackImpl01，对应的包为 cn.sensorsdata.autotrack.demo，然后单击 Finish，如图 7-1 所示。

接着在 MainActivity 的布局文件中定义一个按钮，并且在 MainActivity 中为其设置事件监听，代码如下。

```
public class MainActivity extends AppCompatActivity {

    @Override
    protected void onCreate(Bundle savedInstanceState) {
        super.onCreate(savedInstanceState);
        setContentView(R.layout.activity_main);
        init();
    }

    private void init(){
        findViewById(R.id.btn).setOnClickListener(new View.OnClickListener() {
```

```
            @Override
            public void onClick(View v) {
                Log.i("MainActivity", "Button is Clicked");
            }
        });
    }
}
```

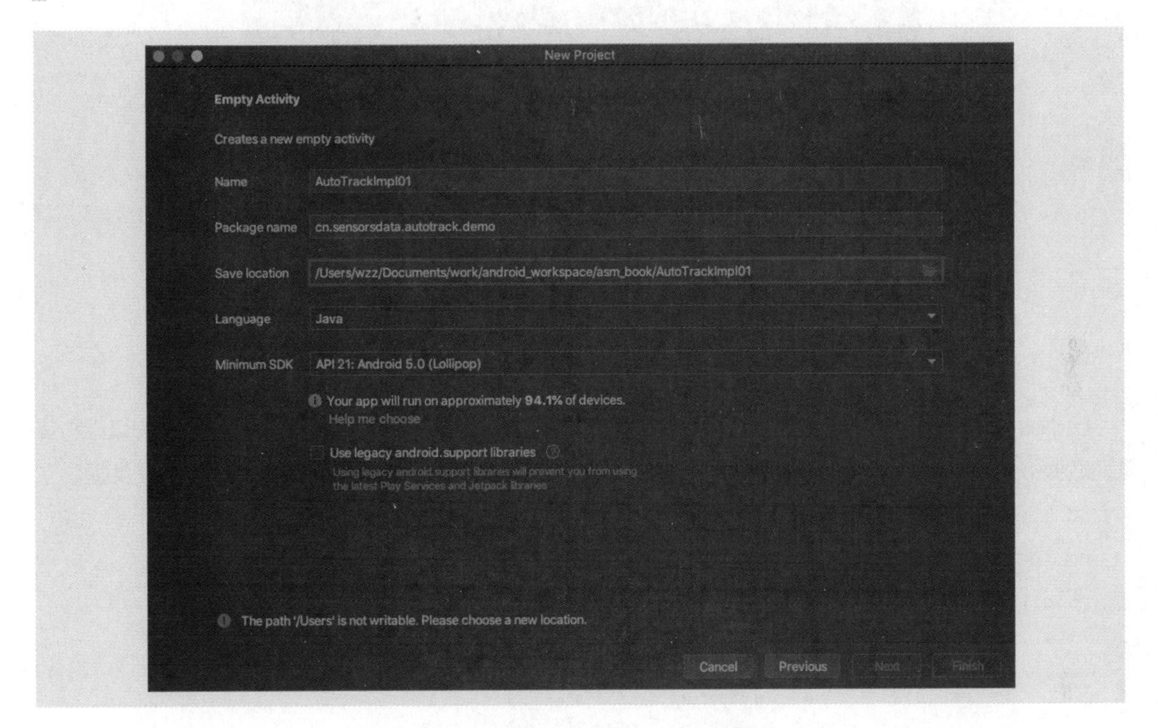

图 7-1　创建 AutoTrackImpl01 工程

至此 Demo 工程已创建完毕，最终对应的界面如图 7-2 所示（注意观察图中按钮内容和页面标题）。

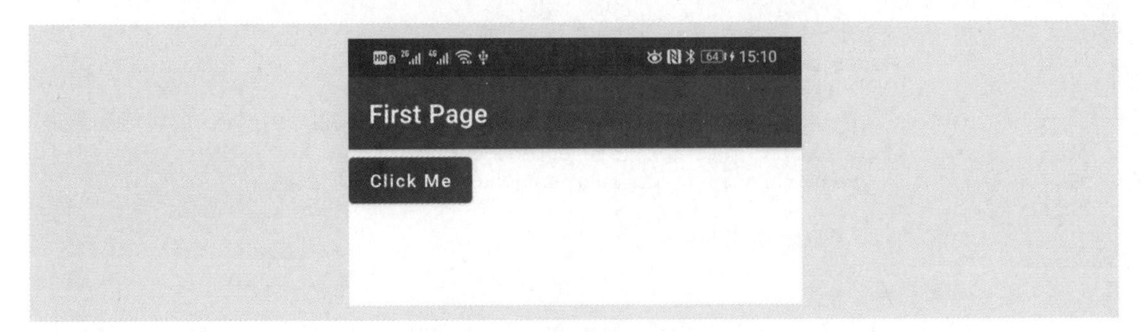

图 7-2　MainActivity 运行界面

接着创建一个名为 autotrack_sdk 的 Android Library 模块，选择 AndroidStudio 菜单 File→New→New Module，在弹出的对话框中按照图 7-3 填写信息。

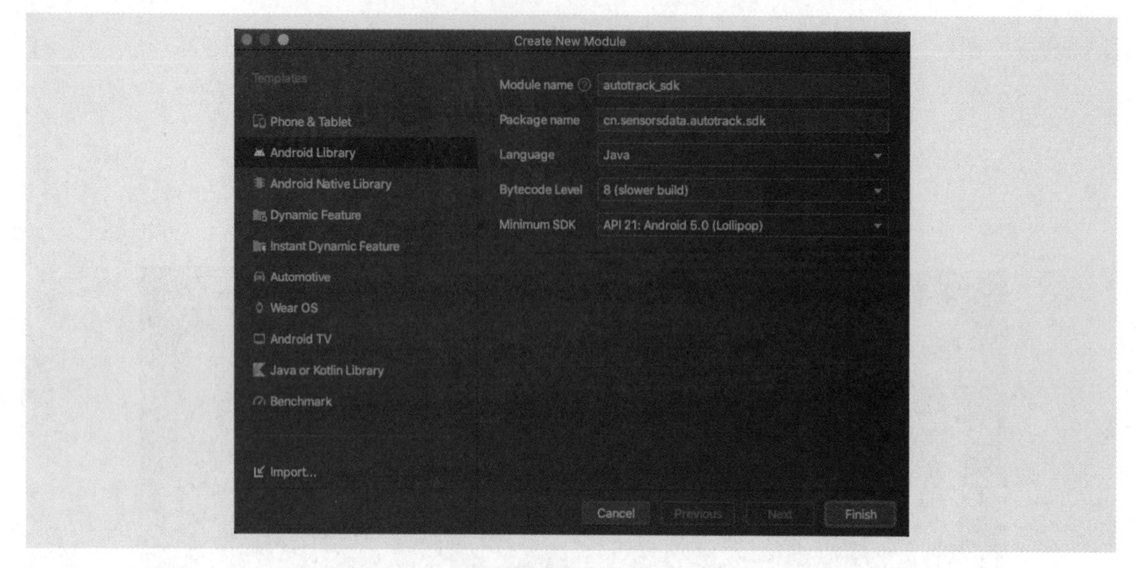

图 7-3　创建 autotrack_sdk 模块

autotrack_sdk 模块用于实现埋点事件的业务逻辑，以及插桩时需要 Hook 的代码。创建好以后将其添加到 app（Demo 工程）模块的依赖中。在 SDK 中创建一个 TrackHelper 工具类，该工具类提供了 trackClick() 方法，在插桩时会用到。该工具类还提供了获取 element_content、screen_name、app_name、time 等内容的方法，其核心代码如下。

```
public final class TrackHelper {
      private static final String TAG = "TrackHelper";

      public static void trackClick(View view) {
          try {
              Log.i(TAG, "track click trigger: " + view);
              if (view == null) {
                  return;
              }
              // 创建一个 JSONObject 用于构建 JSON 数据
              JSONObject resultJson = new JSONObject();
              //1. 获取元素内容
              String elementContent = getElementContent(view);
              if (!TextUtils.isEmpty(elementContent)) {
                  resultJson.put("element_content", elementContent);
              }
              //2. 获取页面名称和标题
              getScreenNameAndTitle(view, resultJson);
              //3. 设置 time
              resultJson.put("time", System.currentTimeMillis());
              //todo 可以扩展其他功能，添加需要的信息

              // 输出结果
              Log.i(TAG, "Final result: \n" + resultJson.toString(4));

          } catch (JSONException e) {
              e.printStackTrace();
          }
```

```
    }

    private static String getElementContent(View view) {
        if (view instanceof Button) {
            Button btn = (Button) view;
            String content = btn.getText().toString();
            if (TextUtils.isEmpty(content)) {
                content = btn.getContentDescription().toString();
            }
            return content;
        }
        //此处可以扩展到其他 View
        return null;
    }

    private static String getScreenNameAndTitle(
        View view, JSONObject jsonObject)
    {
        try {
            Context context = view.getContext();
            Activity activity = getActivityFromContext(context, view);
            if (activity != null) {
                String screenName = activity.getClass().getCanonicalName();
                jsonObject.put("screen_name", screenName);
                String title = getActivityTitle(activity);
                if (!TextUtils.isEmpty(title)) {
                    jsonObject.put("title", title);
                }
            }
        } catch (JSONException e) {
            e.printStackTrace();
        }
        return null;
    }

    public static Activity getActivityFromContext(Context context,
                                                  View view)
    {
        Activity activity = null;
        try {
            if (context != null) {
                if (context instanceof Activity) {
                    activity = (Activity) context;
                } else if (context instanceof ContextWrapper) {
                    while (!(context instanceof Activity)
                            && context instanceof ContextWrapper)
                    {
                        context =
                            ((ContextWrapper) context).getBaseContext();
                    }
                    if (context instanceof Activity) {
                        activity = (Activity) context;
                    }
                }
            }
        } catch (Exception e) {
```

```
                e.printStackTrace();
            }
            return activity;
        }

        ...
}
```

以上只展示了部分代码，完整代码可以参考本章对应的工程代码。为了验证上述代码的正确性，这里修改一下 MainActivity 中的 init() 方法，模拟插桩后运行的结果。修改后的结果如下。

```
private void init(){
        findViewById(R.id.btn).setOnClickListener(new View.OnClickListener() {
            @Override
            public void onClick(View v) {
                Log.i("MainActivity", "Button is Clicked");
                TrackHelper.trackClick(v);// 添加此段代码
            }
        });
}
```

再次运行 app，单击 Button 按钮，输出的日志如下。

```
I/MainActivity: Button is Clicked
I/TrackHelper: track click trigger:
I/TrackHelper: final result:
{
      "element_content": "Click Me",
      "screen_name": "cn.sensorsdata.autotrack.demo.MainActivity",
      "title": "First Page",
      "time": 1633504335964
}
```

在 init() 方法中，手动调用了 TrackHelper.trackClick() 方法，通过输出结果验证了数据采集内容的正确性。假如需要对项目中每一个按钮设置的单击事件都添加这行代码，将会是一个不小的工作量，接下来介绍在编译时自动插入这行代码。

7.2.2 创建插件框架

这一步的目标是创建一个 Gradle 插件，能够自动在按钮设置的事件监听中添加 TrackHelper.trackClick() 代码。

在第 1 章和第 2 章中详细地介绍了如何创建 Gradle 插件与如何使用 Android Transform，根据这些知识创建一个单独工程插件，这样方便后续的插件发布和分享。

首先按照 1.3.3 小节介绍的内容，为 AutoTrackImpl01 工程添加一个 autotrack_plugin 模块，选择菜单 File → New → New Module，在弹出的对话框中选择 Java or Kotlin Library，按照图 7-4 填写信息。

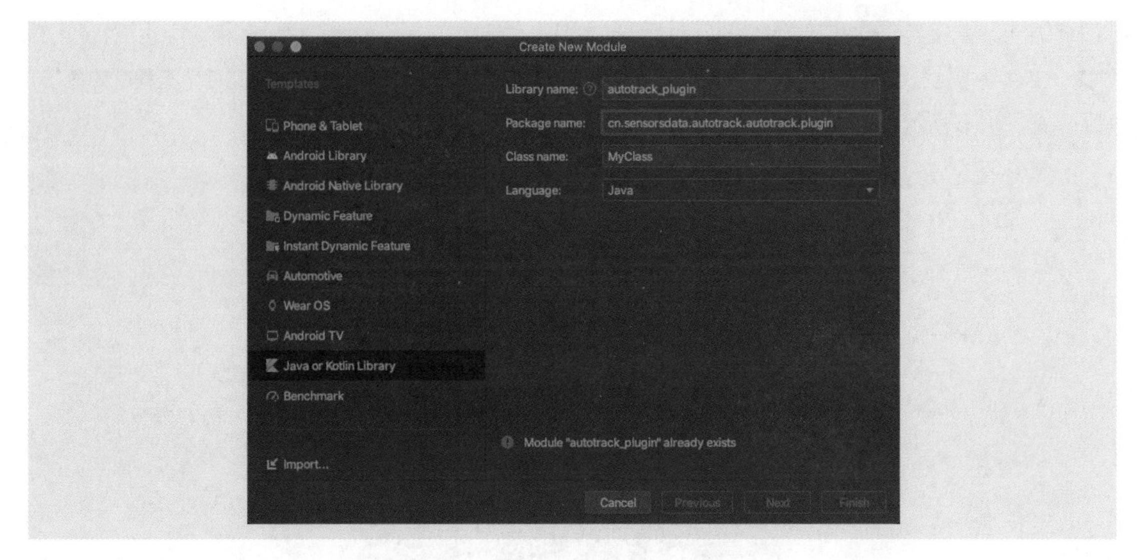

图 7-4 添加 autotrack_plugin 模块

接着修改 autotrack_plugin 模块中的 build.gradle 代码，添加 Gradle API 依赖以及 Maven Publish 功能。

```
plugins {
      id 'java-library'
      id 'maven-publish'
}

java {
      sourceCompatibility = JavaVersion.VERSION_11
      targetCompatibility = JavaVersion.VERSION_11
}

dependencies {
      implementation gradleApi()
}

publishing {
      publications {
          myPlugin(MavenPublication) {
              from components.java

              group = 'cn.sensorsdata.autotrack'
              artifactId = 'plugin'
              version = '1.0.0'
          }
      }

      repositories {
          maven {
              name = 'plugin'
              url = layout.getProjectDirectory().dir("repo")
          }
      }
}
```

添加上述配置后，当执行 ./gradlew publish 命令的时候会在 autotrack_plugin 模块中生成 repo 文件夹，文件夹中包含插件的内容，后面我们将使用这种方式来将插件发布到本地。至此整个工程就建好了，其目录如图 7-5 所示。

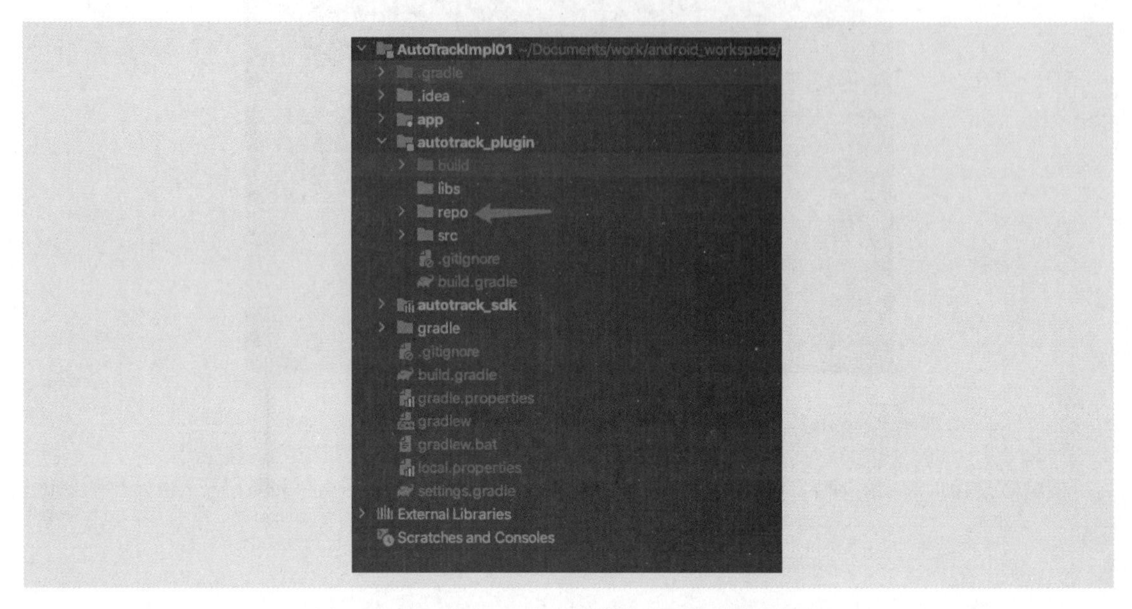

图 7-5　AutoTrackImpl01 工程结构和 repo 文件夹展示

接下来添加插件的实现代码，我们定义 AutoTrackPlugin 类，并实现 Gradle Plugin 接口，在 apply() 方法中添加一句输出日志，具体的实现代码后面再介绍。

```
public class AutoTrackPlugin implements Plugin<Project> {

    @Override
    public void apply(Project project) {
        System.out.println("======plugin=====" + project.getName());
    }
}
```

实现了插件后，还需要在 properties 中添加插件的声明。在 src/main 目录下创建 resources/META-INF.gradle-plugins/cn.sensorsdata.autotrack.plugin.properties 文件，properties 的内容和定义位置如图 7-6 所示。

此时插件的整体框架就已经完成了，还需要在 app 工程中使用插件。在此之前先运行如下命令生成插件对应的 repo。

```
$ ./gradlew publish
```

接着在 project/build.gradle 文件中添加插件的仓库，代码如下。

```
buildscript {
    repositories {
        google()
        mavenCentral()
        maven {
            url(uri("autotrack_plugin/repo")) // 仓库位置
        }
```

```
        }
        dependencies {
            classpath "com.android.tools.build:gradle:7.0.2"
            classpath "cn.sensorsdata.autotrack:plugin:1.0.0" // 依赖插件
        }
}

task clean(type: Delete) {
        delete rootProject.buildDir
}
```

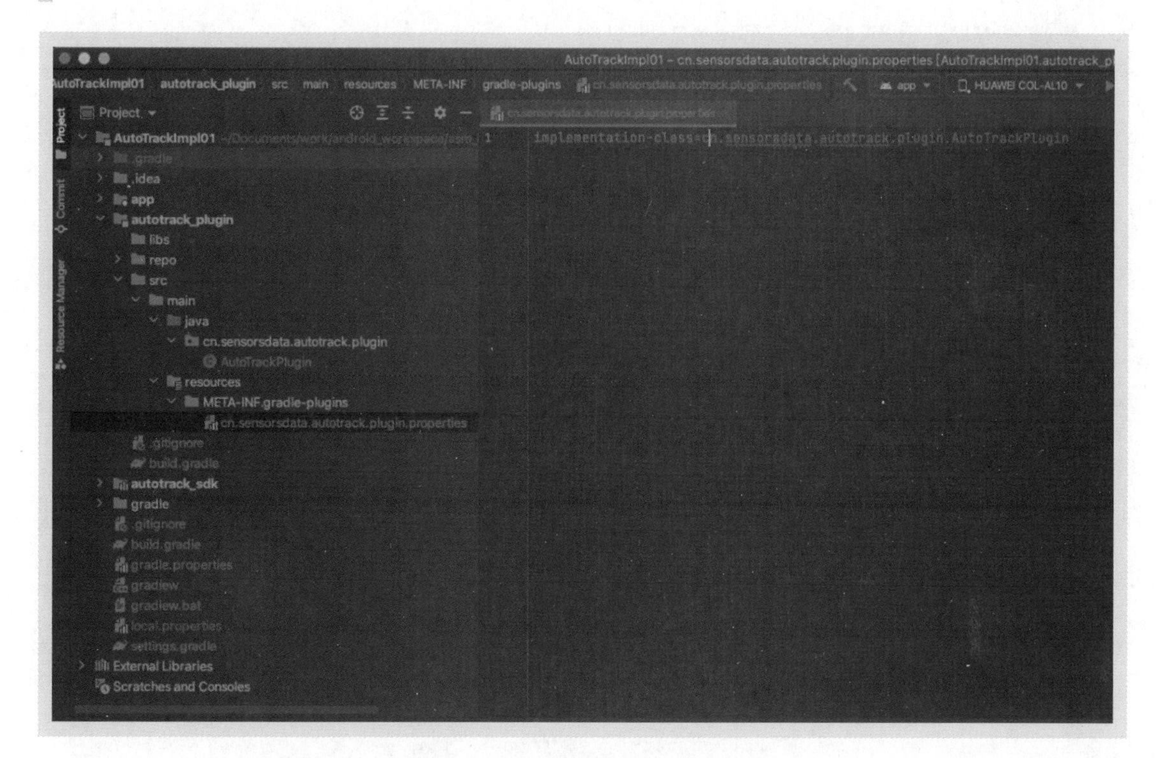

图 7-6　properties 的内容和定义位置

再在 app/build.gradle 中使用插件，代码如下。

```
plugins {
        id 'com.android.application'
        id 'cn.sensorsdata.autotrack.plugin' // 使用插件
}
...
```

注意上述这种使用方式，还需要在 project/settings.gradle 中添加如下配置，如果采用 apply() 方法则不需要。

```
dependencyResolutionManagement {
        //repositoriesMode.set(RepositoriesMode.FAIL_ON_PROJECT_REPOS)
        repositories {
            google()
            mavenCentral()
            maven {
                url(uri("autotrack_plugin/repo"))
            }
```

```
        }
}
rootProject.name = "AutoTrackImpl01"
include ':app'
include ':autotrack_sdk'
include ':autotrack_plugin'
```

最后在 terminal 中输入如下命令验证插件是否生效。

```
$ ./gradlew configure

> Configure project :app
======plugin=====app
```

从输出日志中可以看到插件已经生效了。现在我们的项目中有了 SDK、插件以及 demo 这 3 个模块，并且插件已经起作用了，下一步就是要实现 Hook 功能。

7.2.3 编写插件逻辑

这一步，我们直接使用 2.5 节介绍的 BoilerplateIncrementalTransform 模板，在此模板上进行修改。在此模板中我们已经可以获取到项目中所有的 .class 文件，接下来就是使用 ASM 对 .class 文件进行修改。不过在此之前，我们先来看看 demo 工程中的 MainActivity 经过编译后的产物是什么样的，并确定插入点（hook point）。

先来看看 MainActivity 中的这段代码。

```
public class MainActivity extends AppCompatActivity {
      ...
      private void init(){
          findViewById(R.id.btn).setOnClickListener(new View.OnClickListener() {
              @Override
              public void onClick(View v) {
                  Log.i("MainActivity", "Button is Clicked");
                  TrackHelper.trackClick(v);
              }
          });
      }
}
```

上述代码通过View的setOnClickListener()方法给按钮设置了单击事件的监听，此处setOnClickListener()方法的参数是一个匿名内部类。例子中匿名类在编译后实际上会生成 MainActivity$Xxx 开头的类，可在编译项目后在项目的 /app/build/intermediates/javac/ 文件夹中查看编译后的 .class 文件，图 7-7 所示为本例子中 app 模块的编译结果。

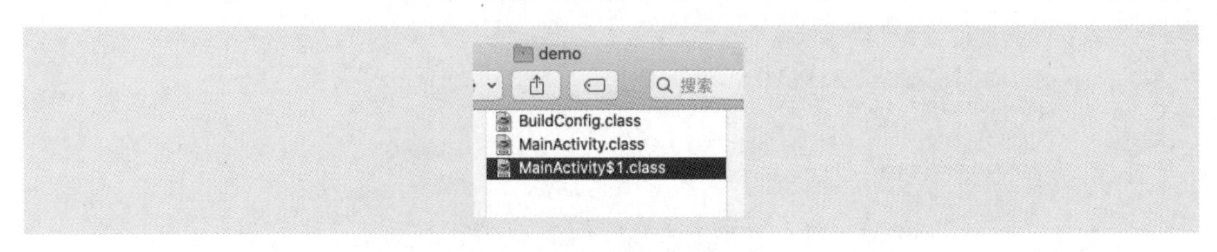

图 7-7 编译生成的 .class 文件

使用反编译工具或者 javap 命令查看 MainActivity$1.class 文件的内容，如图 7-8 所示。

```
//
// Source code recreated from a .class file by IntelliJ IDEA
// (powered by FernFlower decompiler)
//

package cn.sensorsdata.autotrack.demo;

import android.util.Log;
import android.view.View;
import android.view.View.OnClickListener;
import cn.sensorsdata.autotrack.sdk.TrackHelper;

class MainActivity$1 implements OnClickListener {
    MainActivity$1(MainActivity this$0) {
        this.this$0 = this$0;
    }

    public void onClick(View v) {
        Log.i("MainActivity", "Button is Clicked");
        TrackHelper.trackClick(v);
    }
}
```

图 7-8 MainActivity$1.class 反编译的结果

可以看到 setOnClickListener() 方法中设置的匿名内部类，实际上生成了图 7-7 所示的 .class 文件。每个匿名内部类都会产生一个类似这样的 .class 文件，那现在如果想要对 onClick() 方法进行插桩，只需要判断 .class 是否实现了 android.view.View.OnClickListener 即可。插桩的第一步是找准插入点。

找到插入点以后，为了能够修改 .class 文件，还需要在插件中引入 ASM，在 autotrack_plugin 模块的 build.gradle 中添加如下配置。

```
dependencies {
      implementation gradleApi()
      implementation 'org.ow2.asm:asm:9.2'
      implementation 'org.ow2.asm:asm-tree:9.2'
      implementation 'org.ow2.asm:asm-commons:9.2'
      implementation 'org.ow2.asm:asm-analysis:9.2'
      implementation 'org.ow2.asm:asm-util:9.2'
      implementation 'org.apache.commons:commons-lang3:3.12.0'
      implementation 'commons-io:commons-io:2.11.0'
      compileOnly 'com.android.tools.build:gradle:3.4.1', {
          exclude group:'org.ow2.asm'
      }
}
```

上述配置中引入了 ASM 依赖以及 AGP 依赖，现在就可以使用 ASM 对 .class 进行修改了。首先将复制过来的 BoilerplateIncrementalTransform 模板改名为 AutoTrackTransform，并在 AutoTrackPlugin#apply() 方法中注册此 Transform，AutoTrackPlugin 类最终的代码如下。

```
public class AutoTrackPlugin implements Plugin<Project> {
      @Override
      public void apply(Project project) {
          System.out.println("======plugin=====" + project.getName());
          AppExtension appExtension =
              project.getExtensions().findByType(AppExtension.class);
          if (appExtension != null) {
```

```
                appExtension.registerTransform(new AutoTrackTransform());
            }
        }
}
```

然后修改 AutoTrackTransform 类，添加 ASM 处理 .class 的逻辑，其中核心代码如下。

```
package cn.sensorsdata.autotrack.plugin;

// 省略 import

/**
  * 普通写法的样板代码
  */
public class AutoTrackTransform extends Transform {

    @Override
    public String getName() {
        return "myAutoTrack";// 从模板代码中复制过来的代码，记得修改此处的名称
    }

    ...

    /**
     * 用户可以在这里实现具体的处理原始数据的逻辑
     * 例如使用 ASM、Javassits 等工具修改 .class 文件，然后返回处理后的结果
     * 此方法直接返回了输入的值
     *
     * @param data 原始数据
     * @return 修改后的数据
     */
    private byte[] handleBytes(byte[] data) {
        try {
            ClassReader classReader = new ClassReader(data);
            ClassWriter classWriter =
                new ClassWriter(ClassWriter.COMPUTE_MAXS);
            // 为方便阅读和展示
            // 以下的 ClassVisitor 以及 AdviceAdapter 都使用匿名内部类实现
            ClassVisitor classVisitor =
                new ClassVisitor(Opcodes.ASM7, classWriter)
            {
                private String className;
                private String superName;
                private List<String> interfaceList;

                @Override
                public void visit(
                    int version, int access, String name, String signature,
                    String superName, String[] interfaces)
                {
                    super.visit(
                        version, access, name,
                        signature, superName, interfaces);
                    this.className = name;
                    this.superName = superName;
                    this.interfaceList = Arrays.asList(interfaces);

                }
```

```
                @Override
                public MethodVisitor visitMethod(
                    int access, String name, String descriptor,
                    String signature, String[] exceptions)
                {
                    MethodVisitor methodVisitor = super.visitMethod(
                        access, name, descriptor, signature, exceptions);
                    // 当方法名、方法签名和实现了 View$OnClickListener 接口时
                    // 才去处理
                    if ("onClick".equals(name)
                            && "(Landroid/view/View;)V".equals(descriptor)
                            && interfaceList.contains(
                                "android/view/View$OnClickListener"))
                    {
                        methodVisitor = new AdviceAdapter(
                            Opcodes.ASM7, methodVisitor,
                            access, name, descriptor)
                        {
                            @Override
                            protected void onMethodExit(int opcode) {
                                super.onMethodExit(opcode);
                                // 在退出方法时添加 hook() 方法
                                // 加载 hook() 方法需要的参数
                                this.mv.visitVarInsn(Opcodes.ALOAD, 1);
                                this.mv.visitMethodInsn(
                                    Opcodes.INVOKESTATIC,
                            "cn/sensorsdata/autotrack/sdk/TrackHelper",
                             "trackClick", "(Landroid/view/View;)V",
                                        false);
                            }
                        };
                    }
                    return methodVisitor;
                }

            };
            classReader.accept(classVisitor, ClassReader.EXPAND_FRAMES);
            return classWriter.toByteArray();
        } catch (Exception e) {
            e.printStackTrace();
            return data;
        }
    }

    ...
}
```

在 handleBytes() 方法中实现了 ASM 处理 .class 的逻辑，其中在 visitMethod() 方法中进行插入点逻辑判断，当满足条件时在退出方法的位置插入 SDK 中 TrackHelper 类的 trackClick() 方法。至此我们就实现了按钮上的埋点，接下来单击按钮验证结果。

7.2.4 验证

在插件逻辑功能实现后，不要忘了重新对 autotrack_plugin 插件进行本地发布，然后在运行项目或者

编译项目后查看 Transform 后的结果。

运行 Android Demo 项目，并单击按钮，会看到控制台中输出了图 7-9 所示的结果。

```
2021-10-07 16:29:01.460 7828-7828/cn.sensorsdata.autotrack.demo I/TrackHelper: Final result:
    {
        "element_content": "Click Me",
        "screen_name": "cn.sensorsdata.autotrack.demo.MainActivity",
        "title": "First Page",
        "time": 1633595341460
    }
```

图 7-9　输出的结果

通过上述结果可以看到插件生效了。如果不想通过运行项目来查看结果，可以先编译项目，然后在 build/intermediates/transforms/myAutoTrack/ 目录中查看对应的 .class 文件，如图 7-10 所示。

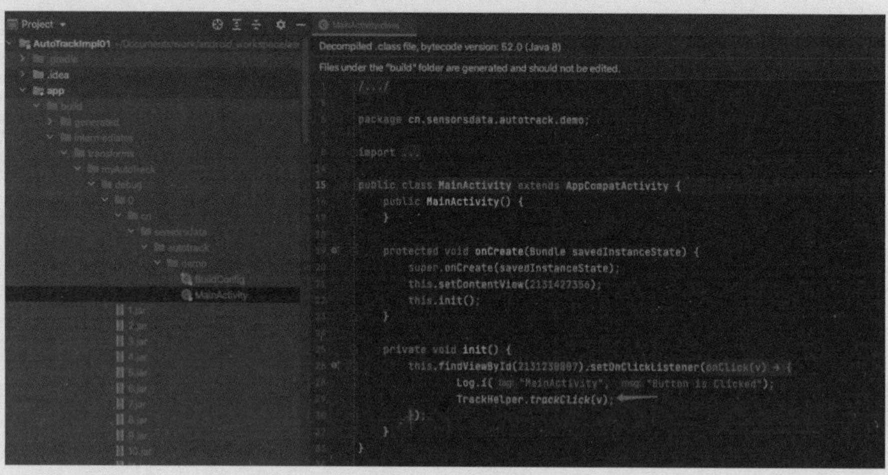

图 7-10　查看结果

因为 app 依赖的 module 工程最终会被编译成 JAR 包的形式依赖，如果想验证 module 或第三方包中插入后的结果，可以借助反编译工具。

7.2.5　发布

完成 SDK 和插件开发以后，下一步就是分享我们的劳动成果，通常是将产品发布到 Maven Central 中。关于如何发布插件已经在 1.6 节中介绍过了，此处就不赘述了。

7.3　小结

为实现按钮单击事件的自动采集，本章详细地介绍了创建 SDK、插件以及实现插件的整个流程，读者务必亲自动手操作。

8. ASM实现全埋点——进阶部分

第 7 章实现了按钮单击事件的自动采集，读者可以按照这种方式实现其他 View 的事件采集。这里指的不仅是 View 的单击事件，还可以采集诸如 CheckBox 的 OnCheck 事件、ListView 的 OnItemClick 事件等。

其实在实际开发中，仅仅做到这一步还远远不够，还有很多技术点要考虑到。示例如下。

- 在开发中肯定会遇到需要忽略一些类，达到不采集其内容的目的，那么它的最佳实践是怎么样的呢?
- 在案例中，给按钮设置单击事件采用的是匿名类，如果换成 Java 8 中的 Lambda 表达式或者方法引用，是否能采集到单击事件呢?
- 在案例中，Hook Point 是在 onClick() 方法退出的时候插入代码，这样做会有什么缺陷，又该如何进行优化呢?

类似的问题还有很多，本章将带着这些实际开发中遇到的问题继续对第 7 章 AutoTrackImp01 项目进行完善。为方便读者参考，复制 AutoTrackImp01 项目，将其副本改名为 AutoTrackImp02。

8.1 黑名单

在实际开发中必然会遇到不想对一些类进行处理的场景。例如对 A 类中的所有 View 自动添加埋点事件或者不处理 Android 官方的支持库，这就需要在插件中建立黑名单机制。通过黑名单可以灵活地配置、忽略哪些类或者包，这个功能很简单，也很普遍。

结合第 1 章中的知识，可以按照如下步骤来实现此功能。

（1）定义一个 Gradle Extension，用于配置黑名单。

（2）注册 Extension。

（3）在处理 .class 文件时，根据 Extension 中的值来忽略 .class。

第一步： 定义 Extension。首先在 autotrack_plugin 模块中创建一个 AutoTrackExtension 类，该类用作插件的配置类，其内容如下。

```
public class AutoTrackExtension {
      public ArrayList<String> exclude = new ArrayList<>(); // 黑名单

      @Override
      public String toString() {
          return "AutoTrackExtension{" +
                  "exclude=" + exclude +
                  '}';
      }
}
```

AutoTrackExtension 中定义了一个 exclude 列表，用于配置黑名单。

第二步： 注册 Extension。为了能在 build.gradle 中正常使用，需要注册配置，具体可以参考 1.4.3 小节。下面的代码是在 AutoTrackPlugin 中注册插件并且在 Project 的声明周期回调中输出其结果。

```
public class AutoTrackPlugin implements Plugin<Project> {

      @Override
```

```
        public void apply(Project project) {
            //注册插件
            AutoTrackExtension autoTrackExtension =
                project.getExtensions()
                .create("autotrackConfig", AutoTrackExtension.class);
            AppExtension appExtension =
                project.getExtensions().findByType(AppExtension.class);
            if (appExtension != null) {
                appExtension.registerTransform(new AutoTrackTransform());
            }
            AutoTrackHelper.getHelper().autoTrackExtension = autoTrackExtension;
            AutoTrackHelper.getHelper().appExtension = appExtension;
            //输出插件的配置
            project.afterEvaluate(proj -> {
                System.out.println("AutoTrack Config: " + autoTrackExtension);
            });
        }
}
```

上述代码中的 AutoTrackHelper 类是一个单例，用于保存插件中公用的数据，其内容如下。

```
public final class AutoTrackHelper {
      public static final AutoTrackHelper INSTANCE = new AutoTrackHelper();
      public AutoTrackExtension autoTrackExtension;
      public AppExtension appExtension;

      private AutoTrackHelper(){
      }
      public static AutoTrackHelper getHelper(){
          return INSTANCE;
      }
}
```

第三步：在项目中配置插件 Extension。注册好插件以后，就可以在 build.gradle 文件中使用。例如需要忽略 androidx.appcompat 包，可以在 app/build.gradle 中按照如下方式配置。

```
plugins {
      id 'com.android.application'
      id 'cn.sensorsdata.autotrack.plugin'
}

android {
     ...
}

dependencies {

      implementation 'androidx.appcompat:appcompat:1.2.0'
      ...
}

//配置插件
autotrackConfig{
      exclude=['androidx.appcompat']
}
```

完成插件 Extension 的创建和配置后，就可以验证扩展是否生效，按照如下命令运行，然后查看结果。

```
$ ./gradlew publish   # 修改后的插件需要重新发布
$ ./gradlew configure  # 执行 configure 任务，查看结果
```

第四步： 更改 Transform 逻辑。为了能让黑名单配置生效。需要对插件中的 Transform 实现进行修改。修改的原则是获取到 Class 的包名加类名，然后与黑名单中的项进行比对即可。可以通过 ASM 的 ClassVisitor 获取到包名和类名信息，但这样就需要先使用 ASM 加载 ClassFile 数据，这显然效率不高。推荐的做法如下。

- 对于DirectoryInput，只需要根据路径即可获得。
- 对于JarInput，则使用JarEntry的name即可。

例如对于 DirectoryInput 的做法如下。

```
private void processDirectoryFile(
      DirectoryInput directoryInput,
      TransformInvocation transformInvocation)
{
    File srcDir = directoryInput.getFile();
    File outputDir = transformInvocation.getOutputProvider() (1)
        .getContentLocation(
          srcDir.getAbsolutePath(),
            directoryInput.getContentTypes(),
          directoryInput.getScopes(), Format.DIRECTORY);
    try {
        FileUtils.forceMkdir(outputDir);

        if (transformInvocation.isIncremental()) {
            ...
        }
        else {
            FileUtils.copyDirectory(srcDir, outputDir); (2)
            FileUtils
                .listFiles(outputDir, new String[]{"class"},true)
                .parallelStream()
                .forEach(clazzFile -> {
                    try {
                        byte[] sourceBytes =
                            FileUtils.readFileToByteArray(clazzFile);
                        //注意此处的获取类名的方式，这里给一个实际路径的参考
                        byte[] modifiedBytes = handleBytes( (3)
                            sourceBytes,
                            clazzFile.getAbsolutePath().replace(
                                outputDir.getAbsolutePath()+ File.separator,      "")
                        );

                        if (modifiedBytes != null) {
                            FileUtils.writeByteArrayToFile(clazzFile,
                                         modifiedBytes, false);
                        }
                    } catch (IOException e) {
                        e.printStackTrace();
                    }
                });
        }
    } catch (IOException e) {
        e.printStackTrace();
```

```
        }
    }

    private byte[] handleBytes(byte[] data, String canonicalName) {
            try {
                // 如果在黑名单中，就直接返回不处理
                if (checkBlackList(canonicalName)) {
                    return data;
                }
            ...
    }
    /**
      * 判断类是否在黑名单中
      */
    private boolean checkBlackList(final String canonicalName) {(4)
            if (canonicalName != null) {
                String tmp = canonicalName.replace(File.separator, ".");
                return AutoTrackHelper.getHelper()
                    .autoTrackExtension.exclude
                    .parallelStream().anyMatch(tmp::startsWith);
            }
            return false;
    }
```

上述代码位置（1）处是获得输出目录，位置（2）处是将源目录中的内容复制到输出目录中，位置（3）处的 handle Bytes() 的逻辑是获取 .class 文件的绝对路径并替换掉目录部分。下面给出 clazzFile 和 outputDir 的绝对路径的一个输出结果，用于参考理解。

```
clazzFile 路径参考：
/Users/.../app/build/intermediates/transforms/myAutoTrack/debug/3/cn/sensorsdata/
autotrack/demo/MainActivity$1.class

outputDir 路径参考：
/Users/.../app/build/intermediates/transforms/myAutoTrack/debug/3
```

上述示例中字符串替换后的最终结果如下。

```
cn/sensorsdata/autotrack/demo/MainActivity$1.class
```

根据这个结果就可以使用位置（4）的逻辑来判断该类是否需要处理。可以看到这种操作不需要使用 ASM 加载类。对于 JarInput，JarEntry 的 name 的结果就与上述字符串替换后是一致的。另外注意 checkBlackList() 方法，这里使用的是 String.startWith() 方法，所以如果设置 androidx.appcompat，那么所有类都不会处理。

8.2 防止多次插入

在实际应用中可能会遇到这样的场景：将插桩后的代码分享给其他项目使用，而其他的项目中可能也使用了同样的插件，如果不对这种情况进行处理，就可能会出现对同一个方法多次插桩的情况，因此需要通过一定的措施来规避此种情况。

具体思路是：使用 ASM 处理完以后，对 Class 添加一个特定注解；当再次处理 Class 时，判断特定注解是否存在，如果存在就表示插件已经处理过，直接忽略后续的处理即可。

首先，在 autotrack_sdk 模块中添加一个注解类。

```
@Retention(RetentionPolicy.CLASS)
@Target(ElementType.TYPE)
public @interface AutoTrackInstrumented {
}
```

注意，此注解类设置为 RetentionPolicy.CLASS，仅用于编译，不能通过反射获取，不过其目的也不是在运行时获取，只是用作标记。接下来需要修改 AutoTrackTransform 类中的 ASM 部分，实现添加 AutoTrackInstrumented 和判断 AutoTrackInstrumented 功能，主要代码如下。

```
package cn.sensorsdata.autotrack.plugin;

public class AutoTrackTransform extends Transform {
      ...


      private byte[] handleBytes(byte[] data, String canonicalName) {
          try {
            ...
              ClassReader classReader = new ClassReader(data);
              ClassWriter classWriter = new ClassWriter(ClassWriter.COMPUTE_MAXS);
              // 为方便阅读和展示，以下的 ClassVisitor 和 AdviceAdapter 都使用匿名类实现
              ClassVisitor classVisitor =
                  new ClassVisitor(Opcodes.ASM7, classWriter)
              {
                  // 注解标记
                  boolean isTracked = false;(1)

                  ...

                  @Override
                  public AnnotationVisitor visitAnnotation(
                      String descriptor, boolean visible)
                  {
                      // 判断类中是否有 AutoTrackInstrumented 注解，并设置标记
                  if ("Lcn/sensorsdata/autotrack/sdk/AutoTrackInstrumented;"
                          .equals(descriptor)) (2)
                      {
                          isTracked = true;
                      }
                      return super.visitAnnotation(descriptor, visible);
                  }

                  @Override
                  public MethodVisitor visitMethod(
                      int access, String name, String descriptor,
                      String signature, String[] exceptions)
                  {
                      MethodVisitor methodVisitor = super.visitMethod(
                          access, name, descriptor, signature, exceptions);
                      // 如果类已经被 AutoTrackInstrumented 标记了，就不再处理方法
                      if (isTracked) {(3)
                          return methodVisitor;
                      }
```

```
                ...
                return methodVisitor;
            }

            @Override
            public void visitEnd() {
                super.visitEnd();
                // 如果类之前没有被 AutoTrackInstrumented 注解标记
                // 就给类添加 AutoTrackInstrumented 注解
                if (!isTracked) (4)
                {
                   this.cv.visitAnnotation(
             "Lcn/sensorsdata/autotrack/sdk/AutoTrackInstrumented;",
                 false);
                }
            }
        };
        classReader.accept(classVisitor, ClassReader.EXPAND_FRAMES);
        return classWriter.toByteArray();
    } catch (Exception e) {
        e.printStackTrace();
        return data;
    }
  }

  ...
}
```

上述代码在位置（1）处定义了一个 boolean 类型的标识符，用来标识当前类是否被处理过。位置（2）处的标识在 ClassVisitor 的 visitAnnotation() 中判断类中是否存在 AutoTrackInstrumented 注解，如果存在，则设置一个标识符 isTracked = true，然后在位置（3）处的 visitMethod() 方法中判断并返回。如果不存在 AutoTrackInstrumented 注解，就在 visitEnd() 方法中给类添加 AutoTrackInstrumented 注解。

以上就是防止重复插入功能的逻辑。用户如果不想处理某个类，也可以在代码中添加此注解。不过这里有一个问题，例如给 MainActivity 添加了注解，但是运行时会发现 Button 的监听仍然会正常插桩，读者可以先想一想这是为什么。示例代码如下。

```
@AutoTrackInstrumented
public class MainActivity extends AppCompatActivity {
      private void init(){
          findViewById(R.id.btn).setOnClickListener(new View.OnClickListener() {
              @Override
              public void onClick(View v) {
                  Log.i("MainActivity", "Button is Clicked2");
              }
          });
      }
}
```

不生效的原因是 Button 设置的 View.OnClickListener 的匿名类在编译时会生成单独的 .class，而 MainActivity 中的 AutoTrackInstrumented 只针对 MainActivity。要解决这个问题可以考虑给方法添加处理完后的注解。

8.3 方法前插还是后插

读者也许发现了，项目中的 Hook Point 是在退出方法的时候进行插桩的，为什么不能放在进入方法的时候插桩呢?

首先就我们目前实现的功能来说，是推荐在进入方法的时候插入代码的。但是有时候因为业务需要，必须要将代码插入方法退出的地方，例如下面代码所展示的用法。

```
findViewById(R.id.btn).setOnClickListener(new View.OnClickListener() {
      @Override
      public void onClick(View v) {
          v.setTag(100,"value");
          Log.i("MainActivity", "Button is Clicked2");
          TrackHelper.trackClick(v);
      }
});

TrackHelper.public static void trackClick(View view) {
      String result = view.getTag(100)
      //do something
}
```

对于上述代码，在退出方法时插入可以获得 View 中的 Tag，而在进入方法时插入就无法获得 View 中的 Tag，对于类似的情况是推荐在退出方法时插入的。不过在退出方法时插入代码会有一个特别严重的问题，要格外的注意，例如下面代码对应的字节码。

```
findViewById(R.id.btn).setOnClickListener(new View.OnClickListener() {
      @Override
      public void onClick(View view) {
         int a = 10;
         // 其他逻辑，但是这些逻辑中没有使用到入参 view
      }
});
```

上述 View.OnClickListener 接口的匿名类中 onClick() 方法对应的字节码如下。

```
public void onClick(android.view.View);
     descriptor: (Landroid/view/View;)V
     flags: (0x0001) ACC_PUBLIC
     Code:
       stack=1, locals=3, args_size=2
          0: bipush        10
          2: istore_2
          3: return
```

从字节码中可以看到变量 a 的值存放在了局部变量表的第三个槽中，此时局部变量中的结果是 this、view、a，这并没有任何问题。但是可以看到在 onClick() 方法的整个过程中并没有使用参数 view，而它却始终占据着局部变量表中的一个槽，这样就白白浪费了内存。假如能够对此种情况下 view 的槽进行复用，就可以减少内存开销。

例如上述字节码可以按照如下方式进行修改。

```
public void onClick(android.view.View);
     descriptor: (Landroid/view/View;)V
     flags: (0x0001) ACC_PUBLIC
     Code:
       stack=1, locals=3, args_size=2
          0: bipush        10
          2: istore_1
          3: return
```

因为引用类型和 int 类型，都使用一个槽来保存数据，所以上述指令是没有问题的（实际上很多字节码的优化工具也确实是这么做的）。此时局部变量表中的数据是 this、a，现在需要在方法退出的时候插入代码，让我们再来看看 ASM 的处理逻辑。

```
methodVisitor =
        new AdviceAdapter(Opcodes.ASM7, methodVisitor, access, name, descriptor)
{
        @Override
        protected void onMethodExit(int opcode) {
            super.onMethodExit(opcode);
            this.mv.visitVarInsn(Opcodes.ALOAD, 1);
            this.mv.visitMethodInsn(
                Opcodes.INVOKESTATIC,
                "cn/sensorsdata/autotrack/sdk/TrackHelper",
                "trackClick",
                "(Landroid/view/View;)V", false);
        }
};
```

可以看到 ASM 的逻辑仍然是获取局部变量表中的第二个槽，此时这个槽中的数据被 int a 替换了，而 TrackHelper.trackClick() 方法的入参是 android.view.View 类型，并不是 int 类型，类型不匹配了。这样就会存在字节码校验也不会通过的情况，这种情况我们称为“参数优化问题”。

当遇到此种情况时可以考虑对需要用到的参数进行备份，具体方式参考 5.10.2 小节的第二种方式。针对 onClick() 方法，对应的 ASM 实现如下。

```
methodVisitor =
        new AdviceAdapter(Opcodes.ASM7, methodVisitor, access, name, descriptor)
{
        int viewPosition = -1;
        @Override
        protected void onMethodEnter() {
            super.onMethodEnter();
            // 使用 newLocal，并根据类型创建一个槽
            viewPosition = newLocal(Type.getType("Landroid/view/View;"));
            // 复制参数 view，并将其保存在新的槽中
            this.mv.visitVarInsn(Opcodes.ALOAD, 1);
            this.mv.visitVarInsn(Opcodes.ASTORE, viewPosition);
        }
        @Override
        protected void onMethodExit(int opcode) {
            super.onMethodExit(opcode);
            // 在退出方法时添加 hook() 方法
```

```
            // 加载 hook() 方法需要的参数
            // 加载新槽中的数据
            if (viewPosition != -1) {
                this.mv.visitVarInsn(Opcodes.ALOAD, viewPosition);
            } else {
                this.mv.visitVarInsn(Opcodes.ALOAD, 1);
            }
            this.mv.visitMethodInsn(
                Opcodes.INVOKESTATIC,
                "cn/sensorsdata/autotrack/sdk/TrackHelper",
                 "trackClick",
                "(Landroid/view/View;)V", false);
        }
};
```

此种方式虽然能避免参数优化问题，但也同样面临更多的内存开销。根据实际情况，可以在 AutoTrackExtension 中提供配置，让用户来决定是在退出方法时插入还是在进入方法时插入。

8.4 支持Lambda和方法引用

Lambda 和方法引用是 Java 8 引入的新特性，这个特性实现的原理就是 Java 7 推出的 invokedynamic 指令。关于 invokedynamic 指令，在前面已多次提及，本节会对此进行进一步的探讨。

首先，使用 Lambda 和方法引用改写 MainActivity 中 Button 设置的匿名类实现方式，代码如下。

```
//Lambda 的写法
private void initLambdaStyle(){
      findViewById(R.id.btn).setOnClickListener(v -> {
          System.out.println(v);
      });
}
// 方法引用的写法
private void methodRefStyle(){
      findViewById(R.id.btn).setOnClickListener(System.out::println);
}
```

按照上述代码改写按钮的单击事件实现后，运行项目并单击按钮时，会发现并不能按照预期触发单击事件，那么产生此现象的原因是什么呢？

8.4.1 原因分析

我们知道匿名内部类在编译时会生成一个新的类，而 Lambda 表达式在编译后是否也会生成类呢？

将 MainActivity 重新编译后打开路径为 AutoTrackImpl02/app/build/intermediates/javac/debug/classes/cn/sensorsdata/autotrack/demo 的文件夹，可以看到并不会产生 MainActivity$X 这样的类（javac 目录下的文件是 Gradle 编译时产生的 app 中的 Java 文件编译后的结果，这个结果会作为 Transform 的输入产物）。

再反编译项目的 APK 文件并查看 MainActivity 的结果，如图 8-1 所示。

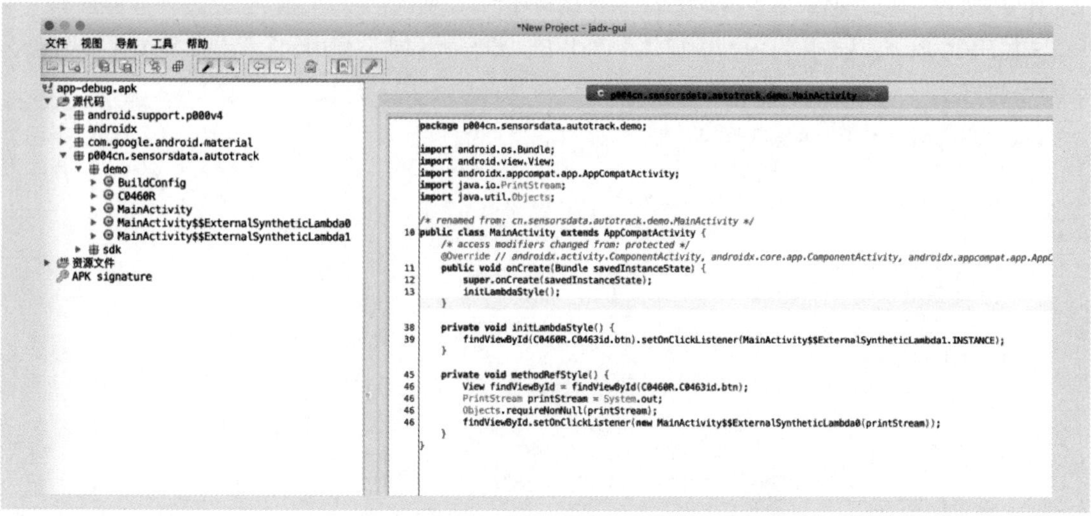

```
package p004cn.sensorsdata.autotrack.demo;

import android.os.Bundle;
import android.view.View;
import androidx.appcompat.app.AppCompatActivity;
import java.io.PrintStream;
import java.util.Objects;

/* renamed from: cn.sensorsdata.autotrack.demo.MainActivity */
public class MainActivity extends AppCompatActivity {
    /* access modifiers changed from: protected */
    @Override // androidx.activity.ComponentActivity, androidx.core.app.ComponentActivity, androidx.appcompat.app.AppC
    public void onCreate(Bundle savedInstanceState) {
        super.onCreate(savedInstanceState);
        initLambdaStyle();
    }

    private void initLambdaStyle() {
        findViewById(C0460R.C0463id.btn).setOnClickListener(MainActivity$$ExternalSyntheticLambda1.INSTANCE);
    }

    private void methodRefStyle() {
        View findViewById = findViewById(C0460R.C0463id.btn);
        PrintStream printStream = System.out;
        Objects.requireNonNull(printStream);
        findViewById.setOnClickListener(new MainActivity$$ExternalSyntheticLambda0(printStream));
    }
}
```

图 8-1　反编译 APK 文件，查看 MainActivity 的结果

从图 8-1 中可以看到，MainActivity 中定义的 Lambda 表达式和方法引用被两个类——MainActivity$$ExternalSyntheticLambda(0/1) 替换了，而在项目中并未定义这样的类。以 MainActivity$$ExternalSyntheticLambda1 这个类为例，查看其反编译的结果，如图 8-2 所示。

可以看到该类实现了 View.OnClickListener，理论上也是符合插桩条件的，那为什么没有正常插桩呢？

```
package p004cn.sensorsdata.autotrack.demo;

import android.view.View;

/* renamed from: cn.sensorsdata.autotrack.demo.MainActivity$$ExternalSyntheticLambda1 */
public final /* synthetic */ class MainActivity$$ExternalSyntheticLambda1 implements View.OnClickListener {
    public static final /* synthetic */ MainActivity$$ExternalSyntheticLambda1 INSTANCE = new MainActivity$$ExternalSyntheticLambda1();

    private /* synthetic */ MainActivity$$ExternalSyntheticLambda1() {
    }

    public final void onClick(View view) {
        MainActivity.lambda$initLambdaStyle$0(view);
    }
}
```

图 8-2　MainActivity$$ExternalSyntheticLambda1 反编译结果

通过 Debug Transform 可以发现，插件能获取到 MainActivity 这个类，但是获取不到这两类，即 MainActivity$$ExternalSyntheticLambda(0/1)，所以无法对其进行插桩操作。

那么这两个类是什么时候生成的，又是如何生成的呢？针对此类情况又该如何进行插桩呢？在回答这些问题之前，先来了解一下 Lambda 表达式的实现原理。

8.4.2　Lambda 表达式的实现原理

关于 Lambda 表达式，我们知道其背后的原理是通过 invokedynamic 指令来实现的，在 4.6.6 小节

中已经介绍了 invokedynamic 指令的结构。现在读者应该明白了 invokedynamic 指令各部分所代表的含义，如果没有完全明白也不打紧，本节会结合实际案例进一步说明。

先抛开 Android 工程（因为 Android 对 Lambda 的处理有所不同），使用一个普通的 Java 工程来举例。举例代码如下。

```
package cn.sensorsdata.lambda;

import java.util.function.Consumer;

public class LambdaMain {
      Object obj = "hello";

      public void inner(){
          Consumer consumer = new Consumer() {(1)
              @Override
              public void accept(Object o) {
                  System.out.println("inner");
              }
          };
      }

      public void lambda() {
          Consumer<String> consumer = o -> {(2)
              System.out.println("lambda1");
          };

          Consumer<String> consumer2 = o -> {(3)
              Object tmpObj = this.obj;
              System.out.println("lambda2");
          };
      }

      public static void main(String[] args) {
          new LambdaMain().lambda();
      }
}
```

上述代码中定义了一个匿名内部类 [位置（1）] 和两个 Lambda 表达式 [位置（2）、（3）]，并且在 main() 方法中调用定义有 Lambda 表达式的方法：lambda()。使用 javac 命令编译 LamdaMain.java，编译后可以看到目录下多了一个 LambdaMain$1.class 文件，这个文件是 inner() 方法中定义的匿名内部类产生的，其反编译结果如下。

```
package cn.sensorsdata.lambda;

import java.util.function.Consumer;

class LambdaMain$1 implements Consumer {
      LambdaMain$1(LambdaMain var1) {
          this.this$0 = var1;
      }

      public void accept(Object var1) {
```

```
            System.out.println("inner");
        }
}
```

从结果中可以看到该类实现了 Consumer 接口，其 accept() 方法中的内容就是匿名内部类中的内容，再使用 javap 命令查看 LambdaMain.class 的字节码。

```
public class cn.sensorsdata.lambda.LambdaMain
Constant pool:
     ...
     #6 = InvokeDynamic      #0:#42 // #0:accept:\
                               ()Ljava/util/function/Consumer;
     #7 = InvokeDynamic      #1:#44 // #1:accept:\
                                        (Lcn/sensorsdata/lambda/LambdaMain;)\
                                        Ljava/util/function/Consumer;
     #8 = Class              #45    // cn/sensorsdata/lambda/LambdaMain
    #11 = Fieldref           #47.#48 // \
                                      java/lang/System.out:Ljava/io/PrintStream;
    #12 = String             #49     // lambda2
    #13 = Methodref          #50.#51 // java/io/PrintStream.println:\
                                                     (Ljava/lang/String;)V
    #14 = String             #52     // lambda1
    #15 = Class              #53     // java/lang/Object
    ...
    #35 = Utf8               cn/sensorsdata/lambda/LambdaMain$1

    #37 = Utf8               BootstrapMethods
    #38 = MethodHandle       #6:#55   // invokestatic \
                          java/lang/invoke/LambdaMetafactory.metafactory:\
                              (Ljava/lang/invoke/MethodHandles$Lookup;\
                          Ljava/lang/String;Ljava/lang/invoke/MethodType;\
                          Ljava/lang/invoke/MethodType;\
                          Ljava/lang/invoke/MethodHandle;\
                          Ljava/lang/invoke/MethodType;)\
                          Ljava/lang/invoke/CallSite;\
    #39 = MethodType          #56      //  (Ljava/lang/Object;)V
    #40 = MethodHandle        #6:#57    // invokestatic \
                      cn/sensorsdata/lambda/LambdaMain.lambda$lambda$0:\
(Ljava/lang/String;)V
    #41 = MethodType          #28      //  (Ljava/lang/String;)V
    #42 = NameAndType   #58:#59   // accept:\
                                 ()Ljava/util/function/Consumer;
    #43 = MethodHandle   #7:#60   // invokespecial \
                       cn/sensorsdata/lambda/LambdaMain.lambda$lambda$1:\
                        (Ljava/lang/String;)V
     #44 = NameAndType     #58:#61   // accept:(Lcn/sensorsdata/lambda/LambdaMain;)
Ljava/util/function/Consumer;
    ...

{
    ...
    public void inner();
           0: new #4      // class cn/sensorsdata/lambda/LambdaMain$1
           3: dup
```

```
        4: aload_0
        5: invokespecial #5   // Method \(1)
            cn/sensorsdata/lambda/LambdaMain$1.<init>":"\
            (Lcn/sensorsdata/lambda/LambdaMain;)V
        8: astore_1
        9: return

  public void lambda();
        0: invokedynamic #6,  0  // InvokeDynamic #0:accept:\
                                              ()Ljava/util/function/Consumer;
        5: astore_1
        6: aload_0
        7: invokedynamic #7,  0  // InvokeDynamic #1:accept:\(2)
                               (Lcn/sensorsdata/lambda/LambdaMain;) \
                               Ljava/util/function/Consumer;
       12: astore_2
       13: return

  public static void main(java.lang.String[]);
    ...

  private void lambda$lambda$1(java.lang.String);
        0: aload_0
        1: getfield      #3 // Field obj:Ljava/lang/Object;
        4: astore_2
        5: getstatic     #11 // Field \
                 java/lang/System.out:Ljava/io/PrintStream;
        8: ldc           #12 // String lambda2
       10: invokevirtual #13 // Method java/io/PrintStream.println:\
                                      (Ljava/lang/String;)V
       13: return

  private static void lambda$lambda$0(java.lang.String);
    descriptor: (Ljava/lang/String;)V
    flags: ACC_PRIVATE, ACC_STATIC, ACC_SYNTHETIC
    Code:
      stack=2, locals=1, args_size=1
        0: getstatic  #11 // Field \
                      java/lang/System.out:Ljava/io/PrintStream;
        3: ldc        #14      // String lambda1
        5: invokevirtual #13 // Method java/io/PrintStream.println:\
                                             (Ljava/lang/String;)V
        8: return
}

InnerClasses:
       #4; //class cn/sensorsdata/lambda/LambdaMain$1
       public static final #74= #73 of #76; //Lookup=class \
            java/lang/invoke/MethodHandles$Lookup of class \
            java/lang/invoke/MethodHandles

BootstrapMethods:
    0: #38 invokestatic java/lang/invoke/LambdaMetafactory.metafactory:\
                     (Ljava/lang/invoke/MethodHandles$Lookup;\
```

```
                        Ljava/lang/String;Ljava/lang/invoke/MethodType;\
                        Ljava/lang/invoke/MethodType;\
                        Ljava/lang/invoke/MethodHandle;\
                        Ljava/lang/invoke/MethodType;)\
                        Ljava/lang/invoke/CallSite;
     Method arguments:
       #39 (Ljava/lang/Object;)V
       #40 invokestatic cn/sensorsdata/lambda/LambdaMain.\
                          lambda$lambda$0:(Ljava/lang/String;)V
       #41 (Ljava/lang/String;)V

   1: #38 invokestatic java/lang/invoke/LambdaMetafactory.metafactory:\(3)
                        (Ljava/lang/invoke/MethodHandles$Lookup;\
                        Ljava/lang/String;Ljava/lang/invoke/MethodType;\
                        Ljava/lang/invoke/MethodType;\
                        Ljava/lang/invoke/MethodHandle;\
                        Ljava/lang/invoke/MethodType;)\
                        Ljava/lang/invoke/CallSite;
     Method arguments:(4)
       #39 (Ljava/lang/Object;)V
       #43 invokespecial cn/sensorsdata/lambda/LambdaMain.\(5)
                            lambda$lambda$1:+(Ljava/lang/String;)V
       #41 (Ljava/lang/String;)V(6)
```

结合inner()方法中位置（1）处的字节码，可以得出编译后的inner()方法的逻辑相当于如下Java代码。

```
public void inner(){
    Consumer consumer = new LambdaMain$1(this);
}
```

这就是 Java 匿名内部类的实现逻辑。其实 Lambda 表达式的实现也与此类似，但是为什么编译后没有看到类似 LambdaMain$1.class 这样的文件呢？这是因为 Lambda 并不是在编译时产生中间文件，而是在运行时。可以在运行的时候添加 jdk.internal.lambda.dumpProxyClasses 参数来将中间文件“dump”出来。

```
java -Djdk.internal.lambda.dumpProxyClasses cn.sensorsdata.lambda.LambdaMain
```

执行上述命令后会发现在目录中多了如下两个 .class 文件，如图 8-3 所示。

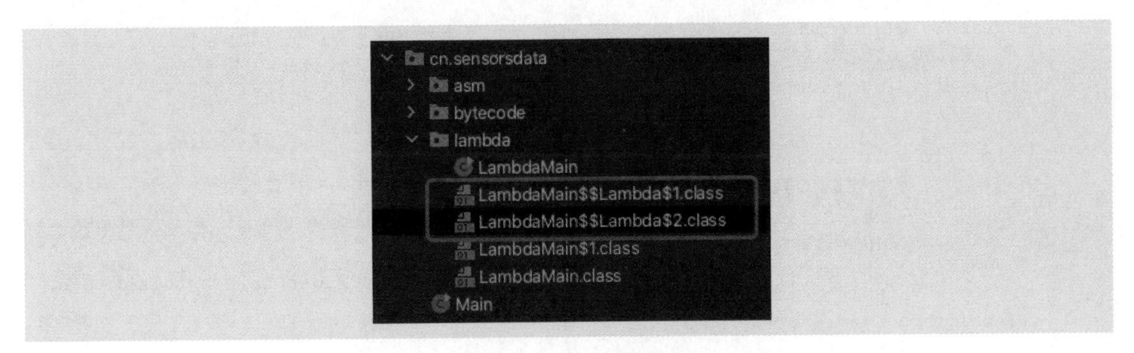

图 8-3 “dump”结果

上述两个 .class 文件的内容如下。

```
//LambdaMain$$Lambda$1.class
final class LambdaMain$$Lambda$1 implements Consumer {
      private LambdaMain$$Lambda$1() {
      }

      @Hidden
      public void accept(Object var1) {
          LambdaMain.lambda$lambda$0((String)var1);
      }
}

//LambdaMain$$Lambda$2.class
final class LambdaMain$$Lambda$2 implements Consumer {
      private final LambdaMain arg$1;

      private LambdaMain$$Lambda$2(LambdaMain var1) {
          this.arg$1 = var1;
      }

      private static Consumer get$Lambda(LambdaMain var0) {
          return new LambdaMain$$Lambda$2(var0);
      }

      @Hidden
      public void accept(Object var1) {
          this.arg$1.lambda$lambda$1((String)var1);
      }
}
```

可以看到它们与匿名内部类的实现类似，也是继承自 Consumer 接口，并且 accept() 方法最终调用的是 LambdaMain 中的 lambda$lambda(0/1) 方法，而这两个方法在源码中是不存在的，是在编译的时候由编译器添加的。再根据 LambdaMain.class 中的内容，使用最基本的 Java 写法来实现匿名类和 Lambda 表达式，最终的效果如下。

```
public class LambdaMain {
      Object obj = "hello";

      public void inner(){
          //使用内部类替换
          Consumer consumer = new LambdaMain$1(this);
      }

      public void lambda() {
          //使用内部类替换
          Consumer<String> consumer = new LambdaMain$$Lambda$1();
          //使用内部类替换
          Consumer<String> consumer2 = LambdaMain$$Lambda$2.get$Lambda(this);
      }

      public static void main(String[] args) {
          int a = 10;
          new LambdaMain().lambda();
      }
```

```
    // 来源于字节码
    private static void lambda$lambda$0(java.lang.String str){
        System.out.println("lambda1");
    }

    // 来源于字节码
    private void lambda$lambda$1(java.lang.String str){
        Object tmpObj = this.obj;
        System.out.println("lambda2");
    }
}
```

上述代码使用 Java 8 之前的写法替换了 Lambda 表达式，这样的操作称为“脱糖”（desugar）。所谓脱糖就是将 Java 中的一些高级的语法糖，在编译时使用普通语法来替换。例如在 Android 中，开发者虽然可以使用 Java 8 的语法来编写代码，但是 Android 手机本身没有 Java 8 的运行环境，所以需要对代码进行脱糖处理。这就是通过反编译工具查看 AutoTrackImpl02 项目的 APK，会发现多出 MainActivity$$ExternalSyntheticLambda(0/1) 这两个类的原因。

以上结合匿名内部类和“dump”出的 .class 文件展示了脱糖后的结果，那么类似 LambdaMain$$Lambda$1、LambdaMain$$Lambda$2 这样的类是如何产生的，以及 invokedynamic 指令又是如何起作用的呢？请继续往下看。

查看 LambdaMain.class 中的 lambda() 方法，方法中定义的两个 Lambda 表达式被 invokedynamic 指令取代了。可以发现 lambda() 方法中定义的两个 Lambda 表达式中的内容分别被放在了 lambda$lambda$0() 和 lambda$lambda$1() 方法中。而且，这两个方法的定义也有些不同：lambda$lambda$0() 方法的访问标识是 private static，lambda$lambda$1() 方法的则是 private，这背后有什么规则可循呢？暂且将此问题放在一边，来先看看当虚拟机执行 invokedynamic 指令时都发生了什么。

以 LambdaMain.class 中的 lambda() 方法，即位置（2）处的字节码为例。

```
6: aload_0
7: invokedynamic #7,  0  // InvokeDynamic #1:accept:\
                         (Lcn/sensorsdata/lambda/LambdaMain;) \
                          Ljava/util/function/Consumer;
```

其中 #7 表示常量池中 CONSTANT_InvokeDynamic_info 结构的位置，即 #7 = InvokeDynamic #1:#44。#1 表示引导方法在 BootstrapMethods 属性表中的索引，#44 表示引导方法需要的动态调用名称、参数和返回类型。#44 常量的结构是 CONSTANT_NameAndType_info，此处描述的是方法名和方法描述符信息。观察其值可知，方法名是 accept，参数类型是 LambdaMain，返回值是 Consumer，这些是引导方法需要的内容。另外可以发现 6: aload_0 指令加载 this 到操作数栈中，这样正好是 #44 方法运行时的参数。

继续看 BootstrapMethods 属性表中索引 #1 的值 [LamdaMain.class 中的位置（3）处]，其中 #38 为指向常量池中 CONSTANT_MethodHandle_info 结构的位置，值是 #38 = MethodHandle #6:#55。由 3.4.5 小节内容可知，#6 表示该 CONSTANT_MethodHandle_info 结构的 reference_kind 的值是 invokestatic，所以 #55 表示它的 reference_index 为指向常量池中的 CONSTANT_Methodref_info 结构，该结构的具体值如下。

```
java/lang/invoke/LambdaMetafactory.metafactory:
            +(Ljava/lang/invoke/MethodHandles$Lookup;
            +Ljava/lang/String;Ljava/lang/invoke/MethodType;
            +Ljava/lang/invoke/MethodType;
            +Ljava/lang/invoke/MethodHandle;
            +Ljava/lang/invoke/MethodType;)Ljava/lang/invoke/CallSite;
```

#38 确定了 invokedynamic 指令引导方法的相关信息，表示当运行 invokedynamic 指令时就会进入该引导方法，LamdaMain.class 中位置（4）处为引导方法的参数信息。接着看看引导方法 LambdaMetafactory.metafactory() 方法的源码。

```
public final class LambdaMetafactory {
    public static CallSite metafactory(MethodHandles.Lookup caller,
                                       String invokedName,
                                       MethodType invokedType,
                                       MethodType samMethodType,
                                       MethodHandle implMethod,
                                       MethodType instantiatedMethodType)
            throws LambdaConversionException {
        AbstractValidatingLambdaMetafactory mf;
        mf = new InnerClassLambdaMetafactory(caller, invokedType,
                             invokedName, samMethodType,
                             implMethod, instantiatedMethodType,
                             false, EMPTY_CLASS_ARRAY,
                             EMPTY_MT_ARRAY);
        mf.validateMetafactoryArgs();
        return mf.buildCallSite();
    }
    ...
}
```

方法各参数说明如下。

- caller：由JVM提供的查询上下文。
- invokedName：被实现的函数式接口的方法名。以LambdaMain为例，此处对应的值是常量#44描述的方法的名字，即accept。
- invokedType：metafactory()方法的返回值是CallSite，invokedType描述的是CallSite中调用方法的描述符信息。以LambdaMain为例，此处对应的值是常量 #44描述的方法描述符(Lcn/sensorsdata/lambda/LambdaMain;)Ljava/util/function/Consumer方法中的参数。
- samMethodType：函数式接口定义的方法描述符。以LambdaMain为例，它的值是(Ljava/lang/Object;)V(这里的sam指的是Single Abstract Method，即函数式接口)。
- implMethod：它的类型是MethodHandle，用于描述运行时应该调用的方法。这里简单回忆一下MethodHandle、MethodType、CallSite之间的关系：MethodType描述的是方法描述符信息；MethodHandle在MethodType的基础上进一步确定了方法名和方法所在的类；CallSite是调用点，它持有MethodHandle对象，并可以调用CallSite#getTarget()方法获取MethodHandle，invokedynamic指令会链接到CallSite，并将调用的信息传给它的target。以LambdaMain.class结果为例，观察位置（5）处的值，对应常量池 #43就是一个CONSTANT_MethodHandle_info结构。
- instantiatedMethodType：函数式接口的具体实现对应的方法描述符。例如Consumer接口中

accept(T t)方法的参数是泛型参数，instantiatedMethodType则是具体实现的描述。以LambdaMain.class为例，观察位置（6）处的值为(Ljava/lang/String;)V。

metafactory() 方法中的参数为一系列 invokedynamic 指令需要的各种输入信息，那么要做的就是根据这些输入信息将它们串联起来，使其调用到特定的方法即可。为搞清楚这部分的逻辑，继续来看看 metafactory() 方法中的内容。方法内部实际上是构建了一个 InnerClassLambdaMetafactory 对象，并且调用它的 mf.buildCallSite() 方法创建最终的 CallSite。下一步一起分析 InnerClassLambdaMetafactory 的构造方法和 buildCallSite() 方法中的逻辑，相关代码如下。

```
/**
  * Lambda metafactory implementation which dynamically creates an
  * inner-class-like class per lambda callsite.
  *
  * @see LambdaMetafactory
  */
/* package */ final class InnerClassLambdaMetafactory
      extends AbstractValidatingLambdaMetafactory
{
      private static final Unsafe UNSAFE = Unsafe.getUnsafe();
      //下面定义常见的方法描述符和一些代码中需要用到的常量
      private static final int CLASSFILE_VERSION = 52;
      private static final String METHOD_DESCRIPTOR_VOID =
                                    Type.getMethodDescriptor(Type.VOID_TYPE);
      private static final String JAVA_LANG_OBJECT = "java/lang/Object";
      private static final String NAME_CTOR = "<init>";
      private static final String NAME_FACTORY = "get$Lambda";

      ...

      // 定义一个累加器
      private static final AtomicInteger counter = new AtomicInteger(0);

      // 定义一个 Class Dumper，用于输出 Lambda 表达式生成的类，通常用于调试使用
      private static final ProxyClassesDumper dumper;

      private static final boolean disableEagerInitialization;

      static {
          //如果用户在运行的时候设置了 jdk.internal.lambda.dumpProxyClasses 参数，
          //就会“dump”出类的内容，用于调试
          final String dumpProxyClassesKey =
                  "jdk.internal.lambda.dumpProxyClasses";
          String dumpPath =
                  GetPropertyAction.privilegedGetProperty(dumpProxyClassesKey);
          dumper = (null == dumpPath) ?
                                  null : ProxyClassesDumper.getInstance(dumpPath);
          ...
      }
      // Name of type containing implementation "CC"
      private final String implMethodClassName;
      // Name of implementation method "impl"
      private final String implMethodName;
      // Type descriptor for implementation methods "(I)Ljava/lang/String;"
```

```
    private final String implMethodDesc;
    // Generated class constructor type "(CC)void"
    private final MethodType constructorType;
    // ASM class writer
    private final ClassWriter cw;
    // Generated names for the constructor arguments
    private final String[] argNames;
    // Type descriptors for the constructor arguments
    private final String[] argDescs;
    // Generated name for the generated class "X$$Lambda$1"
    private final String lambdaClassName;

    // 此构造方法用于初始化一些值，为生成类似 LambdaMain$$Lambda$10 这样的类中的内容做准备
    public InnerClassLambdaMetafactory(MethodHandles.Lookup caller,
                                       MethodType invokedType,
                                       String samMethodName,
                                       MethodType samMethodType,
                                       MethodHandle implMethod,
                                       MethodType instantiatedMethodType,
                                       boolean isSerializable,
                                       Class<?>[] markerInterfaces,
                                       MethodType[] additionalBridges)
            throws LambdaConversionException {
        // 父类中的 super() 方法也用于初始化一些值
        super(caller, invokedType, samMethodName, samMethodType,
              implMethod, instantiatedMethodType,
              isSerializable, markerInterfaces, additionalBridges);
        // 获取宿主类，例如 cn/sensorsdata/lambda/LambdaMain
        implMethodClassName = implClass.getName().replace('.', '/');
        // 获取需要宿主类中需要链接的方法，例如 lambda$lambda$1()
        implMethodName = implInfo.getName();
        // 获取 lambda$lambda$1() 方法的方法描述符，例如 (Ljava/lang/String;)V
        implMethodDesc = implInfo.getMethodType().toMethodDescriptorString();
        //invokedType 就是我们前面提到的常量池 #44 的内容
        //(Lcn/sensorsdata/lambda/LambdaMain;)Ljava/util/function/Consumer;
        // 构造方法就是依靠此内容来构建的
        constructorType = invokedType.changeReturnType(Void.TYPE);
        // 根据宿主类名 + $$Lambda$ + 数字的规则指定匿名类的类名
        // 例如 cn/sensorsdata/lambda/LambdaMain$$Lambda$2
        lambdaClassName = targetClass.getName()
                                .replace('.', '/')
                          + "$$Lambda$" + counter.incrementAndGet();
        // 创建 ASM 的 ClassWriter，该对象用于生成 LambdaMain$$Lambda$2 类
        cw = new ClassWriter(ClassWriter.COMPUTE_MAXS);
        // 根据构造方法中的参数，定义 LambdaMain$$Lambda$2 类中的字段，
        // 例如 private final LambdaMain arg$1
        int parameterCount = invokedType.parameterCount();
        if (parameterCount > 0) {
            argNames = new String[parameterCount];
            argDescs = new String[parameterCount];
            for (int i = 0; i < parameterCount; i++) {
                argNames[i] = "arg$" + (i + 1);
                argDescs[i] =
                   BytecodeDescriptor.unparse(invokedType.parameterType(i));
```

```
            }
        } else {
            argNames = argDescs = EMPTY_STRING_ARRAY;
        }
    }

    //构建CallSite对象
    @Override
    CallSite buildCallSite() throws LambdaConversionException {
        //spinInnerClass()方法是真正创建类似LambdaMain$$Lambda$10内部类内容的地方
        final Class<?> innerClass = spinInnerClass();
        if (invokedType.parameterCount() == 0
                    && !disableEagerInitialization)
        {
            ...

            try {
                Object inst = ctrs[0].newInstance();
                //最终返回ConstantCallSite
                return new ConstantCallSite(
                            MethodHandles.constant(samBase, inst));
            }
            catch (ReflectiveOperationException e) {
                throw new LambdaConversionException(
                            "Exception instantiating lambda object", e);
            }
        } else {
            try {
                //使用UNSAFE初始化内部类
                if (!disableEagerInitialization) {
                    UNSAFE.ensureClassInitialized(innerClass);
                }
                //创建MethodHandle，并返回ConstantCallSite对象
                 return new ConstantCallSite(
                      MethodHandles.Lookup.IMPL_LOOKUP
                       .findStatic(innerClass, NAME_FACTORY, invokedType));
            }
            catch (ReflectiveOperationException e) {
                throw new LambdaConversionException(
                            "Exception finding constructor", e);
            }
        }
    }

    //生成内部类
    private Class<?> spinInnerClass() throws LambdaConversionException {
        String[] interfaces;
        //获取函数式接口的描述符，例如java/util/function/Consumer
        String samIntf = samBase.getName().replace('.', '/');
        //判断是否需要序列化
        boolean accidentallySerializable = !isSerializable
                    && Serializable.class.isAssignableFrom(samBase);
        if (markerInterfaces.length == 0) {
            interfaces = new String[]{samIntf};
```

```
} else {
    // Assure no duplicate interfaces (ClassFormatError)
    Set<String> itfs =
            new LinkedHashSet<>(markerInterfaces.length + 1);
    itfs.add(samIntf);
    for (Class<?> markerInterface : markerInterfaces) {
        itfs.add(markerInterface.getName().replace('.', '/'));
        accidentallySerializable |= !isSerializable &&
              Serializable.class.isAssignableFrom(markerInterface);
    }
    interfaces = itfs.toArray(new String[itfs.size()]);
}
// 使用 ASM 的 ClassWriter 开始创建内部类，
// 例如 cn/sensorsdata/lambda/LambdaMain$$Lambda$10
cw.visit(CLASSFILE_VERSION, ACC_SUPER + ACC_FINAL + ACC_SYNTHETIC,
         lambdaClassName, null,
         JAVA_LANG_OBJECT, interfaces);

// 生成字段内容，这些字段会在构造方法中赋值
for (int i = 0; i < argDescs.length; i++) {
    FieldVisitor fv = cw.visitField(ACC_PRIVATE + ACC_FINAL,
                                    argNames[i],
                                    argDescs[i],
                                    null, null);
    fv.visitEnd();
}
// 生成构造方法 LambdaMain$$Lambda$2()
generateConstructor();
// 生成工厂方法，参考 LambdaMain$$Lambda$2() 中的
//get$Lambda(LambdaMain var0) 方法
// 如果构造方法中没有参数，就不生成，参考 LambdaMain$$Lambda$1()
if (invokedType.parameterCount() != 0 || disableEagerInitialization) {
    generateFactory();
}

// 实现函数式接口中的方法，例如 Consumer 的 accept()
MethodVisitor mv = cw.visitMethod(
                ACC_PUBLIC,
            samMethodName,
            samMethodType.toMethodDescriptorString(), null, null);
mv.visitAnnotation("Ljava/lang/invoke/LambdaForm$Hidden;", true);
// 生成接口中的内容，例如 accept() 中的内容
new ForwardingMethodGenerator(mv).generate(samMethodType);

...

cw.visitEnd();
final byte[] classBytes = cw.toByteArray();

// 如果设置了 jdk.internal.lambda.dumpProxyClasses 参数，
// 此处将类的内容保存到本地
if (dumper != null) {
    AccessController.doPrivileged(new PrivilegedAction<>() {
        @Override
```

```
                public Void run() {
                    dumper.dumpClass(lambdaClassName, classBytes);
                    return null;
                }
            }, null,
            new FilePermission("<<ALL FILES>>", "read, write"),
            // createDirectories may need it
            new PropertyPermission("user.dir", "read"));
        }
        // 最后通过 UNSAFE.defineAnonymousClass 在 VM 中定义匿名类，
        // 并返回 Class 对象
        return UNSAFE.defineAnonymousClass(targetClass, classBytes, null);
    }
    ...
}
```

上述源码中包含详细的注释。通过它可以看到 Java 虚拟机在执行 invokedynamic 指令时进入声明的引导方法 LambdaMetafactory.metafactory() 中，在引导方法中调用 InnerClassLambdaMetafactory 的构造方法，构造方法确定了类似 LambdaMain$$Lambda$1 类名和 lambda$lambda$1() 方法名等的生成规则。在 LambdaMetafactory.metafactory() 方法中，InnerClassLambdaMetafactory 的构造方法执行结束后，接着调用 InnerClassLambdaMetafactory 的 buildCallSite() 方法。buildCallSite() 中先调用 spinInnerClass() 方法，spinInnerClass() 中利用 ASM 构建内部类并使用 UNSAFE.defineAnonymousClass 在虚拟机中定义内部类。如果添加了 jdk.internal.lambda.dumpProxyClasses 参数，则会将其“dump”出来。最后在 buildCallSite() 方法中构建 MethodHandle 并返回 ConstantCallSite，该 ConstantCallSite 声明了如何调用 spinInnerClass() 方法中构建的内部类，虚拟机获取到动态调用点后再继续向后执行。

总结，引导方法 LambdaMetafactory.metafactory() 中会根据输入信息使用 ASM 生成一个具体的类，如 LambdaMain$$Lambda$1，这个类具体定义了如何调用 LambdaMain() 方法中的 lambda$lambda$1 ()。

既然 invokedynamic 指令是在执行的时候创建内部类，那它为什么不像匿名内部类一样在编译的时候就生成对应的 .class 呢？主要原因有如下 3 点。

- 如果Lambda被转换为匿名内部类，那么可能会生成很多 .class文件，又因为每个类文件在使用之前都需要加载和验证，这会影响应用程序的启动性能（加载可能是一个“昂贵”的操作，包括磁盘I/O和解压缩JAR文件本身）。
- 如果Lambda被转换为匿名内部类，由于每个匿名内部类都将被加载，它将占用JVM的元空间；并且需要将匿名内部类中的代码编译为机器码，并存储在代码缓存中。因此，匿名内部类会增加应用程序的内存消耗。
- 最重要的是，从一开始就选择使用匿名内部类来实现Lambda，这将限制未来Lambda实现更改的范围，以及它们根据未来JVM改进而演进的能力。

基于以上 3 点，Lambda 表达式的实现不能直接在编译阶段就用匿名内部类实现，而是需要一个稳定的二进制表示，它提供足够的信息，同时允许 JVM 在未来采用其他可能的实现策略。

8.4.3 Lambda设计参考

8.4.2 小节中提到一个疑问：LambdaMain.class 中的 lambda$lambda$0() 和 lambda$lambda$1() 两个方法的访问修饰符是不同的，lambda$lambda$0() 方法的访问修饰符是 private static，lambda$lambda$1() 方法的则是 private，本小节就来聊聊这背后的原因。

本小节的内容来源于布赖恩 · 戈茨（Oracle Java 语言架构师、《Java 并发编程实战》的作者）的 *Translation of lambda expressions in javac* 一文，我将摘取其中几点知识进行说明，让读者了解 Lambda 表达式到底是如何设计的。希望看完本小节后，读者心中的一些疑惑能够得到解答。布赖恩 · 戈茨的这篇文章主要是介绍编译器遇到 Lambda 表达式的时候生成字节码的策略，以及 Java 语言在运行时如何执行 invokedynamic 指令（8.4.2 小节已经介绍），先来看看转换 Lambda 表达式的策略。

1. 转换策略

在字节码中表示 Lambda 表达式有多种方案，例如内部类、方法句柄、动态代理等，这些方案各有利弊。如何选择转换策略，有两个关键的衡量指标：一是不引入特定策略，以期为将来的优化提供最大的灵活性；二是保持类文件格式的稳定。而 invokedynamic 指令可以同时满足这两个要求，即将 Lambda 在二进制字节码中的表达方式和其运行时的评估机制分开进行，而不是通过生成字节码的方式去创建一个实现了 Lambda 表达式的对象（例如为一个内部类调用构造方法），通过这样的方式在编译的时候将方法需要的静态参数列表和动态参数列表与 invokedynamic 指令绑定，然后在运行的时候链接到指定的方法即可 [此部分可以对比 lambda$lambda$1() 是如何被调用的]。这么做的好处是 invokedynamic 指令使我们可以一直到运行时再去选择转换策略。运行时实现的方式是可以自由地选择转换策略，并且可以动态评估 Lambda 表达式。Invokedynamic 允许这样做，且不需要付出为后续绑定方法可能强加的性能消耗。

具体的的转换策略是：当编译器遇到 Lambda 表达式的时候，它首先会将 Lambda 方法体内容脱糖到一个方法中 [例如 lambda$lambda$1() 这样的方法]，此方法的参数列表和返回值类型与 Lambda 表达式的匹配，可能还会附加一些额外的参数（附加的参数来自外部作用域范围）。同时在遇到 Lambda 表达式的地方会生成一个 invokedynamic 调用点（CallSite 对象），当调用点执行的时候会返回一个函数式接口的实例，这个转换后函数式接口的实现包含 Lambda 的内容（例如实现类 LambdaMain$$Lambda$2）。方法引用也会按照 Lambda 表达式一样的方式进行处理，但是大部分方法引用不需要被脱糖进到一个新方法中；我们可以简单地为一个引用的方法加载一个常量方法句柄，然后将其传给 metafactory。

2. Lambda 脱糖

将 Lambda 表达式转换成字节码的第一步是将 Lambda 方法体脱糖到一个方法中。对于脱糖有以下几个问题需要考虑。

- 将Lambda方法体脱糖到一个静态方法中还是一个实例方法中？
- 脱糖之后生成的方法应该放在哪一个类中？
- 脱糖之后生成的方法的可访问性应该是怎样的？

- 脱糖之后生成的方法的命名应该是怎样的？
- 如果需要一个适配器去桥接Lambda方法体的签名和函数式接口的签名（例如装箱、拆箱、基础类型的扩大和缩小转变、动态参数转换等），那么脱糖的方法是遵循Lambda方法体的签名还是函数式接口的签名，又或者是两者的结合呢？以及谁负责适配呢？
- 如果Lambda从外部作用域（enclosing scope）中获取参数，这些参数应该如何在脱糖的方法体的签名中表示呢？难道是将它们追加到参数列表的前面、后面，或者编译器可以将它们整合在一起，统一放到一个Bean对象里面？

跟脱糖 Lambda 方法体时需要考虑的问题一样，我们也需要考虑方法引用是否需要一个适配器或者桥接方法。

对于以上问题，一般来说，在同等条件下，私有方法优于非私有方法，静态方法优于实例方法，最好的结果是 Lambda 方法体被脱糖在它所在的类里面，脱糖后的签名应该匹配 Lambda 方法体的签名，需要的额外参数应该被添加在参数列表的前面，而且完全不对方法引用进行脱糖。这些准则也不是一成不变的，在某些情况下，我们也不得不偏离这些基准策略。

接下来是关于 Lambda 脱糖的例子。

首先是无状态（stateless）Lambda，所谓无状态指的是 Lambda 方法体没有从外部作用域中捕捉任何状态，例如：

```
class A {
      public void foo() {
          List<String> list = ...
          list.forEach( s -> { System.out.println(s); } );
      }
}
```

这个 Lambda 表达式对应的函数式接口的真实签名是 (String)V［其实这个 Consumer 接口中 accept(T t) 的真实签名，这种情况的签名通常称为 nature signature］，编译器会将 Lambda 方法体脱糖到一个静态方法，静态方法的签名与 Lambda 表达式的 nature signature 相同，然后为脱糖体生成一个方法，脱糖后的结果类似如下。

```
class A {
      public void foo() {
          List<String> list = ...
          list.forEach( [lambda for lambda$1 as Block] );
      }

      // 这个就是脱糖产生的方法
      static void lambda$1(String s) {
          System.out.println(s);
      }
}
```

相比无状态 Lambda，另外一种形式称为有状态 Lambda，所谓有状态指的是 Lambda 方法体中使用了外部作用域的 final 局部变量、隐式是 final 的局部变量，或者外部实例（enclosing instance）的字段（这里可以看作捕获了外部作用域的 this.xx 字段），例如：

```
class B {
      public void foo() {
          List<Person> list = ...
```

```
            final int bottom = ..., top = ...;
            list.removeIf( p -> (p.size >= bottom && p.size <= top) );
        }
    }
```

上面这个例子，Lambda 使用了外部作用域中 final 类型的局部变量 bottom 和 top。脱糖之后的方法将使用的 natural signature 为 (Person)Z，并且在参数列表前面追加额外的参数。编译器有权决定这些额外的参数如何表示：参数可以逐个添加到参数列表的前面，或放在一个 frame class 中，或放在一个数组中。当然，最简单的方式是将参数逐个添加到参数列表的前面，如下面的例子所示。

```
class B {
    public void foo() {
        List<Person> list = ...
        final int bottom = ..., top = ...;
        list.removeIf( [ lambda for lambda$1 as Predicate capturing (bottom, top) ]);
    }

    // 关注这个方法的签名
    static boolean lambda$1(int bottom, int top, Person p) {
        return (p.size >= bottom && p.size <= top;
    }
}
```

以上展示了 Lambda 如何脱糖，那么如何调用脱糖后的方法呢？关于这一部分的内容在前一节已经介绍了，接下来我们主要关注 invokedynamic 指令和脱糖方法之间参数是如何设定的。

3. Lambda Metafactory

先来看看前面提到的例子。

```
//lambda() 方法
public void lambda() {
    Consumer<String> consumer2 = o -> {
        Object tmpObj = this.obj;
        System.out.println("lambda2");
    };
}

// 对应的字节码
public void lambda();
      descriptor: ()V
      flags: ACC_PUBLIC
      Code:
        stack=1, locals=3, args_size=1
          ...
          6: aload_0
          7: invokedynamic #7,  0  // InvokeDynamic #1:accept:\
                    (Lcn/sensorsdata/lambda/LambdaMain;)\
                    Ljava/util/function/Consumer;
         12: astore_2
         13: return
```

上述代码的 lambda() 方法中使用了 LambdaMain 中的成员变量 obj，观察其对应的字节码可以发现在执行 invokedynamic 指令之前先执行了 aload_0，即 this，该值作为参数会在前面提到的 LambdaMetafactory.metafactory() 方法中使用，下面我们看看这部分内容在 Lambda 设计参考中是如何介绍的。

首先看看什么是 lambda metafactory：对给定的 Lambda 来说，这个调用点被称为 lambda factory，lambda factory 的动态参数是从外部作用域中捕获的，lambda factory 的引导方法是一个标准的方法，被称为 lambda metafactory。虚拟机对每个 invokedynamic 只会调用一次这个 metafactory，之后它会链接这个调用点然后退出。调用点的链接是懒加载的，所以 factory sites 不执行就不会被链接。基本的 metafactory 的静态参数如下。

```
metaFactory(MethodHandles.Lookup caller,     // provided by VM
            String invokedName,              // provided by VM
            MethodType invokedType,          // provided by VM
            MethodHandle descriptor,         // lambda descriptor
            MethodHandle impl)               // lambda body
```

前 3 个参数（caller、invokedName、invokedType）是在虚拟机调用链接的时候自动生成的。descripter 参数确定了被转化的 Lambda 对应的函数式接口方法。impl 参数确定了 Lambda 方法，要么是脱糖的 Lambda 方法体，要么是方法引用中的方法名。函数式接口方法的方法签名和实现方法有一些不同，实现方法可以有额外的参数，其余参数也可能不完全匹配。为方便展示，约定用一些符号来替换 MethodHandle、MethodType 与 invokedynamic。

- method handle 常量简写为 MH(引用类型 class-name.method-name)。
- method type 常量简写为 MT(method-signature)。
- invokedynamic 简写为 INDY((bootstrap, static args...)(dynamic args...))，注意这里的参数设定。

对于前面脱糖的类 A，可以使用如下方式来表示。

```
class A {
    public void foo() {
        List<String> list = ...
        list.forEach(indy((MH(metaFactory), MH(invokeVirtual Block.apply),
                              MH(invokeStatic A.lambda$1)( )));// 注意此处的参数
    }

    private static void lambda$1(String s) {
        System.out.println(s);
    }
}
```

因为 A 中的 Lambda 是无状态的，所以 lambda factory 调用点的动态参数是空的。对于例子中的类 B，动态参数并不为空，因为我们必须把 bottom 和 top 的值添加到 lambda factory 中。

```
class B {
    public void foo() {
        List<Person> list = ...
        final int bottom = ..., top = ...;
        list.removeIf(indy((MH(metaFactory), MH(invokeVirtual Predicate.apply),
                          MH(invokeStatic B.lambda$1))( bottom, top ))));
// 注意此处的参数
    }

    private static boolean lambda$1(int bottom, int top, Person p) {
        return (p.size >= bottom && p.size <= top;
```

```
        }
}
```

这就是 LambdaMain 的 lambda() 方法中会有 6: aload_0 的原因。

4. 静态方法还是实例方法

脱糖方法到底是静态方法还是实例方法呢？观察 LambdaMain.class 中的 lambda$lambda$0() 和 lambda$lambda$1() 两个方法，lambda$lambda$1() 是实例方法的原因似乎与 Lambda 使用了 LambdaMain 中的字段 obj 有关，事实上确实如此。总体来说，我们将在 Lambda 中使用 this、super 或者外部实例的成员的情况称为 instance-capturing lambdas，与其相对的是 non-instance-capturing lambdas。

non-instance-capturing lambdas 被脱糖成静态方法，instance-capturing lambdas 被脱糖成实例的私有方法，当捕获 instance-capturing lambdas 的时候，this 会被声明为第一个动态参数。

举个例子，考虑如下 Lambda 表达式中使用了一个 minSize 字段。

```
list.filter(e -> e.getSize() < minSize )
```

我们首先将上面的示例脱糖成一个实例方法，然后把接收者（this）作为第一个捕获的参数。结果如下:

```
list.forEach(INDY((MH(metaFactory), MH(invokeVirtual Predicate.apply),
                    MH(invokeVirtual B.lambda$1))( this ))));

private boolean lambda$1(Element e) {
    return e.getSize() < minSize;
}
```

因为 Lambda 方法体被转换成一个私有方法，所以 metafactory 中的调用点会加载一个常量池中的方法句柄。对示例方法来说，这个方法句柄的类型是 REF_invokeSpecial（CONSTANT_MethodHandle_info 结构的 reference_index 对应的值），而对静态方法来说，这个方法句柄的类型是 REF_invokeStatic。脱糖成为一个私有方法是因为私有方法可以使用所在类的成员。

5. 方法引用

方法引用有多种写法，跟 lambdas 类似，也可以分成 instance-capturing 和 non-instance-capturing 两种。non-instance-capturing 类型方法引用包括静态方法引用（Integer:: parseInt）、未绑定实例的方法引用（String::length）和构造方法引用（Foo::new）。当使用 non-instance-capturing 类型的方法引用时，动态参数列表总是空的，例如:

```
list.filter(String::isEmpty)
```

上面的例子会被转换成:

```
list.filter(indy(MH(metaFactory), MH(invokeVirtual Predicate.apply),
                  MH(invokeVirtual String.isEmpty))()))
```

instance-capturing 类型的方法引用形式包括绑定实例方法引用（s::length）、super() 方法引用（super::foo）和内部类构造方法引用（Inner::new）。当捕获 instance-capturing 类型的方法引用，被捕获的参数列表总是有一个参数，就是 this。

本小节简单地介绍了布赖恩 · 戈茨的 *Translation of lambda expressions in javac* 一文中关于 Lambda 的一些设计思路，有了这些知识后，接下来我们看看到底如何使用 ASM 来实现 Hook Lambda 和方法引用。

8.4.4 Hook Lambda和方法引用

因为Android本身并不支持Java 8的运行环境，而开发者又可以使用Java 8来开发项目，所以Android项目在编译的时候就需要将Lambda表达式脱糖，并将脱糖后的代码与项目中的代码打包在一起。Android的脱糖逻辑主要是在AGP中完成的，AGP早期的版本通过DesugarTransform类来处理，不过最终的脱糖工作是由com.google.devtools.build.android.desugar.Desugar这个项目来完成的；AGP新版本则自己完成了Lambda的脱糖工作，其原理是获取到ClassFile数据后重新生成IL，从语法分析上来脱糖，其过程相对复杂。不管AGP如何处理脱糖，我们暂时没有方法拿到脱糖后的结果，一种可能的方式是去研究AGP的任务流程，对其中的DexBuilder任务添加Gradle任务依赖，在DexBuilder任务结束后执行自定义的任务，通过这种方式去拿到脱糖后的结果。这种方式作者并没有亲自去实践，读者如有兴趣可以研究一下。

项目插件中定义的Transform获取到的Class都是未脱糖的，根据之前的知识，若要对Lambda表达式进行“Hook”，直接“Hook”脱糖方法是一个不错的选择，如Hook lambda$lambda$0()、lambda$lambda$1()，其原理是ASM处理invokedynamic指令时会调用MethodVisitor的visitInvokeDynamicInsn()方法，该方法的定义如下。

```
/**
   * Visits an invokedynamic instruction.
   *
   * @param name the method's name.
   * @param descriptor the method's descriptor (see {@link Type}).
   * @param bootstrapMethodHandle the bootstrap method.
   * @param bootstrapMethodArguments the bootstrap method constant arguments.
   *        Each argument must be an {@link Integer}, {@link Float},
   *        {@link Long}, {@link Double}, {@link String}, {@link Type},
   *        {@link Handle} or {@link ConstantDynamic} value.
   *        This method is allowed to modify the content of the array so
   *        a caller should expect that this array may change.
   */
public void visitInvokeDynamicInsn(
      final String name,
      final String descriptor,
      final Handle bootstrapMethodHandle,
      final Object... bootstrapMethodArguments)
```

此方法的各参数说明如下。

- name：SAM接口的方法名，以View.OnClickLisetener为例，该名字为onClick。
- descriptor：还记得#44吗？对，就是它。以View.OnClickLisetener为例，该值为()Landroid/view/View$OnClickListener;。
- bootstrapMethodHandle：引导方法，即LambdaMetafactory.metafactory，注意这里的Handle并非MethodHandle，它是ASM中的类。
- bootstrapMethodArguments：引导方法的参数，类似LambdaMetafactory.metafactory()方法的后3个参数。

为方便读者更加直观地理解，图 8-4 展示了 visitInvokeDynamicInsn() 方法调试时的截图。

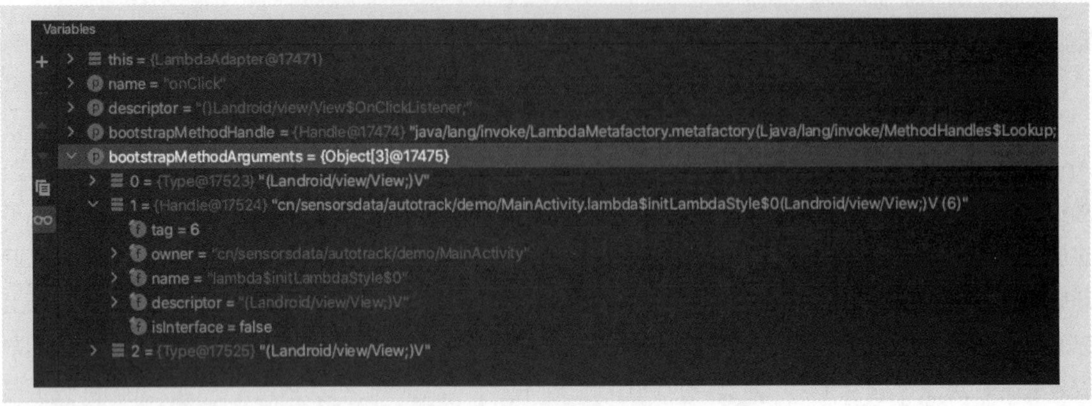

图 8-4　visitInvokeDynamicInsn() 方法调试时的截图

理解了 visitInvokeDynamicInsn() 方法后，接下来利用该方法的参数编写 Hook 逻辑即可。复制一份 AutoTrackImpl02 项目，并将其副本改名为 AutoTrackImpl03，修改 autotrack_plugin 中的 AutoTrackTransform 类，最终的内容如下。

```
public class AutoTrackTransform extends Transform {
    ...
     private byte[] handleBytes(byte[] data, String canonicalName) {
         try {
             ...
             ClassVisitor classVisitor = new ClassVisitor(
                 Opcodes.ASM7, classWriter)
             {
                 private final Map<String, HookType> hookInfoMap
                     = new HashMap<>();(1)
                 ...
                 @Override
                 public MethodVisitor visitMethod(
                     int access, String name, String descriptor,
                     String signature, String[] exceptions)
                 {
                     MethodVisitor methodVisitor =
                         super.visitMethod(access, name, descriptor,
                                             signature, exceptions);
...
                     //处理 Lambda
                     if (hookInfoMap.get(name) != null) {(3)
                         if (hookInfoMap.get(name) == HookType.VIEW_CLICK) {
                             //这里主要是处理 Lambda 有无状态的情况，
                             //以 View.OnCLickListener#onCLick(View) 为例，
                             //descriptor 的最后一个参数是 View，
                             //有状态的 Lambda 的其他参数是追加在方法前面的
                             //假如 descriptor 的值为 (long, int, View)V，
                             //我们需要计算参数 View 所在的槽
                             Type type = Type.getMethodType(descriptor);
                             Type[] argTypes = type.getArgumentTypes();
                             //过滤额外参数的影响，计算 SAM 方法的位置
                             int startPosition = 0;
```

```
        //argTypes.length - 1 实际上需要根据
        //SAM 方法中的值来决定
    //View.OnCLickListener#onCLick(View) 方法只有一个参数，
    // 所以此处减 1
        for (int index = 0;
             index < argTypes.length - 1; index++) {
            startPosition += argTypes[index].getSize();
        }
        // 非静态方法索引 0 的值为 this
        if ((access & Opcodes.ACC_STATIC) == 0) {
            startPosition++;
        }
        methodVisitor.visitVarInsn(
            Opcodes.ALOAD, startPosition);
        methodVisitor.visitMethodInsn(
            Opcodes.INVOKESTATIC,
            "cn/sensorsdata/autotrack/sdk/TrackHelper",
            "trackClick", "(Landroid/view/View;)V", false);
    }
}

methodVisitor = new AdviceAdapter(
    Opcodes.ASM7, methodVisitor, access, name, descriptor)
{
...
    @Override
    public void visitInvokeDynamicInsn(
        String name, String descriptor,
        Handle bootstrapMethodHandle,
        Object... bootstrapMethodArguments)
    {
        // 判断是否需要“Hook”，如果需要就将方法名添加到 Map 中
        if ("onClick".equals(name)(2)
         && "Landroid/view/View$OnClickListener;".equals(
                Type.getMethodType(descriptor)
                    .getReturnType().getDescriptor())
        ) {
            if (bootstrapMethodArguments != null
                && bootstrapMethodArguments.length == 3)
            {
                if (bootstrapMethodArguments[0] instanceof
                    Type &&
                    bootstrapMethodArguments[1] instanceof
                    Handle)
                {
                    Type samType =
                        (Type) bootstrapMethodArguments[0];
                    Handle methodHandle =
                       (Handle) bootstrapMethodArguments[1];
                    if ("(Landroid/view/View;)V"
                        .equals(samType.getDescriptor()))
                    {
                        String hookDesugarMethod =
                            methodHandle.getName();
                        hookInfoMap.put(
                            hookDesugarMethod,
```

```
                                    HookType.VIEW_CLICK);
                            }
                        }
                    }
                }
                super.visitInvokeDynamicInsn(
                    name, descriptor, bootstrapMethodHandle,
                    bootstrapMethodArguments);
            }
        };
        return methodVisitor;
    }
            ...
    };
    classReader.accept(classVisitor, ClassReader.EXPAND_FRAMES);
    return classWriter.toByteArray();
  } catch (Exception e) {
    e.printStackTrace();
    return data;
  }
}
...
}
```

上述代码的逻辑是：首先在位置（1）处创建 Map<String, HookType> 的映射，Map 的作用是记录 Lambda 表达式产生的中间方法名和 Hook 类型的映射关系。它的 key 表示方法名，value 表示属于哪种 Hook 类型，也可以用函数式接口的方法描述符，这里只是做简单的演示。位置（2）处根据 visitInvokeDynamicInsn() 方法的参数信息来判断是否需要进行 Hook，如果需要就将脱糖方法（中间方法）名放入 Map 中，然后在 visitMethod() 方法中的位置（3）处判断方法名是否在 Map 中，如果存在就根据类型添加Hook相关的指令。这种方式利用的是：脱糖方法（中间方法）通常添加在 .class 文件的尾部，ASM 处理 .class 文件时会先处理 invokedynamic 指令，然后才会扫描中间方法。

不过这种方式缺点也很明显。

（1）不能处理方法引用，因为方法引用并不会产生中间方法。

（2）中间方法放在 .class 尾部的行为在未来也可能会改变。

其优点就是简单，最大问题还是在方法引用的支持上，如果不能支持方法引用，那方案就是存在缺陷的、不完整的。

首先，我们看一个方法引用的例子。

```
public class MainActivity extends AppCompatActivity {
    ...
    //方法引用的写法
    private void methodRefStyle() {
        findViewById(R.id.btn).setOnClickListener(System.out::println);
    }
}
```

上述代码是在 MainActivity 中通过方法引用的方式给 View 设置单击事件，对应的字节码如下。

```
public class cn.sensorsdata.autotrack.demo.MainActivity
extends androidx.appcompat.app.AppCompatActivity
{
```

```
    private void methodRefStyle();
            0: aload_0
            1: ldc            #8  // int 2131230807
            3: invokevirtual #9  // Method findViewById:(I)Landroid/view/View;
            6: getstatic      #10 // Field \
                                  java/lang/System.out:Ljava/io/PrintStream;
            9: dup
           10: invokestatic  #11  // Method java/util/Objects.requireNonNull:\
                                             (Ljava/lang/Object;)Ljava/lang/Object;
           13: pop
           14: invokedynamic #12,  0  // InvokeDynamic #0:onClick:\
                     (Ljava/io/PrintStream;)Landroid/view/View$OnClickListener;
           19: invokevirtual #13 // Method android/view/View.setOnClickListener:\
                                          (Landroid/view/View$OnClickListener;)V
           22: return

BootstrapMethods:
    0: #47 REF_invokeStatic
      Method arguments:
        #48 (Landroid/view/View;)V
        #49 REF_invokeVirtual java/io/PrintStream.println:(Ljava/lang/Object;)V
        #48 (Landroid/view/View;)V
```

通过上述字节码 #49 可以看到，方法引用并不像 Lambda 表达式那样将 Lambda 的内容脱糖到一个方法中（实际上也不需要脱糖方法，因为 #49 已经描述了具体需要调用的方法），基于这个因素就无法找到可以插桩的位置。既然没有，那就创建一个 Hook 点。例如下面模拟创建 Hook 点的代码。

```
//方法引用的写法
private void methodRefStyle() {
      findViewById(R.id.btn).setOnClickListener(new View.OnClickListener() {
          @Override
          public void onClick(View v) {
              methodRefStyle$lambda$zw$0(System.out, v);
          }
      });
}
private static void methodRefStyle$lambda$zw$0(
              PrintStream printStream, View view){
      TrackHelper.trackClick(view);
      printStream.println(view);
}
```

其中 methodRefStyle$lambda$zw$0() 方法是我们创建的方法，此方法保留了方法引用的逻辑并且添加上了 Hook 代码。如果能修改原有的逻辑达到这样的效果，就能解决方法引用的 Hook 问题了（同理，Lambda 表达式也可以这么操作）。接下来就来看看如何实现这个操作。

首先复制一份 AutoTrackImpl03 项目，并将基副本改名为 AutoTrackImpl04，在 autotrack_plugin 中添加如下类。

```
package cn.sensorsdata.autotrack.plugin;
...
public class MethodReferenceAdapter extends ClassNode {

      private final AtomicInteger counter = new AtomicInteger(0);
      private List<MethodNode> syntheticMethodList = new ArrayList<>();
      private final String MIDDLE_METHOD_SUFFIX = "$lambda$zw$";
```

```java
    public MethodReferenceAdapter(ClassVisitor classVisitor) {
        super(Opcodes.ASM7);
        this.cv = classVisitor;
    }

    // 判断是否处理 invokedynamic 指令
    // 此处只是单纯地判断了 View.OnClickListener
    private HookType checkShouldHook(InvokeDynamicInsnNode node) {
        if ("onClick".equals(node.name)
                && "Landroid/view/View$OnClickListener;"
            .equals(Type.getMethodType(node.desc)
                    .getReturnType().getDescriptor())
        ) {
            if (node.bsmArgs != null && node.bsmArgs.length == 3) {
                if (node.bsmArgs[0] instanceof Type
                    && node.bsmArgs[1] instanceof Handle)
                {
                    Type samType = (Type) node.bsmArgs[0];
                    Handle methodHandle = (Handle) node.bsmArgs[1];
                    if ("(Landroid/view/View;)V"
                        .equals(samType.getDescriptor()))
                    {
                        return HookType.VIEW_CLICK;
                    }
                }
            }
        }
        return HookType.DO_NOTHING;
    }


    @Override
    public void visitEnd() {
        super.visitEnd();

        this.methods.forEach(methodNode -> {
            ListIterator<AbstractInsnNode> iterator =
                methodNode.instructions.iterator();
            while (iterator.hasNext()) {
                AbstractInsnNode node = iterator.next();
                if (node instanceof InvokeDynamicInsnNode) {
                    InvokeDynamicInsnNode tmpNode =
                        (InvokeDynamicInsnNode) node;
                    // 如果不需要 Hook 的 Lambda 和方法引用，就不处理
                    HookType hookType = checkShouldHook(tmpNode);
                    if (hookType == HookType.DO_NOTHING) {
                        continue;
                    }

                    // 形如 (Ljava/util/Date;)Ljava/util/function/Consumer;
                    // 可以从 desc 中获取函数式接口，以及动态参数的内容。
                    // 如果没有参数，那么描述符的参数部分应该是空
                    String desc = tmpNode.desc;
                    Type descType = Type.getType(desc);
                    Type samBaseType = descType.getReturnType();
```

```
        //sam 接口名
        String samBase = samBaseType.getDescriptor();
        //sam 方法名
        String samMethodName = tmpNode.name;
        Object[] bsmArgs = tmpNode.bsmArgs;
        //sam 方法描述符
        Type samMethodType = (Type) bsmArgs[0];
        //sam 实现方法的实际参数描述符
        Type implMethodType = (Type) bsmArgs[2];
        //sam name + desc，可以用来辨别是否需要 Hook 的 Lambda 表达式，
        // 例如 onClick + (View)V，根据此再加上 samBaseType，
        // 即可判断是否需要 Hook View.OnClickListener
        String bsmMethodNameAndDescriptor =
            samMethodName + samMethodType.getDescriptor();
        // 中间方法的名称 : Lambda 所在的方法名 + $lambda$zw$ + 数字
        String middleMethodName = methodNode.name
            + MIDDLE_METHOD_SUFFIX + counter.incrementAndGet();
        // 中间方法的描述符，例如有状态 Lambda 需要拼接额外参数 + SAM 接口的原本
        // 参数
        String middleMethodDesc = "";
        Type[] descArgTypes = descType.getArgumentTypes();
        if (descArgTypes.length == 0) {
            middleMethodDesc = implMethodType.getDescriptor();
        } else {
            middleMethodDesc = "(";
            for (Type tmpType : descArgTypes) {
                middleMethodDesc += tmpType.getDescriptor();
            }
            middleMethodDesc +=
                implMethodType.getDescriptor().replace("(", "");
        }

        //INDY 指令原本的 handle，需要将此 handle 替换成新的 handle
        Handle oldHandle = (Handle) bsmArgs[1];
        Handle newHandle = new Handle(
            Opcodes.H_INVOKESTATIC,
            this.name, middleMethodName, middleMethodDesc, false);

        InvokeDynamicInsnNode newDynamicNode =
            new InvokeDynamicInsnNode(
                    tmpNode.name,
                    tmpNode.desc,
                    tmpNode.bsm,
                    samMethodType,
                    newHandle,
                    implMethodType);
        iterator.remove();
        iterator.add(newDynamicNode);

        generateMiddleMethod(
            oldHandle, middleMethodName,
            middleMethodDesc,
            bsmMethodNameAndDescriptor, hookType);

    }
}
```

```
        });

        this.methods.addAll(syntheticMethodList);
        accept(cv);
    }

    private void generateMiddleMethod(
        Handle oldHandle, String middleMethodName, String middleMethodDesc,
        String bsmMethodNameAndDescriptor, HookType hookType)
    {
        //开始对生成的方法中插入或者调用相应的代码
        MethodNode methodNode = new MethodNode(
            Opcodes.ACC_PRIVATE + Opcodes.ACC_STATIC,
                middleMethodName, middleMethodDesc, null, null);
        methodNode.visitCode();

        //添加Hook代码
        addHookMethod(
            methodNode, middleMethodDesc,
            bsmMethodNameAndDescriptor, hookType);

        // 此tag具体可以参考： CONSTANT_MethodHandle_info reference_kind
        int accResult = oldHandle.getTag();
        switch (accResult) {
            case Opcodes.H_INVOKEINTERFACE:
                accResult = Opcodes.INVOKEINTERFACE;
                break;
            case Opcodes.H_INVOKESPECIAL:
                //private、this、super等会调用
                accResult = Opcodes.INVOKESPECIAL;
                break;
            case Opcodes.H_NEWINVOKESPECIAL://针对XXX::new的方法引用
                //constructors
                accResult = Opcodes.INVOKESPECIAL;
                methodNode.visitTypeInsn(Opcodes.NEW, oldHandle.getOwner());
                methodNode.visitInsn(Opcodes.DUP);
                break;
            case Opcodes.H_INVOKESTATIC:
                accResult = Opcodes.INVOKESTATIC;
                break;
            case Opcodes.H_INVOKEVIRTUAL:
                accResult = Opcodes.INVOKEVIRTUAL;
                break;
        }

        Type middleMethodType = Type.getType(middleMethodDesc);
        Type[] argumentsType = middleMethodType.getArgumentTypes();
        if (argumentsType.length > 0) {
            int loadIndex = 0;
            for (Type tmpType : argumentsType) {
                int opcode = tmpType.getOpcode(Opcodes.ILOAD);
                methodNode.visitVarInsn(opcode, loadIndex);
                loadIndex += tmpType.getSize();
            }
        }
```

```
            methodNode.visitMethodInsn(
                accResult, oldHandle.getOwner(),
                oldHandle.getName(), oldHandle.getDesc(), false);
            Type returnType = middleMethodType.getReturnType();
            int returnOpcodes = returnType.getOpcode(Opcodes.IRETURN);
            methodNode.visitInsn(returnOpcodes);
            methodNode.visitEnd();
            syntheticMethodList.add(methodNode);
        }
        //添加 Hook 逻辑
        private void addHookMethod(
            MethodNode node, String middleMethodDesc,
            String bsmMethodNameAndDescriptor, HookType hookType)
        {
            if (hookType == HookType.VIEW_CLICK) {
                //这里主要是处理 Lambda 有无状态的情况，
                //以 View.OnCLickListener#onCLick(View) 为例，
                //descriptor 的最后一个参数是 View，
                //有状态 Lambda 的其他参数是追加在方法前面的，
                //假如 descriptor 的值为 (long, int, View)V，
                //我们需要计算参数 View 所在的槽
                Type type = Type.getMethodType(middleMethodDesc);
                Type[] argTypes = type.getArgumentTypes();
                //过滤额外参数的影响，计算 SAM 方法的位置
                int startPosition = 0;
                //argTypes.length - 1 实际上需要根据实际的 SAM 方法中的值来决定
                //View.OnCLickListener#onCLick(View) 方法只有一个参数，所以此处减 1
                for (int index = 0; index < argTypes.length - 1; index++) {
                    startPosition += argTypes[index].getSize();
                }
                node.visitVarInsn(Opcodes.ALOAD, startPosition);
                node.visitMethodInsn(
                    Opcodes.INVOKESTATIC,
                    "cn/sensorsdata/autotrack/sdk/TrackHelper",
                    "trackClick", "(Landroid/view/View;)V", false);
                System.out.println("==== 添加 Hook");
            }
        }
}
```

MethodReferenceAdapter 是 ClassNode 的子类，在其 visitEnd() 方法中获取到类中所有的方法，然后遍历方法中的指令。当遇到 InvokeDynamicInsnNode 时，对该指令进行处理，主要逻辑内容如下。

（1）根据 InvokeDynamicInsnNode 获取 SAM 接口信息以及拼装中间方法，中间方法的方法描述符是 InvokeDynamicInsnNode.desc（暗号为 #44）的值加上 SAM 接口中的参数。

（2）创建新的 Handle，新 Handle 指向了我们创建的中间方法，并拼装新的 InvokeDynamicInsnNode 替换原本的。例如对于一个 Lambda 表达式，这里使用中间方法替换了 lambda&*() 方法，这一步完成了引导方法的转移。

（3）创建中间方法，其关键点是在中间方法中调用原 InvokeDynamicInsnNode 指令中引导方法所指向的方法，也就是说在中间方法中调用原本的 lambda&*() 脱糖方法等。

（4）在中间方法中插入 Hook 逻辑。

以上就是 ASM Hook 方法引用和 Lambda 表达式的主要逻辑，其关键点就是替换原本的 Handle，让 Handle 指向新的方法。下面是发布插件后显示的最终结果。

测试代码如下。

```
package cn.sensorsdata.autotrack.demo;

public class MainActivity extends AppCompatActivity {

      Object obj = new String("sss");
      private void init() {
          //匿名类写法
          findViewById(R.id.normal_btn)
              .setOnClickListener(new View.OnClickListener() {
              @Override
              public void onClick(View v) {

              }
          });
          //无状态 Lambda 写法
          findViewById(R.id.lambda_stateless_btn1).setOnClickListener(v -> {

          });
          //有状态 Lambda 写法
          findViewById(R.id.lambda_state_btn1).setOnClickListener(v -> {
              System.out.println(this.obj);
          });
          //方法引用 new 用法
          findViewById(R.id.mr_style_new_btn)
              .setOnClickListener(MyLayout::new);
          //方法引用静态方法用法
          findViewById(R.id.mr_style_static_btn)
              .setOnClickListener(MainActivity::show);
          //方法引用实例方法用法
          MyLayout layout = new MyLayout();
          findViewById(R.id.mr_style_instance_btn)
              .setOnClickListener(layout::setNewView);
      }

      public static void show(View view) {
      }

      static class MyLayout {
          public MyLayout(View view) {
          }
          public MyLayout(){
          }
          public void setNewView(View view){
          }
      }
}
```

编译后的部分结果如下。

```
...
    // 无状态 Lambda 表达式生成的中间方法
    private static void lambda$init$0(android.view.View);
        stack=0, locals=1, args_size=1
           0: return

    // 我们自己生成的中间方法，并在中间方法中调用 lambda$init$0()
    private static void init$lambda$zw$1(android.view.View);
        stack=1, locals=1, args_size=1
           0: aload_0
           1: invokestatic  #141  // Method
                                   +cn/sensorsdata/autotrack/sdk/TrackHelper
                                   +.trackClick:(Landroid/view/View;)V
           4: aload_0
           5: invokestatic  #143  // Method lambda$init$0:(Landroid/view/View;)V
           8: return
    ...

    // 针对最后一个方法引用生成的中间方法，方法中执行了原方法引用中的逻辑
    private static void init$lambda$zw$5(
                cn.sensorsdata.autotrack.demo.MainActivity$MyLayout,
                android.view.View);
        stack=2, locals=2, args_size=2
           0: aload_1
           1: invokestatic  #141  // Method \
                                        cn/sensorsdata/autotrack/sdk/TrackHelper\
                                        .trackClick:(Landroid/view/View;)V
           4: aload_0
           5: aload_1
           6: invokevirtual #152  // Method \
                        cn/sensorsdata/autotrack/demo/MainActivity$MyLayout\
                        .setNewView:(Landroid/view/View;)V
           9: return
}
...
BootstrapMethods:
    // 无状态 Lambda 表达式的引导方法被修改成了我们自己生成的 init$lambda$zw$1() 方法
    0: #80 REF_invokeStatic ...
      Method arguments:
        #69 (Landroid/view/View;)V
        #73 REF_invokeStatic cn/sensorsdata/autotrack/demo/MainActivity\
                                .init$lambda$zw$1:(Landroid/view/View;)V
        #69 (Landroid/view/View;)V
...
    // 将方法引用中对应的引导方法修改成了我们自己生成的 init$lambda$zw$5() 方法
    4: #80 REF_invokeStatic ...
      Method arguments:
        #69 (Landroid/view/View;)V
        #118 REF_invokeStatic cn/sensorsdata/autotrack/demo/MainActivity\
                        .init$lambda$zw$5:\
                        (Lcn/sensorsdata/autotrack/demo/MainActivity$MyLayout;\
                        Landroid/view/View;)V
        #69 (Landroid/view/View;)V
```

是不是很有意思的一件事情？“偷梁换柱”这样的操作在使用 ASM 操作字节码的案例中真是屡试不爽，前文我们也用到了这样的操作。

8.5 小结

本章介绍了一些实际开发中会遇到的业务逻辑和使用场景，其中重点介绍了 Lambda 表达式的设计原理、运行原理，以及 Hook Lambda 和方法引用的方式。前后涉及的内容较多，读者需要结合 4.7.6 小节的内容多读几遍，并实际练习以加深对这方面内容的理解。

本书使用两章为大家介绍 ASM 如何实现全埋点，搭建了一个最基本的埋点框架，以期起到抛砖引玉的作用。读者可按书中介绍的方法，找准 Hook 点，按实际业务埋入代码即可。

9. ASM实践分享和未来展望

关于 ASM 高级应用相关的知识已经介绍得差不多了，本章再分享一些实际开发中遇到的问题，并展望未来的技术变化和后续深入学习的路径。

9.1 是否可以注册多个 Transform

在我们的开源群里有位朋友问了这么一个问题：插件是否可以注册多个 Transform？

这个问题本身并不复杂，回想一下注册 Transform 的方式，我们是通过 AGP 提供的 AppExtension#registerTransform() 方法注册 Transform 的，registerTransform() 方法的源码如下。

```
public void registerTransform(
        @NonNull Transform transform, Object... dependencies)
{
        transforms.add(transform);
        transformDependencies.add(Arrays.asList(dependencies));
}
```

通过源码可以清晰地看到，registerTransform() 方法是将 Transform 对象添加到一个 List 中，因此是可以注册多个“不同”的 Transform 的。这里的所谓“不同”是指不能重名，即自定义的 Transform 的 getName() 方法返回的值不能相同，这些内容在 2.4.1 小节已介绍。

9.2 插入代码是否会改变行号

同样，有群友询问：使用 ASM 进行插桩后是否会改变行号?

群友问此问题的目的是想确认插桩后是否会影响代码调试。在回答这个问题之前，先了解 LineNumberTable 属性。LineNumberTable 与 LocalVariableTable 以及 SourceFile 属性都是调试时使用的，它们都是非必需的。LineNumberTable 属性的结构如下。

```
LineNumberTable_attribute {
        u2 attribute_name_index;
        u4 attribute_length;
        u2 line_number_table_length;
        {   u2 start_pc;
            u2 line_number;
        } line_number_table[line_number_table_length];
}
```

其中 line_number_table 表示表中每一项的结构；start_pc 表示字节码指令偏移的位置，line_number 表示行号，结合起来就是每个行号对应的字节码偏移位置。举例如下。

```
package cn.sensorsdata.asm;          //linenumber 1
                                     //linenumber 2
public class LineNumberTest {        //linenumber 3
                                     //linenumber 4
    public void foo(){               //linenumber 5
```

```
        String b = "100";        //linenumber 6
        int a = b.length();      //linenumber 7
        System.out.println(a);   //linenumber 8
        int c = 99;              //linenumber 9
    }                            //linenumber 10
}
```

上述代码中的 linenumber *x* 表示 Java 源码对应的行号。foo() 方法对应的字节码如下。

```
public void foo();
   descriptor: ()V
   flags: (0x0001) ACC_PUBLIC
   Code:
     stack=2, locals=4, args_size=1
        0: ldc           #2        // String 100
        2: astore_1
        3: aload_1
        4: invokevirtual #3 // Method java/lang/String.length:()I
        7: istore_2
        8: getstatic     #4 // Field java/lang/System.out:Ljava/io/PrintStream
       11: iload_2
       12: invokevirtual #5 // Method java/io/PrintStream.println:(I)V
       15: bipush        99
       17: istore_3
       18: return
     LineNumberTable:
       line 6: 0
       line 7: 3
       line 8: 8
       line 9: 15
       line 10: 18
  --------------------------
          行号 : 字节码指令偏移
```

说明如下。

- line 6: 0表示第6行的代码的字节码偏移位置是从0开始的。
- line 7: 3表示第7行的代码的字节码偏移位置是从3开始的。

观察源码中的第 8 行代码 System.out.println(a)，可以看到字节码偏移范围 [8,15) 这个区间的字节码都属于第 8 行。以上主要是说明行号与字节码之间的关系，假如在 [8,15) 中间插入一条字节码指令，是否会改变行号呢？从直观上说如果插入一条指令，那么其字节码偏移就会发生变化，这样就无法将行号与字节码偏移进行配对，其实这种看法是错的，或者说角度不对。下面是 foo() 方法使用 ASM 实现的部分代码。

```
methodVisitor = classWriter.visitMethod(ACC_PUBLIC, "foo", "()V", null, null);
...
methodVisitor.visitFieldInsn(
      GETSTATIC, "java/lang/System", "out", "Ljava/io/PrintStream;");
methodVisitor.visitVarInsn(ILOAD, 2);
// 插入一段开始指令
     methodVisitor.visitInsn(Opcodes.DUP);
     methodVisitor.visitMethodInsn(INVOKEVIRTUAL, "java/io/PrintStream", "println", "(I)V", false);
// 插入一段结束指令
```

```
// 此处插入代码
methodVisitor.visitMethodInsn(
      INVOKEVIRTUAL, "java/io/PrintStream", "println", "(I)V", false);
...
methodVisitor.visitEnd();
```

观察上述 ASM 代码，假设在调用 println() 方法之前插入一段代码，代码为上述注释的开始和结束部分。将上述结果保存到 .class 文件中并使用 javap 指令查看。

```
public void foo();
     descriptor: ()V
     flags: (0x0001) ACC_PUBLIC
     Code:
       stack=2, locals=4, args_size=1
          0: ldc           #14  // String 100
          2: astore_1
          3: aload_1
          4: invokevirtual #20 // Method java/lang/String.length:()I
          7: istore_2
          8: getstatic     #26 // Field java/lang/System.out:Ljava/io/PrintStream;
         11: iload_2
         12: dup
         13: invokevirtual #32 // Method java/io/PrintStream.println:(I)V
         16: invokevirtual #32 // Method java/io/PrintStream.println:(I)V
         19: bipush        99
         21: istore_3
         22: return
       LineNumberTable:
         line 6: 0
         line 7: 3
         line 8: 8
         line 9: 19
         line 10: 22
```

可以看到虽然增加了dup和invokevirtual两条指令，但是字节码偏移范围 [8,19) 仍然属于第8行代码，这就证明使用 ASM 插入代码后并不会改变行号。

9.3 是否支持Kotlin

Kotlin 是一种在 Java 虚拟机上运行的静态类型编程语言，它扩展了 Java 编译器，使得 Kotlin 代码能够得以编写、编译和调试。Kotlin 可以将代码编译成 Java 字节码，也可以编译成 JavaScript，方便在没有 JVM 的设备上运行。另外，Kotlin 与 Java 完全兼容，使得 Java 用户可以快速上手 Kotlin 语言，而且 Kotlin 凭借其现代化的语言设计快速吸引了大批的用户。Kotlin 的应用范围也非常广，包括 Android 应用开发、Web 应用开发、iOS 应用开发、Native 开发、Gradle 脚本编写等。另外，许多 Java 开源项目也都纷纷拥抱 Kotlin。可以预见，未来 Kotlin 将会成为一门不错的工作语言。

因为 Kotlin 编译器会将 Kotlin 代码编译成 .class 文件，自然可以使用 ASM 对其进行插桩操作。不过

Kotlin 因其自身语言特性，在选择 Hook Point 时需要一些额外的操作，例如下面的 Kotlin 代码。

```
// 文件名是：commons.kt

package cn.sensorsdata.kotlindemo

fun String.printLength(){
      println("$this 's length is: ${this.length}")
}

fun main(){
      "curious".printLength()
}
```

思考如何对上述 String 类的扩展方法 printLength() 进行插桩。

printLength() 方法是声明在 commons.kt 文件中的顶级函数，Java 中没有对应的语法，不过可以从 .class 文件的角度来分析。上述 commons.kt 代码编译后会产生一个 CommonsKt.class 文件，反编译可以看到如下结果。

```
public final class CommonsKt {
     public static final void printLength(
          @NotNull String $this$printLength)
     {
        Intrinsics.checkParameterIsNotNull(
            $this$printLength, "$this$printLength");
        String var1 = $this$printLength
            + " 's length is: " + $this$printLength.length();
        boolean var2 = false;
        System.out.println(var1);
     }

     public static final void main() {
        printLength("curious");
     }

     // $FF: synthetic method
     public static void main(String[] var0) {
        main();
     }
}
```

这就是 commons.kt 文件中的代码对应的 Java 代码，因此如果要 Hook printLength() 方法，只需要 Hook CommonsKt.class 中方法名为 printLength()、方法签名是 (Ljava/lang/String;)V 的方法即可。这就是在处理 Kotlin 代码时需要注意的事项。

9.4 ASM 如何处理继承关系

思考这样一个场景：假如需要在调用 Android 中 WebView 类的 loadUrl() 方法的地方插入代码，用于获取加载的 URL 地址，应该如何实现这个功能？

可能读者会说这不是一件很容易的事情吗，将 WebView 类中的 loadUrl() 方法作为 Hook Point，在方法开始的地方插桩即可。是的，理论上是这样的，但实际上使用 Android 插件在编译时是获取不到 android.jar 这个包的，自然也就无法对其进行插桩。既然无法直接在 loadUrl() 方法中插桩，换个思路，我们可以在调用 loadUrl() 的时候插桩，比如下面的代码所示的插桩情形。

```
WebView webView = ...;
webView.loadUrl("https://www.sensorsdata.cn");
TrackHelper.loadUrl(webView, "https://www.sensorsdata.cn")// 此处是模拟插入的代码
...
```

分析上述插桩的条件。

（1）必须是调用方法的指令，即 invokevirtual。

（2）方法名必须是 loadUrl，方法的描述符必须是 (Ljava/lang/String;)V。

（3）满足上述两个条件还不够，还需判断调用者是不是 android.webkit.WebView 或者 android.webkit.WebView 的子类。

满足上述 3 个条件，才能进行插桩操作。前两个条件很好满足，在 MethodVisitor 中的 visitMethodInsn() 方法中判断即可，而判断调用者是不是 WebView 的子类应该如何进行呢?

通常有如下两种方式。

- 通过ClassLoader加载Class进行判断。
- 构建Class类图。

9.4.1 ClassLoader方式

此方式的原理是在 Transform 中收集所有编译时的 JAR 包和 classes 文件并构建 URLClassLoader。再通过 URLClassLoader 的 loadClass() 方法获取到 Class 对象，通过 Class 的 isAssignableFrom() 方法来判断继承关系。JAR 包和 classes 文件可以通过 Transform 的 transform(TransformInvocation transformInvocation) 方法中的 transformInvocation 对象获取到。不过通过 transformInvocation 对象并不能获取到 android.jar 包和包中的类，因此还需要获取编译时的 android.jar 包并将其也添加到 URLClassLoader 中才可以。

获取 android.jar 包的方式如下。

```
File androidJar() throws FileNotFoundException {
      File jar = new File(getSdkJarDir(), "android.jar")
      if (!jar.exists()) {
          throw new FileNotFoundException("Android jar not found!")
      }
      return jar
}
private String getSdkJarDir() {
      String compileSdkVersion = appExtension.getCompileSdkVersion()
      return String.join(
          File.separator,
          android.getSdkDirectory().getAbsolutePath(),
          "platforms", compileSdkVersion)
}
```

上述的 appExtension 对象是 com.android.build.gradle.AppExtension Android 扩展的对象，可以在 Plugin 中通过如下方式获取。

```
class SensorsAnalyticsPlugin implements Plugin<Project> {
      @Override
      void apply(Project project) {
           // 获取 Android 扩展
           AppExtension appExtension =
               project.extensions.findByType(AppExtension.class)
      }
}
```

获取上述 JAR 包和 classes 文件以后就可以构建 URLClassLoader 对象，示例代码如下。

```
//Groovy 代码
private void traverseForClassLoader(TransformInvocation transformInvocation) {
      def urlList = []
      // 获取 android.jar 包
      def androidJar = transformHelper.androidJar()
      // 将 android.jar 包添加到 list 中
      urlList << androidJar.toURI().toURL()
      // 通过 TransformInvocation 获取输入源并将它们的 URL 信息添加到 list 中
      transformInvocation.inputs.each { transformInput ->
          //jar 包
          transformInput.jarInputs.each { jarInput ->
              urlList << jarInput.getFile().toURI().toURL()
          }
          //classes
          transformInput.directoryInputs.each { directoryInput ->
              urlList << directoryInput.getFile().toURI().toURL()
          }
      }
      def urlArray = urlList as URL[]
      // 构建 URLClassLoader 对象
      urlClassLoader = new URLClassLoader(urlArray)
      // 将 URLClassLoader 对象保存起来，在其他地方使用
      transformHelper.urlClassLoader = urlClassLoader
}
```

构建了 URLClassLoader 对象后，再来看看如何使用，以前面提到的 WebView 为例。

```
try {
     Class maybeWebView =
         transformHelper.urlClassLoader
         .loadClass("com.curious.demo.MyWebView")
     Class androidWebView = transformHelper.urlClassLoader
         .loadClass("android.webkit.WebView")
     // 判断 maybeWebView 是不是 androidWebView 的子类
     if(androidWebView.isAssignableFrom(maybeWebView)){
         // 是 WebView 的子类
     }
} catch (ClassNotFoundException e) {
       e.printStackTrace();
}
```

以上就是通过 URLClassLoader 来判断继承关系的方法。这种方法相对简单，不过需要注意

NoClassDefError 问题，例如下面的代码。

```
MyWebView webView = ...;
webView.loadUrl("https://www.sensorsdata.cn");
```

假如上述代码存在于 a.jar 中，同时 MyWebView 继承自 XxxView，不过 XxxView 不在 a.jar 中，也不在其他的任何 JAR 中，此时通过 URLClassLoader 来加载 MyWebView 的时候就会抛出 NoClassDefFoundError。出现此错误也很正常，需要捕获异常，不过这里捕获的异常不再是 ClassNotFoundException 或者 Exception，而是 Throwable。这是因为 NoClassDefFoundError 是 Error 的子类，而 Error 是 Throwable 的子类。这里扩大了异常的捕获范围。

9.4.2 类图方式

9.4.1 小节介绍的 URLClassLoader 方式有一个缺点：URLClassLoader 加载 .class 时会去校验 Class 数据格式，如果格式错误会抛出类似 ClassFormatError 这样的异常。例如删除 ClassFile 中的 StackMapTable 属性就会产生此异常。所以在使用 ASM 处理 .class 的时候我们要谨慎地使用 ClassReader.SKIP_FRAMES 配置。

当此类异常发生时自然也会影响继承关系的判断（可以配置 JVM 参数 -Xverify:none 来关闭格式校验），而“类图”就是在此种方式下产生的。简单地说就是在 Transform 中构建一个结构，该结构记录每个类的基本信息，包括类名、父类、接口等内容，判断继承关系是从这些信息中获取即可，例如：

```
class A extends B implements Callback
class B extends
```

假设使用如下结构来保存上述代码的继承关系：

```
class Node{
      String className;
}
class ClassNode {
      Node parentNode;
      List<InterfaceNode> interfaceNodes;
      List<Node> childrens;
}
```

判断 A 是不是 C 的子类，只要判断 A 的 parentNode 是否存在 C 即可。此种方式实现起来相对复杂，并且也需要遍历两遍 .class，但好处是可以规避 URLClassLoader 加载类产生的格式校验问题。类图方式的代表是字节跳动的开源项目 ByteX。

9.5 慎用 static 变量

这是一个比较小的知识点，假设在插件中需要使用一个类字段来保存某种状态，示例代码如下。

```
class TransformHelper{
      static boolean isChecked = false;
}
```

当第一次编译的时候将 isChecked 值设置为 true，读者可以想想在第二次编译时 isChecked 的初始值是多少?

答案可能是 true。原因是 Gradle 在编译时会启动缓存，导致 isChecked 的值还是上一次编译的结果。如果忽略了这一点可能导致项目中与此字段相关的初始设置存在偏差。

9.6 AGP 7的变化

2021 年，AGP 的一个重大变化就是发布了 v7.0.0（新的 AGP 版本号会与 Gradle 版本号保持一致），此版本带来了非常多的特性，具体如下。

- 支持 Java 11：如果想使用 AGP 7 就必须升级到 Java 11。
- lint 行为变更：lint 是一个比较耗时的任务，AGP 7 中的 lint 任务将支持缓存，大大提升构建速度。
- 移除了 Android Gradle 构建缓存：全面使用 Gradle 自身的构建缓存。

不过在 API 层面对我们影响最大的一个点是，Transform API 被标记为废弃了，并且会在 AGP 8.0 中移除该 API。删除的原因是使用 Transform API 的项目会强制 AGP 对 build 使用优化程度不够的流程，从而导致构建时间大幅增加。这意味着这个从 AGP 1.3 一直存在到现在且非常稳定的 API，将退出历史舞台。

按照官方的说明，Transform API 没有单一的替代 API，对于需要处理字节码的工作，官方推荐使用 Instrumentation API，使用这些 API 可以大幅提高 build 的性能。为此，官方也提供了相应的例子，以下是官方插件。

```
import com.android.build.api.variant.AndroidComponentsExtension
import com.android.build.api.instrumentation.AsmClassVisitorFactory
import com.android.build.api.instrumentation.ClassContext
import com.android.build.api.instrumentation.ClassData
import com.android.build.api.instrumentation.FramesComputationMode
import com.android.build.api.instrumentation.InstrumentationParameters
import com.android.build.api.instrumentation.InstrumentationScope
...

abstract class ExamplePlugin : Plugin<Project> {

      override fun apply(project: Project) {

          val androidComponents = // (1)
              project.extensions.getByType(AndroidComponentsExtension::class.java)

          androidComponents.onVariants { variant ->// (2)
              variant.instrumentation.transformClassesWith(// (3)
                                  ExampleClassVisitorFactory::class.java,
                                  InstrumentationScope.ALL)
```

```
                {
                    it.writeToStdout.set(true)
                }
                variant.setAsmFramesComputationMode(// (4)
                    FramesComputationMode.COPY_FRAMES)
            }
        }

        interface ExampleParams : InstrumentationParameters {
            @get:Input
            val writeToStdout: Property<Boolean>
        }

        abstract class ExampleClassVisitorFactory :
            AsmClassVisitorFactory<ExampleParams> {

            override fun createClassVisitor(// (5)
                classContext: ClassContext,
                nextClassVisitor: ClassVisitor
            ): ClassVisitor {
                return if (parameters.get().writeToStdout.get()) {
                    TraceClassVisitor(nextClassVisitor,
                                      PrintWriter(System.out))
                } else {
                    TraceClassVisitor(nextClassVisitor,
                                      PrintWriter(File("trace_out")))
                }
            }

            override fun isInstrumentable(classData: ClassData): Boolean {
                return classData.className.startsWith("com.example")// (6)
            }
        }
}
```

上述代码展示了使用ASM处理字节码的基本方式，其中各位置的说明如下。

- 位置（1）表示获取AndroidComponentsExtension扩展，该扩展是新版本AGP提供的API，通过该扩展可以获取到Variant。
- 位置（2）的onVariant是新版本提供的Variant API。Variant API是AGP中的扩展机制，让用户可以操作各种build时的资源。onVariant回调表示获取已创建的Variant对象。
- 位置（3）的Instrumentation.ItransformClassesWith(Class<out AsmClassVisitorFactory <ParamT>>, InstrumentationScope, (ParamT) -> Unit) 方法表示注册ASM ClassVisitor用于处理给定范围内的classes。该方法有以下3个参数。

① 第一个参数是接口 AsmClassVisitorFactory 的实现，用于处理 classes。

② 第二个参数是指定作用范围，类似 Transform 的 getScopes() 方法。

③ 第三个参数是参数配置，配置的参数影响编译时能否获取到编译产物。

- 位置（4）表示修改ClassFile的行为，与ASM ClassWriter构造方法中的flags相似。

● 位置（5）是实现AsmClassVisitorFactory的createClassVisitor(ClassContext, ClassVisitor)方法。可以看到该方法会提供两个参数。

① ClassContext 包含了 ClassFile 的基本信息。其中 ClassContext.currentClassData 包含了类的名称、注解、实现的接口、父类信息。ClassContext.loadClassData(className:String?) 方法是加载 classpath 中的特定类，利用它可以实现继承关系的判断。

② ClassVisitor 就是 ASM 中的 ClassVisitor。从中可以看出开发者不再需要关注 ClassReader 和 ClassWriter 的创建。

● 位置（6）表示是否对给定的类做处理，如果不处理就返回false。这里很适合处理8.1 节介绍的黑白名单。

以上就是 AGP 7.3 处理字节码的方式，这里之所以选择 AGP 7.3，而不是 7.0、7.1，原因是在 7.3 之前，上述涉及的 API 一直在变动。目前 AGP 7.3 及以后的版本已经相对稳定，对应的 API 变动的可能性较小，建议开发者适配的时候能够以该版本作为参考基准。

从上述介绍可以发现新版本处理字节码的方式与 Transform 相差很大，在实际开发中需要做好两者兼容。

其实这个需求的核心点就是辨别当前使用的 AGP 版本。以 AGP 7.3 为例，当版本号不低于 7.3 时，使用新版本的 Instrumentation API 处理字节码；当版本号低于 7.3 时，使用 Transform API 处理字节码。

AGP 版本可以从如下类的 ANDROID_GRADLE_PLUGIN_VERSION 字段获取到。

```
package com.android.builder.model;

/**
  * @deprecated use com.android.Version instead
  *     <p>TODO: remove (along with the associated version.properties)
  *     once it's no longer used by the gradle build scan plugin
  */
@Deprecated
public final class Version {
      public static final String ANDROID_GRADLE_PLUGIN_VERSION;
      ...
}
```

从上述代码可以看出该类已经被标记为 Deprecated，不过新版本可以使用 com.android.Version 这个类获取 AGP 版本。下面给出一个获取 AGP 版本字符串的代码示例。

```
fun getAGPVersionStr(): String {
        var clazz =
            loadClass("com.android.Version")
               ?: loadClass("com.android.builder.model.Version")
        clazz?.apply {
            return getDeclaredField("ANDROID_GRADLE_PLUGIN_VERSION")
                        .get(this).toString()
        }
        error("Could not find Android Gradle Plugin version.")
}
```

此外读者也可以通过 Chapter9_AGP 项目获取完整的适配示例。

9.7 小结

本章总结了实际开发中一些问题，并简单介绍了 AGP 7 的变化和最新的处理字节码的方式。对于 AGP 的变化，作者认为，不管是 Transform API，还是 Instrumentation API，只不过是获取 Class 的方式不同而已，并不改变对获取到的 Class 使用 ASM 的处理方式。

最后，读者若想更进一步研究，建议深入学习 Gradle 的内容和 AGP 源码。还有一些业界比较不错的开源项目也值得关注学习，例如滴滴公司的 Booster 和字节跳动公司的 ByteX。